"What is important about this book is the clarity, kindness and intelligence of its thesis; what is beautiful about this book is its rootedness in the wisdom of myth."
Jay Griffiths, *author of* Wild *and* Why Rebel

"Jules Pretty has long been a credible and trusted guide on the questions of what we must do, and how we might do it. Now he adds a crucial part of the story: why this will make our lives better, deeper, richer."
Bill McKibben, *founder 350.org, author of* Falter: Has the Human Game Begun to Play Itself Out?

"Brilliant reframing around what living a good life looks like – creative and compelling."
Caroline Lucas, *MP for Brighton Pavilion, Green Party*

"Unputdownable, inspired, hopeful, informed, prescient, millennia and continent spanning, hopeful, cautious, careful, understanding; but above all else, different.'
Professor Danny Dorling, *University of Oxford and author of* Slowdown

"In a truly masterful manner, Jules Pretty weaves empirical fact with myth, data with passion, knowledge with wisdom and poetry with prose. A meticulously argued and wonderfully creative account of how greener, low carbon and happier options can be established for all."
Professor Richard Bawden, *Western Sydney University*

"A fascinating exploration of how we can live happily and sustainably, drawing on sagas and stories, through philosophy and economics, to climate science. This beautifully written book contains a compelling case for a low-carbon good life that can also bring us greater happiness and contentment."

Professor Pete Smith, *University of Aberdeen, Science Director Scotland's ClimateXChange*

"A most valuable ready reference, as we struggle through the paths that lie ahead of us in troubled times."

Robin Hanbury-Tenison, *founder of Survival International, author* Taming the Four Horsemen

"The rare book about global change crafted as story, drawing from Indigenous traditions, philosophy, political science and literature in addressing pathways to increase the proportion of global society with a high quality of life, and the likelihood that such quality of life can be sustained."

Professor Erica Fleishman, *Director Oregon Climate Change Research Inst, Oregon State University*

"This book is almost unprecedented, in its scope, scientific foundations, inter-disciplinary reach and contemporary relevance. It is simple and vivid, a personal conversation. It brings the climate challenge back home, and with hope. This book offers practical paths for enriching lives and saving the planet along the way."

Dr Geoff Wells, *Director, Rural Communities Australia*

"A skilful blend of science, metaphysical poetry and a manifesto for good low carbon living. The result is a book about the climate emergency that is full of hope. A joyful, playful and engaging read."

Right Reverend Roger Morris, *Bishop of Colchester*

"A terrific and unique book. It fills gaps and links subjects that have previously been separate. If we are to find a good path through the next few decades for us as individuals, for communities, societies and for biodiversity and climate, this should be the book that world leaders put at the top of their reading lists."

Professor Lloyd Peck, *British Antarctic Survey*

THE LOW-CARBON GOOD LIFE

The Low-Carbon Good Life is about how to reverse and repair four interlocking crises arising from modern material consumption: the climate crisis, growing inequality, biodiversity loss and food-related ill-health.

Across the world today and throughout history, good lives are characterised by healthy food, connections to nature, being active, togetherness, personal growth, a spiritual framework and sustainable consumption. A low-carbon good life offers opportunities to live in ways that will bring greater happiness and contentment. Slower ways of living await. A global target of no more than one tonne of carbon per person would allow the poorest to consume more and everyone to find our models of low-carbon good lives. But dropping old habits is hard, and large-scale impacts will need fresh forms of public engagement and citizen action. Local to national governments need to act; equally, they need pushing by the power and collective action of citizens.

Innovative and engaging and written in a style that combines storytelling with scientific evidence, this book will be of great interest to students and scholars of climate change, sustainability, environmental economics and sustainable consumption, as well as non-specialist readers concerned about the climate crisis.

Jules Pretty is Professor of Environment and Society at the University of Essex. His books include *Sea Sagas of the North*, *The East Country*, *The Edge of Extinction*, *This Luminous Coast* and *Agri-Culture*. He advises governments, has won awards and prizes, chairs the Essex Climate Action Commission, is host of a podcast, and received an OBE in 2006.

Routledge – SCORAI Studies in Sustainable Consumption

Series Editors: Halina Szejnwald Brown
Professor Emerita at Clark University, USA.

Philip J. Vergragt
Emeritus Professor at TU Delft, The Netherlands; Research Professor at Clark University, USA.

Lucie Middlemiss
Associate Professor and Co-Director of the Sustainability Research Institute, Leeds University, UK.

Daniel Fischer
Associate Professor for Consumer Communication and Sustainability, Wageningen Research and University, The Netherlands.

This series aims to advance conceptual and empirical contributions to this new and important field of study. For more information about The Sustainable Consumption Research and Action Initiative (SCORAI) and its activities please visit www.scorai.org.

Power and Politics in Sustainable Consumption Research and Practice
Edited by Cindy Isenhour, Mari Martiskainen and Lucie Middlemiss

Local Consumption and Global Environmental Impacts
Accounting, Trade-offs and Sustainability
Kuishuang Feng, Klaus Hubacek and Yang Yu

Subsistence Agriculture in the US
Reconnecting to Work, Nature and Community
Ashley Colby

Sustainable Lifestyles after Covid-19
Fabián Echegaray, Valerie Brachya, Philip J Vergragt and Lei Zhang

Sustainable Products in the Circular Economy
Impact on Business and Society
Edited by Magdalena Wojnarowska, Marek Ćwiklicki and Carlo Ingrao

Narrating Sustainability through Storytelling
Edited by Daniel Fischer, Sonja Fücker, Hanna Selm and Anna Sundermann

The Low-Carbon Good Life
Jules Pretty

For more information about this series, please visit: www.routledge.com/Routledge-SCORAI-Studies-in-Sustainable-Consumption/book-series/RSSC

THE LOW-CARBON
GOOD LIFE

Jules Pretty

LONDON AND NEW YORK

Designed cover image: Hornstrandir, Iceland

First published 2023
by Routledge
4 Park Square, Milton Park, Abingdon, Oxon OX14 4RN

and by Routledge
605 Third Avenue, New York, NY 10158

Routledge is an imprint of the Taylor & Francis Group, an informa business

British Library Cataloguing-in-Publication Data
A catalogue record for this book is available from the British Library

ISBN: 978-1-032-38817-5 (hbk)
ISBN: 978-1-032-38820-5 (pbk)
ISBN: 978-1-003-34694-4 (ebk)

DOI: 10.4324/9781003346944

Typeset in Bembo
by Apex CoVantage, LLC

For Gill, Freya and Theo

"The best things in life,
Are not things."
[*Tao Te Ching*, ch 33, transl. Gary Snyder]

"**The Low-Carbon Good Life**: there are many good lives, ways of living that bring happiness, contentment and longevity. Across the world today and throughout history, these are characterised by healthy food, connections to nature, being active, togetherness, continuing personal growth, a spiritual and ethical framework, and sustainable consumption. A good life emphasising these components is also low-carbon, and can help to reverse the ecological, economic and social ills of affluence. This in turn requires collective action on an unprecedented scale."

"There is sweet music in that pine tree's whisper,
There by the spring."
[Theocritus of Syracuse, *Idyll I*, 308–260BCE]

CONTENTS

BOOK STRUCTURE AND CONTENTS

This book commences at Transgression and closes at Transformation. The world faces great interlocking crises on climate, inequality, nature loss and abundant food systems that make us ill. Material affluence has betrayed advocates, and now much will have to change: habits, comforts, norms and collective action. A trickster or two will help with transgressions. Eventually, too, will come transformation to low-carbon and more equal ways of living. These will be happier and promote healthy longevity. Nature and land, it seems, might recover and regenerate.

This book is about how to get from here to there.

What tales should be told, and who will be willing to set off on heroic journeys? Who else might join this great game afoot?

Recently, things went wrong, quite badly, even as affluence seemed to grow. How to make them better, or perhaps, how to find something about ways of living we used to know, so very well. The good life is a common phrase and ethic in many cultures. The things that make us happy and contented are not things, after a certain point. The good life is also low-carbon. It tends to add to natural and social capital, not subtract. It centres on togetherness, nature contact, healthy food, being active, personal growth, an ethical or spiritual framework and sustainable consumption.

Chapter 1: Dust and Air: A Dangerous New Economic Worldview

And now today, dust again. And the heat of consumption. An Icelandic author asked, how do you say goodbye to a glacier? This chapter sets the scene by reviewing the emergence of a new economic worldview that has changed so much. Things have got better for some, worse for the planet. Some celebrated selfishness and said wealth would trickle down. Yet, since 1990, the last year of human safety when atmospheric carbon stood at 350 parts per million (ppm), has come an unravelling of ecosystems. But now a slowdown beckons: human populations are heading for stability. It is consumption patterns that drive planetary threat. And we are finding this: acquiring things is not freedom. An era of slower living and stable economies awaits.

Chapter 2: Ten Thousand Good Lives: Sustainable and Kind Ways of Living

There were once ten thousand good lives. Our human cultures knew how to live well and lightly on the land. Now a perilous journey lies ahead. Yet, it always did. It is the monomyth of Joseph Campbell, the experience of living that is common to us all. This chapter tells stories of the Timbisha Shoshone in the desert, the Koyuko and Haida people, of Oðinn from the North, of the Amish in Ohio, and Tuvans on the grassy steppe. Human cultures have centred on togetherness and belonging, the foundation for a good life. Now we know, too, that ever more GDP does not bring more happiness, after the essentials. Data is presented from 150 to 200 countries: over time and across countries, more stuff does not bring more happiness. This is another common human experience. At the same time, though, populations are growing older, and non-communicable diseases, the ills of affluence, have rolled on stage. The domains of the good life are discussed. Most are low carbon. And here we have a hint of a path ahead. "Cling if you want, to things," wrote Meister Eckhart, "but where then will your heart be?"

Chapters 3 to 6

These four chapters now address the four great crises affecting health, happiness and the fabric of the planet: the climate, inequality, nature and food crises.

Chapter 3: The Climate Crisis: The Safety of One Tonne Each

This chapter sets out the details of the crisis of climate change. It has been coming a long time, and then suddenly accelerated. Since the pre-industrial era, world temperature has warmed by +1.2°C (into the early 2020s), over land by 1.5°C, and over oceans by +0.75°C. Each season, new records are set for high temperature, deeper drought, rainfall and flood, wildfire, melting permafrost, shrinking glaciers. This is not weather; these are signatures of a greater malice. Our world of

7.8 billion people produces some 53 Gt (billion tonnes) of CO_2eq (carbon dioxide equivalent). A safe place is about 1 tonne per person, around 10 Gt. Data on 222 countries is presented: GDP against carbon emissions. A large number of countries and people emit on average less than 1 tonne per person: they need to increase consumption to escape poverty. The majority need to reduce inequality of carbon emissions, which also plays out within countries. The top 1% wealthiest in Europe and North America emit 55 tonnes of CO_2eq per person. The world average is 7 tonnes. And a new twist. This transition to 1 tonne may not be hard. "An era is ending," has written Danny Dorling. This chapter concludes by setting out a schedule of 30 personal behaviours and choices that show how to reduce personal emissions. Start somewhere, and the race to net zero is on.

Chapter 4: The Inequality Crisis: Togetherness Is Better Than Selfishness

This chapter centres on the inequality crises and the importance of togetherness to all humans and cultures. "There is only one true form of wealth," wrote Antoine de St Exupery, "that of human contact." When dusk falls, that is when we gather, to share food and stories. Cooperation stuck people together. And yet, the modern neoliberal project seems to have focused on celebrating selfishness. The advocate Ayn Rand called altruism, "The poison in the blood of western civilisation." This chapter reviews two of the ills of selfishness: rising inequality and loneliness. The evidence is clear. Equal societies with strong connections extend life expectancy for everyone. They keep all people happier. But because social capital has fallen, often dismantled deliberately, a grand project lies ahead. We will want to celebrate community activities and sports, integrated schooling, volunteering, taking collective action in landscapes, the formation of millions of new social groups, especially in emerging economies. Participatory and public engagement help create this groundswell of actions.

Chapter 5: The Nature Crisis: Regaining Earthsong and Attentiveness

This chapter tunes to Earthsong and the nature crises. What shall remain, if we carry on this way? Nature is our home and more than that. "Attention is the beginning of devotion," wrote poet Mary Oliver. Look closely, and you have to sing. And be humble, for nature knows way more than we do. Here nature is discussed as sacred place, alongside the values and ethics of care, the grim norm of exploitation. And yet, a pilgrimage walk, a swim in the river, tending the garden, these are low carbon activities. The term nature-based solution is becoming a common currency. These "expand the solution space," said the International Panel on Climate Change in 2022. They also offer opportunities for attentiveness and immersion, and these, as we shall see in Chapter 7, unlock the quiet and contentment all people seek in one way or another. Nature is an essential infrastructure; environments structure behaviours; and we receive health services from nature. The chapter closes with nine trickster questions: what are your best moments in life? If you listen carefully and are attentive for long enough, you will hear.

Chapter 6: The Food and Agriculture Crisis: It's Nourishment, Not Calories

This chapter lays out a long table of the food and agriculture crises. How did the world turn from nourishment through apparent plenty to the new silent costs imposed on public health? It was, it seemed, a wondrous revolution, more food from the land, more food per person worldwide. Then in supermarkets, thousands of new foodstuffs each year, all remarkable invention and packaging of desire. Yet, we all know, the right amount is best, as a Swedish proverb puts it. Food that brought people together became a world of sorrow. It imposed huge and hidden costs on nature and the environment, and it provoked the obesity and type 2 diabetes crises. The world agriculture and food systems contribute 25% of global greenhouse gases. We could do good by eating well, but this is a big burden of responsibility. Help could come with policies and practices for regenerative agriculture, slow food, adoption of new diets, and celebration of local and distinctive foods. No country worldwide has been able to reduce obesity rates once they rose. Two countries managed to keep levels very low: Japan and South Korea. Put a rainbow on your plate, say people in Okinawa. Eating less meat will favourably change food systems worldwide. "Every meal should be one of the best meals of your life," writes Claire Greenwood. The chapter closes with a famed story of giving: three grapefruits.

Chapters 7 to 9

These next three chapters now take us from individual experience of the world to the collective action needed to create sustainable futures of enoughness. Immersion and flow help create new habits, enoughness leads to slowth rather than growth, and public engagement can create the new expressions of power necessary for changing social and economic structures.

Chapter 7: The Best Things in Life: How Immersion and Flow Make the World Feel Better

This chapter now comes inside to explore the mechanisms and structures of brain and mind that are common to every human. This legacy made us in a certain way, ready to respond to natural and social environments around us. We find: the best things in life are not things. We all start life in the same place, with a beginner's mind. And then it all starts to develop. The chapter commences with an analysis of the goods and bads of habits. Habits mark us as human; they automate behaviours leaving space and time for cognition. Yet, they can be hard to change. It takes about 50 hours of hard effort to break or make a habit. Flow activities help and are central to the contemplative and immersive states that are common across all cultures. Flow is the state when we feel and perform our best, when the self and ego disappear. Problems vanish, and the inner chatter is quietened. Flow activities tend, too, to be low carbon. They create the state of transient hypofrontality: the mental chatter that is a significant source of ill-health is switched off, for a while. The chapter discusses the structures of blue and red brain, and how attentiveness and flow can help create calm. Evolution gave us no off-switch for the mind. We have to choose body behaviours to press the switch. Be attentive, become immersed, be focused and become in flow. The green mind created in this way is pro-social and more connected to nature.

Chapter 8: Enoughness: Creative Slowth Is Better Than Infinite Growth

Well. What now? Four crises, the keys to habit change and quietened inner chatter, the use of flow activities to reduce carbon. This chapter centres on the concept of enoughness, the sufficiency that creates slowth. The corporate model, for many, is always for more: more growth, more sales, more desire, more growth again. Yet, we always have lived within enoughness, the space between not-enough and too-much. Overshoot means going too far, wrote Donella Meadows, one of the authors of *The Limits to Growth*. The bad life has sharp teeth. "How big should it be?" asked Hermann Daly, a big question referring to the world economy. This economy is anyway a wholly-owned subsidiary of the environment. In stable-state economies, interesting things happen, as they did and do in every indigenous culture worldwide. The Edo period in Japan was 260 years of slowth, and gave rise to an age of pilgrimage, poetry, painting and printing, to new temples and festivals of belonging, to great libraries. But worldwide, growthism brought rising inequality and environmental harm. Tim Jackson commented, "There is no growth on a dead planet." There are many ways to engage in personal and collective forms of growth: these are the infinity games. It is through contemplative crafts and art, the flow of activity and skill, that we find a good life in which work is about living well and long. These slow the climate crisis and also create divergence and diversity.

Chapter 9: Public Engagement and New Power: The Race to Net Zero

Now for the new power of public engagement. We cannot solve these crises alone. Yes, please do your best, goes the farewell to pilgrims in Japan. But the dilemma is this. We know the problems, we have identified some solutions. Now how to get more than 200 countries serious about the race to net zero? How, too, to create the social movements that move themselves? This chapter sets out the principles and mechanisms of engagement, the changes that come from new rather than old power. Public engagement means giving up some authority and control, so this is going to be hard for some. A new typology of public engagement is discussed: how to go from passive and consultative modes to co-creation and transformation? Yet, there is good news. Many novel institutional models have been created, rather quietly and forming new structures of social capital. Devolved health and care institutions, climate action groups, farmer field schools, participatory civic budgeting, crowdfunding, social prescribing, web-sharing platforms, microfinance for women's groups: all are models of new power than can help build regenerative cultures. Five guidelines are set out for slowth and regeneration.

Chapter 10: Transformation: Achieving the Low-Carbon Good Lives

The book concludes on transformation. The end is not near, if new ways to organise are found, new stories told, new aspirations for the good life shared and promoted. Low carbon behaviours produce the good life, not just giving up the high material consumption that caused these crises. So come now, let's dance. Peace is not freedom from problems; they will always come. It is a wide space and quiet mind. So put a stone in your pocket, a reminder to be more present, and slow down and reconnect. And tell stories, for this cannot be done alone.

Coda: Let's Dance, Together

And so. Let's dance. Together. There has never been such a great collective project of deliberate change. It could fail. But then the current economic system is already failing both nature and people. The many ways of living a low-carbon good life could deliver more happiness, greater health, longer lives, a whole and recovered natural world. Change is a snare, permanence is a snare, wrote Joseph Campbell. A gate, a threshold, lies ahead. Beyond: immersion in something new.

Supporting Information, Terminology and Sources

You will find endnotes for each chapter at the end of the book, and a combined list of all references. The endnotes are linked to the appropriate segment of the chapter through bold text.

There is a separate list of sources for the data used in the figures and tables. Data tends to run to 2019 inclusive, as so much shifted during 2020 and 2021 owing to COVID and lockdown responses.

I have allowed weights and measures to be mixed and flexible. Mostly distances are reported in kilometres, but also miles and leagues; on a smaller scale, variously feet and metre; areas by hectare, acre and square kilometre; weight in kilogramme, tonne and pound. A billion is a thousand million (a 1 with nine zeros). One Gt (gigatonnes) is a billion tonnes.

The terms for Common Era (CE) and Before Common Era (BCE) are used throughout the book as alternatives to the Dionysian BC (Before Christ) and AD (Anno Domini). The two notation systems are numerically equivalent.

The illustrations at the start of chapters are all monochrome photographs with white backgrounds digitally created. These images are intended to echo the simple designs of ink calligraphy on white paper. Locations for the images are listed at the end of the book. All original photographs were taken and treated by the author.

PREFACE

Transgression

Let's start this tale like this.

The parting will be sweet, eventually.

Some say now is that time to whisper in the wind. All will have to change. There will have to be a great goodbye, an adieu and so long. Deep down there will be worries. But this farewell to many ways of living will be a threshold, a ferry crossing to somewhere better.

Yet, there is this too. When you come home from the daring journey, the demons slain and the elixir cradled in your palm, what do you find: the old and ordinary world is indifferent. It does not know it needs your potion. Now the work begins. You might find yourself saying, that was the old way, now we must do this thing instead.

"To make an end is to make a beginning," wrote T. S. Eliot in *The Four Quartets*. "The end is where we start from."

And what a place to start, at the Earth's great interlocking crises. The crushing of the climate, the deepening of inequality and loss of contentment, the loss of biodiversity and species, the illness from food, all driven by a relentless pursuit of material consumption. In the modern world of affluence, many things have been getting better, but some suddenly became much worse. Once upon a time, all kinds of people knew what a good life could be, and yet somehow it slipped from grasp. We might well ask, how could greener, low-carbon and happier options for people and economies emerge?

When you enter the forest at its darkest point, wrote Joseph Campbell, there is no path. If you find one, it is probably someone else's. The idea is to make your own way. It's over there, the start line. We just need to get in the game, to gather up a staff and enough food and possessions. And start walking. For this has to be done together.

So here we are. This is one way a story starts.

The chief's wooden hut was in a palm grove, set back in the shadows. Inland it had been humid, yet here a breeze blew in from the South Seas. The lace curtains stirred. Nearby a *bure*-longhouse was being repaired, the steep thatched-roof damping sounds of saw and hammer. The visitors passed across yaqona as a gift and sat on pandanus matting facing the Fiji village elders. The guardian of the bowl sat to one side. The yaqona was pounded and mixed with water, sieved and served in half-coconut shells. There were echoes of the old patterns, stories and ceremonial clapping. Permission was sought to bring in specialists from across the Pacific to listen to local people, and the chief smiled and opened his arms. This would also take the time it should, with words drawn up from memory and stillness.

The blue sea rolled against the shore, chickens and children ran outside in the dappled sunlight. Time expanded around each story. When Arthur Grimble wrote *A Pattern of Islands* from Kiribati, there were islanders who could murmur into sea winds, to transmit news to distant atolls. When James Cook pulled in, his *Endeavour* was shorter than local Fijian canoes that could carry 250 people. His ship's top speed was 6 knots, the canoes cut the waves at 22. Sometime later, I took to the ferry around the southern isles. As the boat passed from lagoon to ocean, the crew fetched guitar and ukulele, and leaned against the superstructure singing to the flying fish. Passengers laughed and danced, the sun sparkled on the sea.

So let us hear why it was the world became weary.

Today, unprecedented affluence in many parts of the realm should have meant more happiness. Instead, it fetched new problems, helped them spread. Its side-effects tolled loud for ill-health of planet and people. The ancients often said, when gold dust gets in your eyes, it causes blindness. This was why dragons of old came to help, guarding hoards in cave and barrow, and why it seems serpents at the end brought flood to wash away the noise.

Long ago, the night air had been warm, and dawn light washed the northern hills. A wren scolded from the bushes. It was a day for calling. The deep mine was hot as Hades. The grinning men and boys were shirtless, what else could they do? It was only 15 years after the Aberfan disaster, when colliery spoil had swamped a south Wales school. The science group had been testing grasses that could grow on heavy metals, a way to stabilise the heaps of slag. A time would come when those polluted lands would turn green, and then all of a sudden the jobs be lost. But first, this crossing into darkness and hidden heat. All shift, the miners hewed inside the coal seam half a mile wide and two feet high. Hydraulic rams held up the rock.

All went in one end and crawled. There was no room to turn. A rotating cutter ground the raw coal, the miners shoveled it back, blue-black and glistening in the torchlight. When a yard had been cut, the rams were moved forward, one by one, and a mile of rock above fractured, closing down with groan and crack and thump. The men laughed and dug again the coal. The visitors crept and slithered, and from just this trip it took a week to shift the ingrained dust from eyes and ears. It lined the lungs and blackened skin. This very coal that also scarred the sky. Up on top, the western sky was briefly bright and watery.

Yet, it was a grievous error, letting the old extractive economy die without replacement. One day, the mine owners just locked the gates, and no one sought to create something new for the community. Stopping bads looks hard, but is the easy bit. Creating goods is an altogether harder grail.

Call for a route map. All good stories contain instructions for living. The characters show what is possible, if they open up to such a journey. "What I know is that a good life is one hero journey after another," said Joseph Campbell in *The Hero with a Thousand Faces*. This will be the greatest journey ever undertaken, the one that lies before all of us. The economic system adopted and promoted by the affluent countries in the past couple of generations has breached Earth boundaries. Infinite material growth seemed a clever idea, but on a finite planet was it ever going to be possible? Much will have to change. More rain is due, more record heat too.

It might now be asked: could low-carbon good lives reverse the climate, biodiversity and inequality crises, and at the same time make us happier?

On a golden afternoon, a girl follows a rabbit down a hole. She is transformed by potion and cake, meets a beaming cat and talking mouse, has tea with a mad hatter and a hare; a queen threatens to take off her head, all are playing by the strangest of rules. After more adventures, she returns to the old world above ground.

The Monkey King goes on a Journey to the West, from China to India, after being asked to guard the Peaches of Immortality. He finds he longs to eat them, to know their secret taste, so he feasted, and was pinned under a mountain for 500 years. He then went west on a pilgrimage and suffered 80 ordeals; in the end, he found peace, his mind was clear at last.

Another girl works all hours in a hot and dirty kitchen, chained there by the wicked stepmother. She knows she cannot go to the ball and cries. A fairy godmother appears and with a touch of magic gives her a beautiful dress, a coach and horses, so she dances with the prince and forgets her old life of trouble. Suddenly she recalls she must run away, and at midnight leaves behind a slipper.

A girl walks alone in the dark woods, and the path leads her to a handsome cottage in a glade. The door is open, and there are three chairs, three bowls of porridge, and three beds. She eats until full, falls asleep, and when the bears return from their morning walk, they threaten . . . to eat her up.

Well, it could easily have happened that way. You go on the journey, and it goes badly wrong. A voice inside says you shouldn't even have tried. Stay at home, don't stray from those safe habits. On the other hand, if the story comes to a close and suggests you will now live happily ever after, that's when the real work begins.

The trickster helps when it comes to these essential tilts and transgressions. They long have revealed the world's workings, they tell you how to think and act. They challenge ingrained habits, they indicate the voyage will be worth it. Everyone goes on journeys through life. Many are hard, some are scary. You learn to walk, go to school, leave school; you learn to drive, leave home, travel alone across countries where no one speaks your language; you try to diet or give up smoking, you retire. And now fast comes the advance of death. The aim of the trickster story is to improve lives, wrote William Hynes and William Doty of the Trickster Myth Group. And everywhere you look: "A shear richness of trickster phenomena." They are ambiguous, deceiver, shape-shifter, messenger, imitator of gods, situation inverter, loud and always ingenious. Solutions emerge when tricksters walk on stage.

Koans are types of trickster stories. They are short tales that unmake and subvert. You don't realise you live in a small room, and then abruptly you find you are in a wide meadow of flowers thrumming with bees. A change of heart and mind has occurred, if you let it work, and work some more.

A thousand-year-old koan begins: "Does a dog have Buddha nature or not? Zhaozhou said, "No." [Actually, he said "*mu*."]

In any situation, you could use *no* to open a door and set you on a journey. There could be *no* sun and you'd not worry, *no* cool breeze, *no* fear. You could say *no* to habits, *no* to the noise of inner chatter. This koan also says, *no* might also be a *yes*, depending on how we look at it.

There were 700 koans composed by Mumon Ekai, Setchō, Engo and Xutang Zhiyi a thousand years ago, contained in *The Gateless Gate*, *The Blue Cliff Records*, and *The Record of the Empty Hall*. These are not riddles, you are not asked to find a correct answer: they are myths intended to be keys to unlock habits of thought and behaviour, to open up an ecosystem that you cannot currently imagine.

Each is a call to adventure.

Karen Armstrong has written that myths force us to go beyond our experience to a place never seen. Myth is about the unknown, saying to us, go on, look into the heart of silence. You might find a guide to behaviour. You might make a path.

I will be the Sun-god, declared Coyote to the Salish people of modern Idaho, and they agreed to let him try. He took the Sun-lodge across the sky; he saw the secret meetings of people, what they were hiding from each other, and so he shouted down to them, laughing at their embarrassment. The people were glad when that day was over, and took Coyote away from the Sun-lodge. But Coyote always has the last word, and there is something instructive if we can discern it.

The Yoruba trickster Èṣù-Eshu came walking between fields, wearing a cloth hat of the four world-colours, one side red, the other white, green in front and black behind. He saw two farmers who were best friends and wanted to test them. They came to argue on the way home. One had seen an old fellow with a white hat, the other from his field was convinced it was red. They disagreed so much that they drew knives and fought, and the headman was at a loss to know where justice lay. From the crowd, Eshu spoke, "Both of you are right, for there is not one correct view of the world." The old vows of friendship should have been strong enough, and now could recover.

Story is currency in a stable economy. When the world slows and stops material growth, and distributes the benefits more evenly, when population stabilises, when less is spent on curing ills and more on enjoying health, then interesting things will happen.

A world in which each person produces no more than One Tonne of carbon would be a safe and more equal place for humanity. Stability and slowth will create new opportunities for personal growth.

Yes, there will be hesitations. The track curves up steeply, and clouds are heavy.

So pause. It has always been this way: a common set of behavioural choices make up a good life that is happy and long: attention to healthy food, contact with nature, physical activity, social connections, personal growth through lifelong deployment of skills and crafts, and a coherent spiritual or belief framework. When these create a state of flow, then happiness can follow.

Now, look around. There it is.

A path. And another. Many options for low-carbon regenerative lives. All this needs is the grandest ever project of collective and worldly action.

~★~

THANKS AND ACKNOWLEDGEMENTS

Many people helped on the journey out and back for this book. Some accompanied voyages on the land; others were key collaborators on research that contributed to the book; others were fellow Commissioners on the Essex Climate Action Commission and the Essex Renewal Project; others were members of staff and collaborators at WWF-UK. I am very grateful to Hemant Kumar Badola, Caroline Barratt, Richard Bawden, Ian Budge, Chris Collins, Steve Cornelius, Danny Dorling, Barbara Durham, Níels Einarsson, Pauline Esteves, Erica Fleischman, Jay Griffiths, Robin Hanbury-Tenison, Jason Hickel, David Kline, Emily Kline, Martin Laird, Caroline Lucas, Simon Lyster, Andri Snær Magnason, Jacquie McGlade, Bill McKibben, Roger Morris, Tero Mustonen, Nagasiddhi, Astrid Ogilvie, Umesh Patel, Brad Rendle, Brenda Dardar Robichaux, Johan Rockström, Archie Ruggles-Brise, Dorothy Schwarz, Jo Smith, Pete Smith, Srivati, Tanya Steele, Roma Tearne, Graham Underwood, Geoff Wells, Chris West, Meik Wiking and Rob Wise. A special mention for the late John Devavaram, a good friend for decades and leader of rural transformation in southern India. Lloyd Peck provided comprehensive and hugely helpful comments on a late draft. I am very grateful to four anonymous reviewers for the wise comments and insights on an earlier manuscript. Thank you too to Annabelle Harris and Kitty Liu.

I am appreciative of the following for kindly checking the translations of the term "good life" into a range of languages: Ann Waters Bayer, Hemant Badola Kumar, Nick Dickson, Yuelai Lu, Femke Tonneijek, Puyun Yang.

I am grateful to the following for wise advice and suggestions about public engagement principles and practice as deployed in the university, higher education and research sectors: many thanks to Richard Bawden, Peter Beresford, Emily Burns, Edd Codling, Laura Cream, Sheena Cruickshank, Uwe Derksen, Heather Doran, Alex Flores Quiroz, Tracey Johns, Liz Kuti, Paul Manners, Niamh Nic Daéid, Guy Poppy, Vanessa Potter, John Preston, Steve Scott, Julian Skryme, Graham Underwood, Beverley Wilkinson, Rich Yates and Vippy Yee.

I would like to thank the Secretariat and fellow Commissioners for the Essex Climate Action Commission. This was set up by Essex County Council chief executive, Gavin Jones, supported by Council leader, Kevin Bentley, and all other elected councillors (of all parties). The secretariat is led by Sam Kennedy. I became Chair of the Commission in 2021.

There have been many fabulous contributors on public engagement to the *Louder Than Words* podcast, launched in 2021 and produced by Martha Roberts, Ali Walker and Luke Fitch. The pod can be found on the normal pod platforms.

I especially thank those respondents to the good life survey who provided further wise observations and comments on what the good life meant to them: Javed Ahmed, Glenn Albrecht, Simon Amstutz, Gerald Assouline, Hemant Kumar Badola, Caroline Barratt, Jo Barton, Ann Waters Bayer, Richard Bawden, Mike Bell, Dani Broitman, Tom Brown, Eric Brymer, Fabian Bush, David Butcher, Ricky Carline, Naomi Carmon, Nicholas Charrington, Angus Clark, Catherine Clarke, Chris Collins, Nicholas Colloff, Nigel Cooper, Marta Blanco Cordero, Nick Dickson, Martha Dixon, Tom Dobbs, Frands Dolberg, Helen Doran, Caroline Drummond, Nigel Dudley, Níels Einarsson, Paul Ellis, Edward Ellison, David Farrow, Elske van der Fliert, Robert Erith, Erica Fleischman, Cornelia Butler Flora, Bernhard Freyer, Theodor Friedrich, Tom Fyans, Ann Gallagher, Kevin Gallagher, Dennis Garrity, Chris Gibson, Robert Golden, Dave Goulson, Ros Green, Benny Haerlin, John Halsey, Christina von Haaren, John Haward, Maddy Haughton-Boakes, Keeley Hazelhurst, Fiona Hearn, Robin Hanbury-Tenison, Jenny Harpur, Alice Hughes, Haruka Imai, Andrew Impey, Caroline Jessel, Brandon Keim, Jan Willem Ketelaar, Matt King, Martin Laird, Howard Lee, Jonathan Lichtenstein, Ulrich Loening, Yuelai Lu, Simon Lyster, Daisy Malt, Georgina McAllister, Tom McMillan, Amelie Michalke, Jan Middendorf, Jo Millar, Ron Milo, Kotra Mohan, Susie Morgan, Stephanie Morren, Nagasiddhi, Alix Nadelman, Tapan Nath, Astrid Ogilvie, Tim O'Riordan, Umesh Patel, Lloyd Peck, Abbot Xavier Perrin, Avi Perevolotsky, Roberto Peiretti, Linda Phelan, Ruth Philo, Hattie Phillips, Jo Phillips, Nigel Poole, Vara Prasad, Qing Li, Dave Reay, Jake Ricks, Nick Robins, Johan Rockström, Gwen Roland, Sasha Roseneil, Asaf Shwartz, Hila Segre, Joe Sempik, Jafar Shah, Sawako Shigeto, Andrew Simms, Ranjay Singh, Jan Paul Smit, Pete and Jo Smith, Srivati, Julia Steinberger, Anna Stenning, Kirsten Stevens-Wood, Emma Stock, Richard Stock, Roma Tearne, Vincent Thompson, Janet Thorogood, Femke Tonneijek, Gideon Toperoff, Mardie Townsend, Katherine Trebeck, Matthis Wackernagel, Sue Waite, Paul Wake, Songliang Wang, Olly Watts, Geoff Wells,

Janie Wilson, Yunita Winarto, Mike Winter, Hannah Wittman, Puyun Yang, Rich Yates and Yeulai Lu.

A subset of contributors were from low-impact communities and ecovillages in the UK, and I thank them for their contributions to both the good life and how they were making transitions to low-carbon living. Many thanks to Juan Sebastian Giraldo Armilla, Rosemary Betterton, Fran Blockley, Alan Brown, Emma Close, Roger Douda, Philippa Draper, Simon Fairlie, Maren Freeland, Leone Graham, Sarah Kleppsattel, Gill Lowing, Diana Martin, Graham Meltzer, Liz Neat, Annabeth Orton, Andrew Pring, Sophie Rackham, Jane Riach, Richard Sanders, Torbørn Schei, Marjorie Sikkal and Chris Thompson.

Many other astute observations on related or earlier work have been made on social media platforms and at readings or talks: to you all, again, much thanks.

THE ILLUSTRATIONS

The illustrations at the start of chapters are all monochrome photographs with backgrounds digitally removed. These images are intended to echo the simple designs of ink calligraphy on white paper. Locations are listed below. All photographs were taken by the author.

Summary: Fulmar, Iceland
Enso: Drawn by the author, ink on paper
Contents: Green Man, roof rafters, Nayland Church, Suffolk
Book structure: Embroidered tapestry, Lofotr Viking museum at Borg,
    Lofoten Islands, Norway
Preface: Dragon, Aysgarth church, Yorkshire
Chapter 1: Car on seashore, Dengie peninsula, Essex
Chapter 2: Mayan heads, Guatemala
Chapter 3: Pollution, Kings Lynn, Norfolk
Chapter 4: Hand prints in ink
Chapter 5: Bristlecone pine, Timbisha/Death Valley, California
Chapter 6: Metal tea cups, Japan
Chapter 7: Esrum monastery, Sjæland isle, Denmark
Chapter 8: Victorian bottles, from badger diggings, Northamptonshire
Chapter 9: Tinners rabbits (hares), stained glass window, Long
    Melford church, Suffolk
Chapter 10: Pink-footed geese, Essex marshes.
Coda: Enso, drawn by author
Thanks and acknowledgements: Humpback whale, Húsavík, Iceland

1

DUST AND AIR

A Dangerous New Economic Worldview

Along the trail's edge, beside a sparkling river,
In the willow shade, I lingered to take a nap,
Lingered, and I'm still here.

[Saigyō Hōshi, Kyoto, 1116–1190]

Abstract: And now today, dust again. And the heat of consumption. An Icelandic author asked, how do you say goodbye to a glacier? This chapter sets the scene by reviewing the emergence of a new economic worldview that has changed so much. Things have got better for some, worse for the planet. Some celebrated selfishness and said wealth would trickle down. Yet, since 1990, the last year of human safety when atmospheric carbon stood at 350 parts per million (ppm), has come an unravelling of ecosystems. But now a slowdown beckons: human populations are heading for stability. It is consumption patterns that drive planetary threat. And we are finding this: acquiring things is not freedom. An era of slower living and stable economies awaits.

DOI: 10.4324/9781003346944-1

The Dust of Dark Times

Dust. Noise.

Hubbub everywhere, inside and out.

It happened like this: all threads are connected. The way we live, the nature of this blue-green marble of a planet. The way we eat and move, the rising heat in air. The things we buy, the fossils and forests that have been burned. Our happiness, or often not.

There once was a god called Enlil, who sent drought and famine, and finally a deluge. The people had hauled clay in baskets, built fine temples, hitched plough and diverted water to the desert fields. Yet, they seemed to want more, and so new city-states grew and the minds of gods and people were filled with clutter. He who saw the deep, Gilgamesh was the prince of the sheepfold Uruk, all its date groves and cool gardens, all the herds of antelope beyond the fields of spelt. In those days, scorpion men guarded the sun, and Gilgamesh wandered the wilds with his friend Enkidu in search of wisdom.

For this is the oldest written story in the world. Its lesson was simple.

In many a city, in civilised settlements, the teeming markets drowned the peace. By specialisation, as merchant and shepherd, as cook and carpenter, potter and weaver, some people could decorate their homes with shiny goods. And Gilgamesh at first cut down the ancient cedars guarded by Humbaba, and killed the very last fiery bull with wings. In the presence of nature, he found he should have been glad of life, but it was too late. A piece of sky fell down. In the deluge, mayflies floated on the water and turtles flitted fast. Gilgamesh could beat neither sleep nor death and realized all the people's labours had been in vain. He travelled to the underworld and found the flood had washed away the dust.

This epic was written on tablets of clay 5000 years ago. It contains other clues. For years before, the traders of those first cities may have travelled north and west, carrying seeds of corn and leading tamed cow and pig, and found at the shores of western Europe there was a sea that had once been land. And on those steppes once called Doggerland, bands of hunter-gatherers had roamed far. There had been mammoth and giant elk, broad-browed auroch, on the marsh a million whistling waders. There had been no better place, said the coastal people to the traders. Yet, as the air had warmed and waters risen, their homes and hills had been drowned. The dry plains had been taken by the sea, and they had to dig up ancestral bones and take them to the higher land.

And now, today, dust again.

Another era when nature's assets will strand, by storm and flood and fire.

Looking back, a century on from now, the children of our grandchildren will hear the tales and may be puzzled. What were you thinking back then? How did you let it happen? Andri Snær Magnason has recently asked, how do you say good-bye to a glacier, when the first of Iceland's 270 had melted? The Okjökull was to the north-west of Vatnajökull, the second largest ice cap in Europe. Now at Ok there is only rock and ash. There is a memorial plaque, on which he composed a letter to the future. Was this farewell or a funeral for such eternal nature, this white giant of frost?

Andri Snær wrote, "This monument is to acknowledge that we know what is happening and what needs to be done. Only you know if we did it."

"What kind of times are they," wrote Berthold Brecht, "When a conversation about trees is almost a crime, because it implies silence about so many horrors." We people have again entered a dark time, in the heat and horror, the fire and melted ice. Brecht wrote in 1939:

> In the dark times,
> Will there also be singing?
> Yes, there will also be singing.
> About the dark times.

Carpets take time to weave. Yet, a heavy modern tread has shaped new ills of affluence: the climate crisis, yawning gaps of inequality, the sixth episode of species extinction, ill-health within richer countries, the persistence of poverty and hunger, the anxious hopes of economic and conflict migrants, the loss of half the world's languages. And yet, at about the same time, world-spanning public and human assets have been created: 15% of the world's land area has become protected, 90% of girls and boys worldwide completed primary schooling in 2020, 80% of all children are immunized from common infectious diseases, drinking water is available and safe for hundreds of millions, whales are increasing in number.

It Seems Things Can Both Be Bad and Getting Better.

From their safe vantage point a century on from now, those great-grandchildren might concur: it all happened so fast. Certainly, seeds were sown early, the soil prepared, coalitions of interest created, all such provisions do take time. Yet, the worst occurred across one fateful generation. It was fast, and what a contrast with the slowth that followed. Someone had trashed the planet, and afterwards, well, many came to learn all the components of the good life had been there all along.

Our descendants may mutter in the future, they may point and say we had been hoodwinked by an orthodoxy. It had swept the planet. There were shadows on the shining sea. People wept and mourned, but at long last came applause. They will know the climate had become safe and that human numbers had long ago stabilised, and mostly people had found ways to live that improved the planet and spread some happiness. They will understand better how mind and body were linked to land and planet, they will know the value of community and social connectedness.

Yes, well, some said, we've learned our lesson.

> Fools put all their faith in things,
> And so become angry and bitter.
>
> *[Yoshida Kenkō, Kyoto, 1283–1350]*

The Emergence of a New Economic Worldview

Let's go back about a generation. It is 1990. Atmospheric carbon had been rising since the advent of the industrial revolution but was still at levels safe for humanity. The concentration had not yet slid past 350 parts per million (ppm). The world had warmed a little, but not by much. It was only 45 years since the end of a world war, yet the peace brought a stand-off by superpowers with different dogmas. Then the Berlin Wall fell.

Farmers worldwide had already increased food production sharply, post-war shortages had receded and the world's 5.3 billion people had more food on average than their mothers and fathers. Obesity in affluent countries was rare, as was type 2 diabetes. A hundred countries had become newly independent of colonial control. In some parts of the world, there still were people who had lived stably on their terrains for 2000 generations, but they were increasingly seen as relics.

The children born in 1990 are in their thirties now. One generation has passed. It was not their fault.

What was this wondrous worldview, bringing good and grim? The one that later ended hastily as wind tore in and storms crushed the coast.

Everything was connected. This world ideology had its roots in that European industrial revolution. Technological leaps brought infrastructure and factory, and empires with global reach accessed natural resources at low cost, built systems of cheap labour. Companies kept down costs by not paying to prevent air and water pollution. In Britain, 3500 Enclosure Acts were passed by a parliament in the 1700s and 1800s, when only men could vote, and lands held in common were converted to private ownership. The 1834 Poor Law made criminals of the feckless, broke up their families, put adults and children to work for only food. Between 1830 and 1914, a tenth of British people were incarcerated in a workhouse. The economy grew fast.

The era of consumption spread through the 20th century, there was bubble, over-reaction and collapse. Yet, in the longer run, all was fine: this was a pinnacle of progress. All cultures had been steps on a long ladder, and there were many rungs. Yet, in a spirit of benevolence, those other peoples who were assumed to be primitive should be granted, it was felt, access to the modern ways. At last, there was convergence. Some economists knew what good looked like. Some would later even say it was the end of history.

This era of individualism was called a new liberalism. It was not cloying, it would not be constrained by leaden government or border tariff. It would work for all. And sure enough, this form of capitalism marked its own homework by using a single progress measure. Gross domestic product (GDP) put an economic value on all the goods and services produced in a country. Looking back, advocates could see this could grow sharply; looking forward, they predicted more growth. This would go on forever, provided people bought more goods and services, and then ensured those in other countries would do the same.

There was no sign of any dust, at this point. The air, too, was clear and the views to every summit were fine.

Kate Raworth describes in *Doughnut Economics* how in those early days economists built models using the design principles of engineers. Pull this lever, and over there the system grows. But there were fatal flaws. Externalities of economic subsystems were not measured as bads, and so environmental and social side-effects were ignored. Before long, the results were supersonic, fast growth could happen year on year. When you have compound growth, a few percent more on a few percent previously, then the graph of GDP rises very fast. The arithmetic seemed a physical law: 2% growth means a doubling in size in 34 years. And 2% per year did not seem like much.

Yet, as Kate Raworth also whispered, now give a pencil to an economist and prompt: draw the curve after this period of growth. What happens next?

Does growth go on forever, does it slow, does it fall back to zero? These classes of questions had been avoided. And how in hell, or in heaven's name too, could anyone envisage infinite economic growth on a finite planet?

Yet, the underlying ideology was convincing. Individualism was good, and when freed from the constraints of government, each person could be so creative and effective that they soon would become wealthy. Growth was possible everywhere. Environmental and social externalities, the pollution and inequality, well, these were at worst just temporary phases. Once economies grew past a certain point, then of course the wealthy would invest back to solve those passing problems. Free trade was good within countries and across borders. Some places are suited to certain economic advantage, so could trade freely with other places with their different specialisms. We'll grow pineapple and trade for your mobile phones. Everyone will be a winner.

And so the global economic machine spins smoothly, and wealth itself comes to trickle down. And if you do not join the game, if you cannot get on your bike to find a job, well the fault is clearly yours.

An advocate of this emerging neoliberalism was a 1920s Russian émigré to the USA. Ayn Rand was the pen-name of Alice O'Connor, and she became a novelist, screenwriter and activist. She wrote two works of fiction, *The Fountainhead* and *Atlas Shrugged*, around the time that Rachel Carson was gathering the post-war evidence on the unravelling of the natural world. Rand had been purged from university for being from a bourgeois family, and later was not permitted to bring her parents from St Petersburg. She came to take on a mix of stances: she was pro-abortion rights and opposed the Vietnam War, she called people in civilised countries heroes for fighting savages, she opposed all gay rights. Her Objectivism emphasised individualism and opposed all state action.

Rand's books are widely read today. Her institute purchased 4 million to send to every school in the USA (30 copies per school). In the year 2020, members of the White House and Britain's cabinet said they were readers, as did a billionaire leader or two from the silicon corporates. Some have said, proudly, that they have them by their bedsides. You can see, perhaps, these are not just any old novels. They had become tracts, words of orthodoxy that pointed to a promised land. In 1974, Alan Greenspan was appointed by President Gerald Ford as chair of the US Council of

Economic Advisers. He was sworn in with a copy of an Ayn Rand book in his right hand.

To be kind, we should say Ayn Rand was a single-minded philosopher who had suffered much. A web of acolytes was attracted by a worldview that said all government was bad, all selfishness good, solidarity a trap and taxes always bad. She wrote that those at the top were creators of wealth, often threatened by the mob. She said, why must the rich be made to suffer? Why were the producers of wealth being made victims of confiscatory taxes?

Selfishness had now become a thing of merit, and by the 1980s, the Randians had grown influential. Now appeared the Atlas Society, John Birch Society, Freedom Movement, Mont Pelerin Movement, Cooler Heads Coalition, Global Climate Coalition, each built on the shoulders of the Chicago School of Economics led by Frank Knight. These organisations and advocates promoted self-interest and repeatedly stated altruism was a poison. In *Evil Geniuses*, Kurt Anderson recently concluded of the US, "This was a quite deliberate reorganizing of our economy and society by a highly rational confederacy of the rich, the right and big business." Yet, they slept without troubling dreams.

The foundations were in place.

The sun went down, and all the ways were about to stay dark.

Changes Since 1990

What was it, then, with 1990?

A country seemed still to regret the loss of empire, another was about to triumph in the cold war, and both had been led for a decade by market ideologists. Margaret Thatcher and Ronald Reagan appointed factions to their cabinets, and they themselves strode the world stage. They disparaged natural and social assets, thinking only of national economic growth.

Reagan said of the ancient American forests, "Once you've seen one redwood, you've seen them all."

Thatcher observed of Britain, "You know, there is no such thing as society. There are individual men and women . . . it is our duty to look after ourselves."

You might recall this: the selfishness of the neoliberal project was untested, fragile even. It might have failed. And this too: in the 1980s, the prevailing popular view had been that an ice age was on the way, and average obesity levels were so low that no one thought too much food could possibly become a problem. The oxygen-producing climate-cooling Amazon rainforest, home to one-quarter of the world's species, was 400,000 square kilometres larger than today.

All hail utopia, now a new world can be made. The philosopher John Gray called this free-market liberalism an unofficial civil religion. The foundations of prosperity were positioned behind walls of high tariffs and government subsidies, public investments in New Deals and Marshall Plans, and lately in vast public support of private businesses through furlough schemes and tax breaks. These ideologists also assumed welfare systems created poverty by incentivising not working, even as free-marketeers were creating more poverty and social exclusion. Now, within the borders of the affluent countries, there quickly came the effects of climate change and social inequalities.

Some had climbed a ladder all their life, only to find it was propped against the wrong wall. You could feel sorry for them.

Except, perhaps, for this: we find, in the USA, wildfire, heat, flood and storm, but also low-intensity civil conflict over cultural values and identity, endemic racism, the detention of a tenth of the world's prisoners. We find the breakdown of families, 12 million people infected by tropical hookworm parasites after deregulation of water pollution laws, and the building of walls and gated communities to keep out the others. This quotient of sorrow is hard to bear.

This might not have occurred if neoliberalism had set out to produce net benefits for people and the planet. But it did not. Growth in GDP should have made people happier. It did not. Inequalities should have been temporary as affluent counties grew. They have grown. Environmental harms should have been temporary stages that could be endured, for a little while, to get to somewhere better. Now comes the unravelling climate crisis. Consumerism brought fabulous choices of accessible foods. Now food is killing people with non-communicable diseases (NCDs).

"Affluence," as Tim Jackson put it in *Prosperity Without Growth*, "has betrayed us."

Yet, things can be different. In the social-democratic countries of Scandinavia, taxes are high, and some common rights are universal: paid parental leave, care for the elderly, welfare systems. People approve public investment in development aid for poorer countries and support domestic transitions to non-fossil fuel futures. Denmark has many months each year when all its electricity is produced by wind; Norway's electricity is now all renewable. In *Factfulness*, Hans Rosling said this about his home country, Sweden: the country invests in public schooling and free university education, and produces, amongst others, notable volcanologists. Yet, there are no volcanoes in Sweden.

Why do we do this, Rosling asked. Well, because we feel we are commoners of the whole world, not just of one particular country.

We all have such choices. There are ten thousand proven ways of living well. Human cultures have been good at creating the good life, founded inside stable environments and economies, often egalitarian, always creative and offering opportunities for personal growth. Looking back, our descendants again will observe, wise as they will be, how did those ideologists not see it all a little sooner? How were they blinded by this single way, the one above all others?

There is great wisdom, after all, in every culture.

Happiness Awaits

This book will show there are choices available that can bring happiness and avoid planetary disruption.

But these are often hard to make as individuals. Despite the neoliberal ideology lauding individualism, we are not entirely free to make choices that might make us healthy and the economy environmentally sound. We are shaped and conditioned by our surroundings, by each of our lives lived to this very point. If you dwell in a forest, you will come to know trees very well and respond to them. If you live in a consumerist society, you will come to know advertising, packaging, political and economic signals, and you will respond to them, without necessarily knowing. Habits make us and are hard to break. This is going to be tough.

One theme for this book is that the green mind has the ability to make choices, develop healthy habits, and live well and long. Yet, we as individuals need the help of others, of collective action and movements, of new policies and institutions.

There are routes to both opportunity and revolt. If you decline to buy meat from animals raised in crowded shed and feedlot, fed antibiotic and hormone, then the producers will get the message soon enough. If you refuse to buy energy generated with fossil fuels, then the fossil fuel industry will hear too. If you install solar panels, buy an electric car, and refuse to fly so often? Well, you cut your carbon footprint, and perhaps people in the poorest 75 countries could increase their consumption to meet basic needs.

Where might this be heading? Towards a regenerative kind of stability, into an era where we have more time to invest in other people, in family, friends and community; more time to restore nature; more time for storytelling and cultural celebration. We could produce caring economies where people work less and invest more in others, where material consumption was low and only green. And the evidence shows, people would be happier.

If this all sounds too fanciful, from a distance, then caution might indeed be right.

But don't take too long. Fossil fuel companies knew in the late 1960s that their products caused heating of the atmosphere. Their playbook, shared with the tobacco giants, was to create doubt and fund opposition. In the early 2020s, mining companies were still blasting sacred Aboriginal sites in order to mine the iron ore by opencast methods. Tar sands still were being quarried, and new rigs were being sent to the Arctic for one last dash for oil.

Lest you think the fossil fuel industry and petro-states dependent on oil and gas exports have got the message, and that they are keen to go green, lest you are drawn into making this mistake, then consider this tale of the anti-solar-panel movement in Arizona. Bill McKibben established the 350.org movement to help keep this number of 350 ppm at the forefront of our minds. It is a safe place for humanity. He wrote in *Falter* of Arizona, where the sun shines for nearly 300 days each year. The fossil fuel industry lobbied to prevent the fitting by individuals and companies of solar panels. They successfully persuaded the state government to prevent solar installations, for they in business faced a death spiral.

Which is, some might say, the whole point.

And now Bangladesh, a recently poor nation, is seeking to become a country powered only by solar. It soon will have a female solar engineer trained in every village in the country. Uruguay is already 100% renewable for electricity, Costa Rica at 95%. Arizona, or Bangladesh and Costa Rica and Uruguay, which looks to be the more progressive?

And yet, what's not to like about modern affluence? Why be worried now? Material consumption has brought great good, offered escape from disease and drudgery, created cheap communication platforms accessible to almost every person, ringed the planet with space vehicles, created accessible food systems with many thousand new products each year.

Naomi Klein has observed that the timing of this crisis is bad. In many affluent countries, social capital and collective capacity have been wrecked just at the time it is needed most. Organisation will have to come in other forms and from new entrants. With the Friday schools-out campaign, young people across the world have been eager to speak out and act. A remarkable 8 million social groups have been formed worldwide in just 20 years to advance collective forms of agricultural and forest management: they have a membership of a quarter of a billion self-organised people improving livelihoods and nature. There are thousands of communities in ecovillages and transition towns; there are companies signed up to Race to Zero; others to Lab B as corporations intended to produce side benefits.

This is the kind of growth we need, in ways of living and working together.

There have also been a small number of political concessions to indigenous cultures, who already know how to live happily and well. In Nunavut of Canada, in Tuva of Siberia, and in Santa Marta of Colombia, indigenous people have political control. In Ecuador first (1991) and then Bolivia (2009), the concept of *buen vivir* (the good life, derived from the indigenous concept of *sumac kawseng* of the Quechua people, sometimes written as *vivir bien*) has been signed into national constitutions. These are deliberate attempts to celebrate difference and divergence. Other experiments are happening in equally fundamental ways on Native American/Indian sovereign reservations in the USA. Casinos are controversial, but income has allowed previously poor tribal nations to create sufficient resources to pay each member an annual basic income. Progressive countries elsewhere are looking at the option of paying their citizens such a universal basic income. The strange thing about these income guarantees, as Ian Budge has put it in *Kick-Starting Government Action*, is that free marketeers think people (especially poor people) don't deserve support as it will make them lazy, and yet basic income payments give choice to individuals, they extend freedom of choice. Such ideas could change everything.

This is the topic of this book: act now to slow and halt a runaway train. Is it too late to save the climate? Nearly, but not quite. There are ways to save us from ourselves.

We have three sets of ten years available. Johan Rockström of the Potsdam Institute for Climate Impact Research has called for a new kind of Moore's Law. We need to cut net emissions by a half in the first ten years, by half again in the next ten, and then half again. In this way, we would get to a roughly tenth, from 1.00 to 0.125, from 100% to 12.5%. It is not quite zero, but this would pretty much do the job alongside carbon capture. The greatest changes thus have to occur in this very decade. This mission is going to have to succeed without significant reliance on geoengineering. This may help, but will take too long to sequester enough carbon to stop the crisis. Such approaches might suggest, to some, that we could do this by not cutting emissions.

Smoke still clings to the hills. There is a stark choice.

Allow the climate crisis to change everything, or change everything first to avoid this fate. Yes, this will spell extinction for fossil fuel and extractive industries, for clear-felling and agricultural expansion. But be sympathetic and generous. The green future offers much more for all cultures and people, at every location.

> Take good care, of what is good,
> In your life, use it well.
>
> *[Meister Eckhart, The Book of Secrets, 1260–1328]*

Different Forms of Growth

It happened fast, the weightless form of capitalism. First two centuries of groundwork, philosophy and political testing, acceleration during the cold war, emerging maturity in the 1980s, and then one generation since 1990 at full hustle. Here are five popular phrases and underlying ideas that seem to float up repeatedly.

You might choose to be suspicious if you hear them.

We have never had it so good: plain wrong. Ten thousand ways of living have already shown how people have had it so good. Of course, medical advances, technology and communications are features of the modern world we'd like to keep and make sure they are run on renewable sources of energy.

There is a ticking demographic time bomb: this is a convenient diversion, suggesting the problems with the world lie with people over *there*, amongst *those* kinds of people, and not over here. Numbers of people are not the problem; it is how we consume that matters. In affluent Europe, the richest 1% of people have carbon footprints eight times the world average. And anyway, the world population is about to stabilise. When it falls, that's when we might start to worry.

When it comes to the commons, there is always tragedy looming: wrong again. Garrett Hardin's notorious 1968 essay seemed to suggest that the commons would always be inefficient and subject to destruction. Yet, most commons are well-governed by rules, obligations and reciprocity. People knew how to manage collectively. We always have.

There is no altruism in the natural world, and so humans are right to be selfish: wrong again. This derives in part from conclusions drawn from Richard Dawkins' 1970s' book, *The Selfish Gene*. This played into misunderstandings of natural selection and evolution, and appeared to suggest that culture and altruism were irrelevant to human experience.

Nothing will change, it's not worth the effort: our bodies and brains are fixed when inherited, and the mix of poor and rich countries will always stay the same. Both are wrong. We now know of the power of neuroplasticity. Our brains and body respond to outer environments and can be changed by deliberate action. And the state of countries and cultures can also change, very quickly, if the political and social will is present.

So if anyone says it is just human nature, just say this: the problem really is *not*. It is not human nature to be greedy, to disrespect other cultures, to think government and regulations are evil, to over-consume. It is only a current ideology that makes these acceptable norms. It has done this in a short period of time, so could change again.

Human nature is this: the commonalities between people, our shared genetic legacy, the common brain–body–mind structures, the common pasts and legacies. Friendliness has been vital to our evolution. Look carefully, and you will see the world is already more caring, less cruel, more prosocial, less poor.

But it has not yet learned how to control the greedy.

In future, we might have fewer possessions, and be happy for it. We will eat well and healthily. We will be connected in communities with high trust and mutual support. We will be connected to natural places, be cognitively engaged for life, and thus live long and well. Planet-busting ideologies will be cast away.

When he was alive, Hans Rosling often used this story: many people, especially the older, have simply missed a system upgrade. It was not their fault. We commonly see the world as we learned it when at school and college. Rosling posed this question: what country were you born in? He was born in Sweden in 1948, when Sweden looked like Egypt does today. Setting aside the differences in ecology and climate, Egypt is on track to look like Sweden in one more lifetime. The country you were born in has changed dramatically. But have you checked the data? The assembled folk are waiting.

Population is one area where some people seem to need an upgrade.

Those people who say, "They have too many children," referring to some distant *other* group, are ignoring the recent data. Their knowledge is outdated. At the time of the US moonwalks and its own national independence, Bangladesh was a country of extreme poverty. In 1972, women had on average 7 children, life expectancy was 52 years, and infant mortality was 200 per 100,000 births.

Today, women in Bangladesh each have on average 2.4 children, life expectancy has jumped to 73 years, and child mortality fallen to 30 per 100,000 births.

> Have a cup of tea. Wake up.
> The pure land is right beneath your feet!
> [W S Wilson, *The One Taste of Truth*, 2013]

It's Not Population

Before we come to responsibility and targets for cutting carbon emissions, it might help to observe: the problems with pollution and externalities, with the climate crisis and non-communicable disease, these are not to do with population alone. They are about consumption. Not numbers of people, but what each one of us does.

The world population increased substantially during the 20th century, from 1.6 billion in 1900 to 3 billion in 1960, then to 5.3 billion by 1990. By the start of the

2020s, it had grown to 7.8 billion. Alarmists have in the past called this a population bomb, even today some sensible people still write that we may exceed 10 or 12 billion people on the planet.

It is the worldwide data that tells us this is incorrect. It also feeds a racist stereotype: there are too many people, other sorts, who are having too many babies, and the world cannot support them. Hans Rosling was once asked by an audience participant in an affluent country, should people over there be prevented from living like us? He chided, "*We* cannot live like us."

And if you meet anyone who has fallen into this trap of population bomb-and-explosion, then their current knowledge is probably wrong, for the days of fast population growth are almost over. Stability beckons, quite soon.

Figures 1.1 and 1.2 tell a remarkable story. They show the outcome of successful implementation within countries of programmes to reduce poverty, some with the support of international development assistance by governments and charities. There have been increases in the rates of female education and literacy, for both mothers and children, increases in access to contraception, organisation of women into microfinance groups for saving and lending. Families have been ensured access to enough food and clean drinking water, better housing, electricity services and security.

Figure 1.1 shows the changes in fertility rates from 1960 to 2019 for more than 200 countries, roughly over two generations. The number of children per woman has been falling fast. Fertility is measured by the average number of children each woman bears in a lifetime, and 2.1 is taken as the replacement rate (2.0 for the parents, 0.1 for mortality up to the age of reproduction).

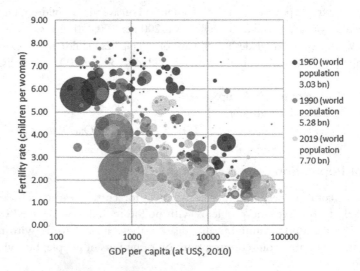

FIGURE 1.1 Changes in fertility rates by per capita GDP, 1960–2019 (population by bubble, n=207 countries)

In 1960, the average woman worldwide had 4.98 children; by 1990 this had fallen to 3.25, and by 2019, another generation passing, to an average of 2.40 children (Table 1.1). By the year 2020, there were 95 countries with fertility rates below 2.1. In all other countries, rates were falling towards 2.1.

There is still population growth to come, in a declining number of countries. This data suggests that world population could stabilise at about 9.0 billion, perhaps less, perhaps a little more. Not 10, and not 15 billion. Another billion more than now is not a small number, but look again at the changes in fertility since 1990. Families are better-off, have confidence about their children, who are likely also to have smaller numbers of children. And of those children, more likely to survive infancy, both girls and boys can go to school, and can themselves imagine every kind of hopeful future.

Figure 1.2 then shows the infant survival rates against fertility since just 1990: survival has been increasing while fertility has fallen. All countries are heading

TABLE 1.1 Changes in human fertility rates by country, 1960–2019

Average fertility rates in each country	Number of countries 1960	Number of countries 1990	Number of countries 2019
>3.00	161	121	64
2.10–3.00	35	36	42
1.50–2.10	5	36	71
<1.5	0	8	24
Total countries	201	201	207
World average fertility	4.98	3.25	2.40

Note: Data from World Bank and United Nations

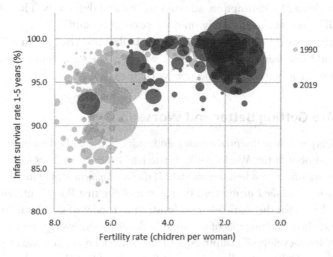

FIGURE 1.2 Changes in fertility rates and infant survival by country, 1990–2019 (population by bubble, n=207 countries)

towards the top right, where there is low mortality of children under the age of 5 years, and fertility is at or below 2.1.

Let this news sink in. And then think, if nearly half the countries in the world have women bearing fewer than 2.1 children, then their populations will fall. In 2021, the fertility rate in South Korea had fallen to 0.98, less than one child per two adults, the world's lowest for a country. This *Slowdown*, as Danny Dorling has wisely called it, changes everything. It changes the way we think about the future.

There could be a stable world population by 2050, possibly sooner. When it stabilises within countries, it is likely then to fall.

Some countries have already started to respond. Japan's peak population was 130 million; it is now 125 million, and predicted to fall to 80 million by 2060 on current trends, if nothing else changes. The government of Japan recently announced a scheme to recruit economic migrants. It wishes to recruit 350,000 doctors, nurses, engineers and teachers to migrate permanently to Japan. The OECD has calculated that between 2020 and 2060, the USA and Europe will need to attract 50 million migrants, for without such a workforce, the tax base shrinks so much that states risk going bust.

Will attitudes to migrants change, you might ask. The poet Warsan Shire in her *Bless the Daughter* says no one leaves home unless it's become a mouth of a shark: "You only run for the border when you see the whole city running as well."

There might be this kind of speculation: will walls and barriers come to look even more absurd when countries begin the search for people? Meanwhile, a new dread could come, as analysts with pencils extrapolate lines on graphs, showing where human numbers turn to slip downward, heading for zero.

Set that concern aside, and think what happens to economies relying on infinite material growth. There will now be fewer consumers, and they will understand the world's problems of consumption, so will buy less and differently. They will realise the good life does not need to rely on only consuming stuff.

The current world economy will fall over. Unless it is changed first.

The world has been getting better and worse, at the same time. Now ways are needed to make it better for all and not worse for some.

Things Are Getting Better and Worse

Herman Daly was an influential economic insider in the early days of the Environment Department at the World Bank in the late 1980s to mid-1990s. It felt like a moment of significance when the annual *World Development Report* for 1992 focused on *Development and the Environment*. In the year of the first Rio Conference, something had shifted for the world's largest lender and funder of development projects. At the time, though, concepts of sustainable development were seen to apply only to poorer "less-developed" countries, and the material growth mindset was not to be challenged. Daly and colleagues prodded and pushed, but he later observed, "The devotion to growthmania dies hard."

Herman Daly could see what was happening. He was close enough to appreciate a paradigm in action. One economic adviser to a US President, Pat McCracken, had quipped in 1975, "Unless the abstemious rich make the sacrifice of consuming more, the poor will not be able to sell their resources." A decade later, the influential left-leaning Oxford economist, Wilfred Beckerman, stated, "There is no reason to suppose that economic growth cannot continue for another 2500 years."

Daly set out the principles for a steady state economy, in which there would be, "Sufficiency for a good life." If we destroy natural assets, he wrote, then the human-made economy goes too. What good is a sawmill without a forest, a fishing boat without fish, an irrigated farm without an aquifer or glacier? In *Beyond Growth*, he wrote, "We should strive for sufficient per capita wealth, efficiently maintained and allocated, and equitably distributed." The goal should be enoughness, not maxima.

The key words are these: steady state, sufficiency, enough, good life.

He also observed, "Most classical economists dread the steady state as the end of progress." Twenty-five years later, many still do. The docks lie in ruins, yet many think there still are fish in the sea.

And yet, those wedded to the ideals of neoliberalism might have missed something greater. In a steady state economy, there still is growth. Just not the unending growth of material things, but personal and social growth instead. Human experience comes from many compass points, and living within limits has long been our normal way. It is entirely possible to live a good life without breaching personal, community and global boundaries. If you tell me a story, we both benefit. If you listen to the birds singing at a spring dawn, at the end the birds are still there. If you walk attentively, you bring home only memories.

Meanwhile, we find ourselves in a state of emergency, and not just around the impacts of pandemic and climate. The ills of affluence are spreading, not only ills of health but of inequality and racism and poverty. This is as human fertility is falling, infant survival rising, education opportunities spreading, and world population coming towards stability. We have remarkable medical technologies and practices, new social movements, communication platforms for all. The cargo cult of consumerism has not completely driven out all that is good.

There still are large parts of our daily lives not governed by endless growth and the search for profit: the public sectors of health, social care and education, the cooperatives and non-profits, the charities and volunteers, the cultures of community and family. It feels as if an era could be ending. The super-rich may survive slowth, when world population stabilises, consumption growth ceases, economies stop growing, and debt no longer borrows from the future. But selfishness cannot survive the high connectedness of people across the world, their increased capacity to act together, the emerging worldviews, the collective action especially among younger people.

Stability will bring wisdom. We are circling back to constancy, the old normal human condition, perhaps also influenced by stable environments and ecosystems. Prosocial and contemplative economies in a restoration era will be more caring, less cruel. They will be more connected and less hierarchical.

Yet, such a regenerative era will require social organisation and action. It will not happen by accident, nor by individuals acting alone.

The foundations are already strong in many countries, especially in rural communities supported in recent years by progressive external organisations. In India, 4 million social groups have been formed for rural collective action over the past 20 years, in Bangladesh nearly 2 million. In one corner of south India, the non-government organisation SPEECH had for years been forming women's groups in poor dryland villages. One March day in the mid-1990s, I joined a walk with 1500 women from 40 village sanghas. The sun had dropped towards rice fields and rows of coconut palm, and the groups formed into a kilometre-long procession to mark International Women's Day. Three hundred men and boys walked on the outside for protection, and the marching crowd cried, "No more dowry, equal wages, unite–unite, share the burden between men and women."

In the town, some local people smiled, many frowned. But there was no attack. "Just look," said John Devavaram, "There is only one cop. We don't need anyone to control us. We are well-organised."

But this centre for agricultural regeneration was also location for Victorian-style child labour. It has a dry climate, and so was favoured by the match and explosives industry. Children have small fingers, perfect for packing matches into boxes. Let's go to look at a factory, said John one day, and you'll see. The daily wage then was Rs. 15–20 for adults, Rs. 10 for children. This group bluffed a way in, the four of us, into a hall with about 500 children seated on a concrete floor. At one end was a furnace that mixed the potassium chloride for the match heads. It was fiercely hot. The youngest child was perhaps five years old. They were bused from villages in the hours before dawn, each day was a 12-hour shift. Not only were their childhoods stolen, but their futures too. Many became partially blind from the chemical mix.

Over the years, I watched young village girls grow up and have their own children. I played cricket with John's boys, and they too grew up to have families. I have seen how village wells were dug by hand, trees grew, and watersheds became enhanced to save soil and rainfall. The dry and sandy land of 1990 has since turned green.

Thus, you see, things have been both better and bad, getting worse and being good.

Shunmyō Masuno is head priest and garden designer at the 500-year-old temple Kenkō-ij, and wrote:

> Acquiring lots of things isn't freedom,
> In the world of nature, every day is a new day,
> The garden is never the same, from one day to the next.

This World Worth Saving

There is no country outside of conflict zones in the world where fertility rates are not falling, where infant mortality is not falling too. Equally, there is no country in the world where high levels of obesity have been reduced, though there are notable examples where it has remained low. We might ask, what about those who say this idea of crisis around climate and modern living is all a hoax? Well, if green new deals and actions create new jobs and detoxify the environment, as Naomi Klein observes, who really cares?

Three options are going to be important: change your own habits, behaviours and choices; join organisations and build social capital; speak loudly to politicians and corporate leaders. But also, be attentive.

Look closely at this world worth saving.

Let's be blunt. Some economists lost a piece of moral core. Parts of the discipline became unhinged, lost sight of what is important to people. It measured the wrong things, was obsessed with waged labour, and seemed to ignore other forms of human contribution to societies. It underplayed the negative environmental and human health effects. Here's a mission statement, a variant on the Hippocratic Oath of *do no harm*. How about do more good, not just less harm? Or better still, do only good. Andrew Simms and Joe Smith of the New Economics Foundation wrote that good lives do not have to cost the earth, and in *Post Growth*, Tim Jackson has set out the terms for *life after capitalism*.

A proposal for the first half of the 21st century, for the remaining three decades, could feel like this: redefine what we mean by progress. What should our goal now be, now that it is not infinite wealth, material consumption and economic growth?

Something far bigger and better awaits: happiness, well-being, stability, egalitarian structures, wisdom and flourishing. And, as later chapters will show, we humans have done this before, many thousand times. We know what the good life looks and feels like. We just have not learned to reject a recent project so blinding in its certainty.

In both *Slowdown* and *On Fire*, Danny Dorling and Naomi Klein have each observed that neoliberal orthodoxy can only exist as long as states prevent political expression. This is an era of individual communication platforms available to almost everyone. Expression and organisation are possible in ways quite impossible to imagine in former eras. And what happened when COVID struck, a rehearsal for the climate crises? People worldwide celebrated social goods, public medical services, schooling, protection from crime, conserved nature, and relied on governments to intervene in private and corporate spheres with unprecedented financial support for businesses.

Another angle is to consider how we humans value collective actions that seem to make little or no contribution to GDP.

Why would rationalists with only self-interest devote time on pilgrimages to sites of particular cultural and spiritual significance? All this activity that appears to have no instrumental or guaranteed benefit. Each year, at least before the pandemic, 40,000 people came to the Sinkara Valley in the Andes for three days of spiritual renewal. In the western Himalaya, many thousands climbed 5000 metres to Mount Kailash and walked around the sacred mountain that brings together Buddhist, Hindu and Jain pilgrims. More than 2 million people visited Mecca each year, walking round the sacred pillar; 6 million travelled to Lourdes in southern France.

Even the moonshot programme, you might say, was not solely about economic growth and winning a race, for it brought back 400 kilogrammes of rock and an image. Apollo astronauts took that first stunning photograph of our blue-green planet, a jewel all alone in space. And a number felt the deep cognitive shift of *the overview effect*, a profound awakening and sense of awe.

Astronaut Don Lind said, "There is no way you can be prepared for the emotional impact," and William Anders added, "We came all this way to explore the moon, and found we discovered the Earth."

This Earth is not infinite, but it is very large, Bill McKibben said in *Falter*. Yet, since the Apollo programme we have warmed it by +1.2°C, created extremes of weather pattern, burned too much forest, ruined many a species, bleached coral reef, brought heat dome, melted ice cap and shrunk glacier. Some people still say climate change is a falsehood, that it is a narrative designed only to stop ceaseless material growth of economies, and thus to undermine basic freedoms for a certain way of life. Yet, insurance companies are now saying basements from New York to Miami and Mumbai will be uninsurable. Many of the privileged and elite seem oblivious, yet when the fall comes, it will hurt them hard. The greatest danger comes from fondness for the current ways and habits.

And there is this: the modern financial and personal weight of environmental pollution and non-communicable diseases could be prevented, perhaps almost eliminated. And that resource could be spent in other ways. Living in a new way could move billions of dollars from treating obesity, diabetes and other ills of affluence, it would mean not needing to spend to mitigate and adapt to climate change. Resources could come to be diverted to ensure businesses are protecting environmental and social goods. We might not need to spend billions to treat all dementias, a half of the incidence of which are preventable by behaviour and choices.

Now we could instead spend to create renewable energy sources, community social capital and accessible natural places. Just as a start. And thus ensure people have the chance to be happier and more content.

We know what a good life looks like.

A new era of slower living and stable economies may now await.

> From the South Sea, did a guest appear,
> And give to me, a mermaid's tear,
> And in that pearl, there were cloudy signs,
> Wrapped it away, in a little box,
> Opened and saw, it had turned to blood,
> Alas today, I have nothing else!
>
> [*Tu Fu, From the South Sea, 712–770*]

~*~

2

TEN THOUSAND GOOD LIVES

Sustainable and Kind Ways of Living

A hundred thousand sparrows, descend on my empty courtyard,
A few gather atop the plums, chatting with clear evening skies,
Suddenly they all startle away, and there's silence: not a sound.

[Yang Wan-i, *Cold Sparrows*, 1127–1206]

Abstract: There were once ten thousand good lives. Our human cultures knew how to live well and lightly on the land. Now a perilous journey lies ahead. Yet, it always did. It is the monomyth of Joseph Campbell, the experience of living that is common to us all. This chapter tells stories of the Timbisha Shoshone in the desert, the Koyuko and Haida people, of Oðinn from the North, of the Amish in Ohio, and Tuvans on the grassy steppe. Human cultures have centred on togetherness and belonging, the foundation for a good life. Now we know, too, that ever more GDP does not bring more happiness, after the essentials. Data is presented from 150 to 200 countries: over time and across countries, more stuff does not bring more happiness. This is another common human experience. At the same time, though, populations are growing older, and non-communicable diseases, the ills of affluence, have rolled on stage. The domains of the good life are discussed. Most are low carbon. And here we have a hint of a path ahead. "Cling if you want, to things," wrote Meister Eckhart, "but where then will your heart be?"

DOI: 10.4324/9781003346944-2

The Perilous Journey

The forest is gloomy, the night desert cold. Beyond the blue lagoon, there is a heartless ocean; a blizzard blows on the mountain pass.

Joseph Campbell called it a monomyth, the experience of living that is common to us all.

Many a time across life, we are called to adventure. You leave your place, pass across a threshold to a new realm, have to fight demons, thieves and thoughts, you find the elixir, come to a new and deep realization, and return to the cool fields of normal life. Something called, and we took to this new game. But often, we refuse, for everyone fears the journey ahead.

And then when we return, bringing back something essential for the world, no one seems to care. They don't know they need it yet. After the ecstasy, said Jack Kornfield, it's time to do the laundry.

The perilous journey was one thing. What are you now going to do next? Pick the mask up again, tell the story.

Joseph Campbell said life was one hero journey after another. And by this, he meant: learning to walk as an infant; going to school, to big school, to bigger college; learning to drive; quitting one job to do what you really desire; offering your song, painting or poem to an unknown gatekeeper; taking up tai chi or open-water swimming; going into hospital for an operation; adopting a low-carbon way of living. And for many, the most heroic of great cross-continental journeys to escape conflict. Habit is this great journey too.

Once you get started, it feels fine. You succeed or fail, but you're on the journey now. The problem is getting going, waving goodbye. Mostly we prefer to stay, even if that habit was causing harm. "Over and over again," said Campbell, "You are called to the realm of adventure, to new horizons. Each time, the same problem: do I dare?" There is the possibility of fiasco, but also of bliss.

In some ways, it is just a matter of time. One way or another, the selfish brand of late-stage capitalism will fail, and be replaced. It could happen intentionally, by heroic choice. It could happen by mistake, the outcome of rising sea, extreme heat, drought and melted glacier, shore-side assets stranded and abandoned. Expect many journeys. Ways of good living do await. We just need a map or two and a good mythology to help us on the way.

The monk and mystic of the early 1300s, Meister Eckhart of Thuringia, wrote this:

> There are many ways,
> Some are crowded, others less travelled,
> But every way, is the right one.

So, a start might be best in two great libraries. The first is longitudinal: how the many human cultures crossed eternal time still to live in particular ways today. The second is latitudinal: how people living under the rule of affluence have today already sought the good things in life.

Yes, like that. We have choices.

And this much is known: a good life never damaged a whole planet. It exists within a boundary of physical stability and sufficiency, there's enough food, communities are bound together, ethical behaviours with a moral core ensure no long-term harm to nature. In these stable places, people have the time to create stories, telling why things happen, how they live meaningful lives. And nature itself, the Earth system, is safe. At each dawn, you can see a red thread stretched across the horizon.

How should we think of these other ways of living? Some tricksters are ready to help. They will open a swinging gate.

One-eyed Oðinn had two ravens, Huginn-Thought and Muninn-Memory, and they flew out at dawn over the whole realms of Midgard and Utgard. At each eve, they perched on his shoulders and whispered the tidings, and Oðinn was wise, for the ravens helped him see the world in depth. His mind had stretched far, as it does for seers and shamans. The ravens were guardians and shape-shifters. They were also tricksters, they showed the way.

When thought and memory go away for a bit, of course, there is silence in the mind.

For the north-western Athapaskan and Haida cultures, Raven is a powerful deity and mischief-maker, clown and guide. Raven has guile and genius. When the world was covered in flood in Distant Time, he supervised each pair of species onto the raft, when the sun went missing, he flew up to bring it to the sky. The Koyukon, as Richard Nelson has recorded, were guided by animals and their stories, so knew what behaviours were right and wrong. One of the great Haida sagas is *Raven Travelling*, some 1400 lines long.

Raven is ingenious, ever-watchful from high spruce and red cedar. "Bring us good luck," people called out. Says one old man, "He always fools everybody, so he gets by easy."

For the south-western peoples, Coyote shares these traits with Raven. He is so clever, knows much, is prone to showing off. He is an instructor and teacher through behaviours, and always appears before a story. And the story itself is intended to work on us, burrowing deep and one day provoking us to cross a threshold. In the desert, the silence is long and deep. Yet, in ancient times, animals and humans could speak the same language.

The culture of the Timbisha Shoshone was built on persistence in the deserts. They hunted and gathered, shared ceremonies and rituals to celebrate nature and place, took care of water and trees, and had no need for centralized political organisation. They were given back a small reservation in the year 2000, fully 150 years after settler contact, in the heart of the land the maps call Death Valley. They know it as Timbisha. Walking on the way over to see tribal elder, Pauline Esteves, there was Coyote, sitting on her haunches with ears pricked, at the boundary of the playa and lush green grass of the motel golf course.

Pauline has lived nearly a hundred years, and said, "When we see Coyote crossing the road, we say, there he goes. He supposed to be our father, but what a father! He is always teaching us." One day, Coyote was looking down into water, even though you should not adore your own image. Something terrible could happen. "When you come to a dry lake," Pauline said, "You say, I wonder if this is where Coyote passed through?" The story is always about attentiveness, creating something inside for us.

> That dry place was called Clay Camp, but we elders always call it The Brains. Coyote's mind was like clay, then became clear. But he had been mocking a bird, saying he could fly better. So Sun helped him fix wings, but in his pride he flew so high where all was quiet. Then Sun's heat melted the pine resin, and he fell to Earth. He started to vomit, and a richness started draining out of his mouth and ears and nose. It was his brains coming out. He ate his brains. When we think of that place, or mention its name, we remember how to behave.

Coyote had revealed something else. How to consume the self, how to silence the world and become transcendent. The Early Morning Song, *Imaa Hupia*, was danced and sung by Timbisha people as dawn rose over the desert valley:

> How beautiful is our land/how beautiful is our land,
> Forever, beside the water, the water/how beautiful is our land,
> How beautiful is our land/how beautiful is our land,
> Earth, with flowers on it, next to the water,
> How beautiful is our land.

Thousands of Good Lives

More rain is due, it seems. And more heat.

Let's put it this way. All things always were connected. How promising that we might gaze away from the ills of affluence, and see the assembled folk gathered. Now there is a hush in the life of the hall and longhouse.

Today there are some 500 million indigenous and tribal peoples in some 90 countries. There are about the same number of small farms, and together they bring to this century knowledge of the watchful world, traditions and truth-stories, ethical and moral obligations, respect towards nature. Terrains are seen as nourishing, a healthy country is a good country.

Modern ideology cast a veil over these ten thousand other ways of living. It has called them pre-modern, the people savage, they were barbarians and uncivilised. Those cultures were placed at a prior evolutionary stage. At the top was an extractive economy, with endless goods and things and foods, seeking infinite growth and freedom from history. Oddly this philosophy still drenches people with unhappiness and inequality, and deepens environmental harms.

What the tricksters say is that there are many ways to live. Over time, cultures diverge. There never was one model better than all the others. A city-based culture, the Mayans of Central America, developed advanced mathematics and astronomy, built temples higher than the rainforest canopy, and developed regenerative methods of agriculture. They knew well of the wheel and metal, yet rejected both. There were 1000 riverine settlements of Harappa in the Indus basin, spanning 2000 years, yet even today no one has found a way to unlock their script. They had sophisticated urban planning and drainage, hydraulic engineering, and brick housing and metallurgy. But not one temple or palace, though at one city an enormous bath has been found, it seems for communal and spiritual bathing.

And sure enough, indigenous people today wish to sustain their ways and access some of the benefits of affluence. I sat once with shamans of the Adyg-Eeren bear society in a wooden hut in Siberia. On the walls were three bearskins pinned out, the skulls of ox and boar, eagle-wings, a bow with arrows, prayer flags and many braided ropes and beads. The room was silent, there was no sign of any other world. Stated headman Dopchun-ool Kara-ool, "The main task of shamans is to protect nature. We act for the mountains, rivers, taiga and different landscapes of our homeland."

A mobile phone rang, and one of the female shamans stepped out. Kara-ool waited a moment, "To be a shaman, you have to know the voices of nature and the language of spirits. This is a time of bad fortune, and our goal is the renaissance of a traditional culture."

"And we are also modern people, using mass media and the internet." On yurts across the steppe were solar panels, to charge the phones and TVs of the nomad families.

Under the same broad sky, many small farmers say their goal in life is contentment, and so see time and money in distinct ways. Gene Logsdon called himself a contrary farmer in Ohio, not at all like his big-tech neighbours, and wrote, "To understand a meadow, you mostly need to sit down in one a while. Maybe like for twenty years."

He asked a fellow shepherd, "Why do you raise sheep when there's no money in it?"

And she replied, "Well, there's a little money in it, but the main reason is that my sheep make me happy."

Also in Ohio, the Amish have a term called worldliness. To be worldly is to be a little on the fast side, a tilt away from established norms. The Amish have some solutions to the contradictions of affluent society, and are curious about the idea of progress. They say you should think before you speak, take a long and slow stride, and brake deliberately to slow things down. They ask, rhetorically perhaps, why do modern people deposit their elderly in bleak retirement homes, why do they move house so often and lose touch with family and friends, why do they sit on ride-on lawn mowers and then drive to the gym?

I first travelled to Holmes County with David Orr of Oberlin College. We walked around David Kline's 120-acre organic farm under wet occluded skies. David Kline's elegant book, *Great Possessions*, documents the gentle rhythms of Amish organic farms and their abundant wildlife. He was once asked by a magazine editor to write about the advantages and disadvantages of their way of life.

"This bothered me all summer," David said. "Quite honestly, I couldn't think of any disadvantages."

These different ways of living also represent forms of resistance. They have no wish to abandon their ways. I sat once on the iron boulder slopes of the Karratha peninsula in north-west Australia, where had lived the *Pilbara* people for 50,000 years, perhaps a bit more. They had chiseled and carved a community of a thousand petroglyphs on rock, there was emu and turtle, thylacine and fat-tailed macropod. These were the very spirits of the animals. Yet, those settled people who had lived so long, who knew so much, were killed on colonial contact.

In 2020, a multinational mining corporation set explosives at nearby sacred caves, and blew that ancient place and knowledge into skies. Their plan was for opencast iron extraction: their leader received a bonus.

Back in Siberian Tuva, every place has an *ee*, a spiritual guardian. People feel awe in the beauty of the light upon the steppe, and at every place you show respect. The rivers are alive, they sing. The winds of the steppe ease through the yurt and sheepfold, and nomads are said to have a wide stride. They have freedom to pass an entire lifetime of days and nights, dawn and dusk, watching the turning of the seasons.

Author and musician Ted Levin called Tuvans, "Among the subversive heroes of modernity."

He quoted *khomei* throat-singer, Sayan Papa:

> You let things happen and evolve, and you don't destroy the nomadic way of life, you put up your yurt and you haven't harmed anything, not the grass, not the sunlight; when you play music, you've taken from nature and you give back to it; everything passes through you, and in the end, you're back to zero; a big and beautiful zero.

The Tuvan throat-singers sing on the steppe:

> Let my people live well,
> Let their work go well,
> Let the children live well,
> Let life be without obstacles.

Friendliness and Belonging

In 1968, Marshall Sahlins famously spoke of hunter-gatherer-foragers as, "The original affluent society."

It was a phrase that made people sit up and notice. Those cultures have few things, have never heard of financial markets, and seem not to want any of the new food-like products generated yearly for our supermarkets. They work few hours, spend time lying around and being, well, apparently lazy.

Just as someone might today wish a holiday in the sun should be, except they never call it being idle.

How then shall we define the good life? The core elements seem to be the same worldwide, even as they exist in strikingly different cultures emergent from their particular places. This should not be a surprise. The brain, body and mind works in identical ways within and across human cultures. The structures and internal chemical communications are the same, and we all start life at the same place and with the same amount of knowledge and language. We each diverge and develop as we live and grow, responding to the social and ecological environments around us.

The phrase *the good life* has often come to mean in popular culture some kind of dipping out, a rejection actually of the many things in life. This suits consumerists, who will be saying, over and again, you need to buy our stuff to be happy. We shall see shortly this contention is not supported by evidence from the affluent and high-emitting countries. Material consumption above a certain point has not brought everlasting happiness and contentment for all. Growth advocates would never want you to say you have enough, to hear that you will now sit and listen to the earth sing, or that you will gather vegetables grown in the garden and cook a dish, and eat with family and friends, all together sitting at the kitchen table.

The striking thing, when considering each way of living, is what seems almost absurd to mention. Every human culture and society was founded on sharing. In *Survival of the Friendliest*, Brian Hare and Vanessa Woods observed that "No folk theory of human nature has done more harm than the *survival of the fittest* meme." They pointed out that the best way to win the evolutionary game was to maximize friendliness, as this reduced social stress, increased the speed at which innovations could spread, and brought new forms of cognition and social networks. Humans and bonobo apes gather food and bring it back to camp to give to others to eat. They do this because it brings benefits to the whole group.

About 100,000 years ago, *Homo sapiens* coexisted with five other species of humans. *H. erectus* had been using tools for more than a million years; Neanderthals had culture, art and jewelry. By 25,000 years before present, *H. sapiens* had become the most numerous species and had spread the furthest. A larger cortex brought self-control, a less reactive red brain and amygdala, and thus the capacity to think before we acted. Friendliness became an evolutionary advantage.

And with the passing of time at particular places, cultural divergence became the common state of human affairs. We know today of the existence of 7000 living languages, half of which are under threat. We know every continent save Antarctica has been inhabited for several thousand generations by modern humans, as hunter-gatherer-foragers and lately as agriculturalists and then city-dwellers. What we also know, from recent study by archaeologists, anthropologists and ecologists, is that those cultures retained similar worldviews and spiritual frameworks centred on respect and reciprocity for people and nature.

Over time and across these places, you will find that people rarely claim theirs is the only truth. You will not hear assertions from one that they have the only answers. Over there is another country, so of course those people know best. These principles have profound implications. The world cannot wholly be altered to suit only one people.

Thus, stable worlds, their economies and environments, are not boring or uninteresting. Things happen, and acute attentiveness reveals the workings of nature, a tiny sound or scent, a shift in the wind, the source indeed of all life, the thread of grace inside us all.

It is also a common trait: people give to each other, and to nature. And as gifts move, their value increases as they acquire layers of story. The more a thing is shared, the greater is its worth, and so gifts bring obligations and responsibilities. An old rural aphorism from eastern England says, "You should farm as if you are going to farm the land forever; you should live as if today were your last."

By these are regenerative priorities, things get better. Or put another way, the capital assets of nature and society can be grown.

Robin Wall Kimmerer in *Braiding Sweetgrass* tells of wild food and sweet grass, they are gifts of the earth. She braids sweetgrass to give it away, never to sell. Braiding is an original instruction, passed on by generations of storytellers. She sets out guidelines for the honourable harvest: take only what you need, share, leave some for others, give thanks, give a gift. She says you should put such gifts of the earth in one bowl, and share them with a single spoon.

The Kula Ring is a cultural structure as wonderful and impressive as any pyramid, as any glass skyscraper, more than any moonshot or satellite array. This is why we sail, said Pacific voyagers to anthropologist Wade Davis, because when we voyage we are creating new stories. In the western Pacific, island cultures are tied together by reciprocity and gifts, by objects of no apparent intrinsic value. The Trobriand islanders sail huge journeys to pass on red shell necklace that move clockwise, and armshells that move anticlockwise, and these are passed to others to gather and grow magical power. Sailors might be stranded for months by unfavourable winds, but they are patient. It can take 2–10 years for one item to circulate. Reputations grow by keeping people connected through the ring of gifts.

The Kula sailors know prestige and status comes to those who give well. Both receivers and givers are made happy.

In Tibet, similar value is put on coral buttons and peacock feathers by religious and political leaders. Objects grow when they give them to others. Patrick French said this, despite the prevailing political and economic conditions,

> This is the good life: a group of men and women sitting, laughing on the edge of a field of barley after a morning's harvest, wearing wide-brimmed hats to keep off the sun, each one spinning wool on a spindle, or carving a peg, telling a joke, passing the day.

In the 1950s, Lorna Marshall gave Kung San! women with whom she had been living a gift of cowrie shells. These had originated in the Pacific. On her return a year later, she found the cowries distributed in families and groups across the whole land. Their high value as a gift meant they were also given away, and away again. At the same time, you would expect, stories travelled alongside the shells.

Gifts that have not used up the planet are also builders of togetherness. Not long ago, Larry Summers was a man of great influence, President of Harvard and then the World Bank. He once said, "There are no limits to the carrying capacity of the earth that are likely to bind at any time in the future. The idea that we should put limits on growth because of some natural limit is a profound error."

Deborah Bird-Rose quotes in *Nourishing Terrains* a Bilinara man, Anzac Muang-anyi, who said, "White men just come up blind, bumping into everything."

This is a good description for modern economic philosophy, it could be said. Over 50,000 years or more of deep time, the Dreamings are the Law, they cannot change, are passed from generation to generation. Our job, say surviving people, is to keep it going, keep the land and people safe. "The winding path is just how the path is," said Tyson Yankaporta in *Sand Talk*, in sharp comparison with the "impossible physics of civilisation."

So, how might affairs play out? No previous culture left permanent scars on the Earth, none permanently damaged the planet itself in pursuit of a worldview. And though cultures diverged, moral worldviews did not. In common across the earth, people have behaved ethically towards land and nature, both by doing good and doing no harm. We built foundational social ties and connections and created stable environments. We composed stories and narratives to describe events, code behaviours and explain origins. And nature itself had intention, provided the people were respectful and showed humility. These common pillars produced ways of living founded on reciprocity and obligation, and a sense that surroundings are sensate, aware and personified.

The 17th-century Chinese author of *Master of the Three Ways*, Hung Ying-Ming, wrote, "There is a natural equality of all people and all things in nature."

In the 1930s, Genzaburō Yoshino wrote *How Do You Live*, a book that became a national classic about the good life, and that came to be read by almost every child in Japan. Gentle things happen, they go wrong and right, to Copper and the other young characters, and the voice of his uncle the commentator adds a narrative layer to their unfolding worlds. "The things you feel most deeply, from the bottom of your heart, will never deceive you in the slightest," says the uncle to Copper.

Suddenly, a cold day comes along, and then on another, how nice it was in the warmth and silence. The uncle asks of Copper, "What will you create? What will you give to the world, when people are so used to taking? People built our world only by working together, it is said." Behind each item, thing, piece of food, are so many people, the boy is told, so many they cannot be counted. But they are connected, from field to factory, to driver and to us in the home. Poverty, he said, makes you feel inferior. When you are wealthy, you should show restraint and modesty. Poverty means people are unable to live their lives as humans.

By contrast, some modern commentators have been deliberately provocative. Richard North wrote in the 1980s, "A world where cooperation is stressed over competition may well be so boring." He scoffed, "Dolphins are all very well, but build no cathedrals." Yet indigenous, native, tribal and local, all these people caused little damage, and generally lived stably. As do dolphins, even though they also argue and sometimes fight.

I once sat on a conference stage in the American Museum of Natural History, in central New York City. Like many a museum, it was filled with artefacts that belonged elsewhere. My Finnish fisherman-friend, Tero Mustonen, elected mayor of his sub-polar village, stood up in the audience and said: "Tell me this, where is the escape route for my people, how do we get away from this destructive modern world?"

All was silent in the hall. You could hear the traffic from the nearby busy roads.

The Earth is a place where there is a dance of a thousand snowflakes. "What makes the corn field happy, I'll now begin to relate," wrote Virgil in *The Georgics* in 30 BCE.

Saichō was the founder of Tendai Buddhism in Japan, and composed this verse in 766 CE, long famed and learned by every child in Japan:

> What is the treasure of a nation?
> A person with a mind set on the Way . . .
> The ancients have said that a nation's wealth,
> Does not consist of a heap of precious stones.
> One virtuous individual who illuminates a thousand leagues,
> Is a national treasure.

More GDP Does Not Bring More Happiness, After a Point

There is no sleeping with untroubled dreams, not now.

There are 1.3 billion people living in poorer countries who should be con-suming more to escape poverty, hunger and water-borne disease. In the richest countries, there are 3.0 billion people who need to consume materially much less. If these people could adopt new habits, replace material consumption with a combination of sustainable consumption *and* choose non-material consumption activities that increase well-being, then a fiendish knot might be untied. They will need help from governments, they will need guiding policies. They will also need to organise into movements.

Freeing individuals from the constraints of culture and history, and creating ever-growing demand for goods, should have, it was said, grown the economy for all, thus bringing greater life satisfaction and happiness. This is why world's econo-mies annually spend US$500 billion on advertising and another US$400 billion on gambling. Buy a thing, join this service, become a winner. You will be happier and life will be more meaningful.

It is promised: each thing you buy will be a charmed vessel. And everyone can win.

These days, drought comes and all the world looks bleak. There are two problems. Consumerism needs you to keep buying, or how else will there be continued economic growth? And second, there is compelling evidence that populations have not become happier as economies have grown. After a low point, roughly US$25 per day of income, happiness no longer increases. It stays bunched around happiness scores of 7 to 8 (on a scale of 1–10). This has been called the Easterlin paradox: how could it be that increases in GDP, achievements so sought-after, do not go on bringing more well-being and satisfaction?

Well, there really is no paradox. The stuff that makes us happy is not stuff, after a certain point.

There are two ways to look at the data: across countries at the same moment in time, and within countries over time. First, as countries escape poverty defined by lack of essentials, people do become happier (Figure 2.1). At low annual GDP per capita, below about US$10–15,000 per person, largely in C1 countries where carbon emissions are less than 1 t C per person per year, people are much less happy. They lack food, clean water, electricity, transport, free education and health care.

The graph has two features: a *consumption cliff* showing a steep rise in happiness as more income delivers basic needs, and then broadly a flattening out on

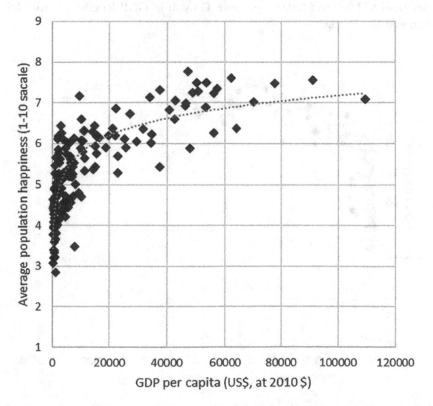

FIGURE 2.1 GDP per capita (average 2015–2019) and happiness (2018) (n=151 countries)

the *affluent uplands*. Put another way, after a certain point, the marginal returns to income are low to zero. The headwinds are strong. What was the point of trying to earn more, then, you might well ask?

In these affluent uplands, there are 18 countries with a similar average happiness, yet their per capita GDPs range widely from US$25,000 to US$120,000 per year. At happiness scores above 7, GDP is lowest at $9500 for Costa Rica; is in the $30,000s for Israel and New Zealand; between $40,000 and $60,000 for the UK, Finland, Austria, Iceland, Canada, Netherlands, Australia and Sweden; and above $60,000 for Denmark, Ireland, Switzerland, Norway and Luxembourg.

GDP does not buy us love or happiness, in the affluent uplands. But policies do matter. Five of the top ten happiest countries in the world are the Nordics and Scandinavians: Denmark, Finland, Iceland, Norway and Sweden. These are countries with relatively high levels of trust, generosity, support for others and social capital. They have chosen to use their income in particular ways that benefit their people's well-being.

A similar cliff and flat uplands is found in the cross-country relationship between GDP and life expectancy, the marginal changes with rising GDP even flatter above $20,000 per capita (Figure 2.2). Again more GDP makes a great difference in the poorer low-carbon countries, but not much after a threshold. There are 29 countries with life expectancies between 80 and 85 years, and their per capita GDPs vary from $22,000 to $120,000 per year. Growth in GDP lengthens a poor life, then stops delivering.

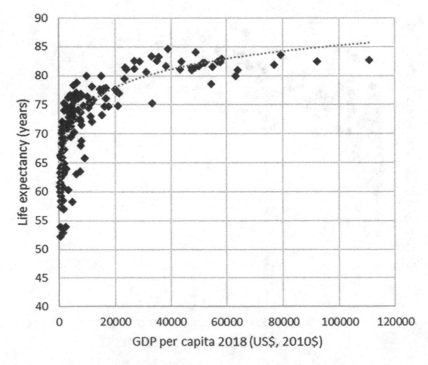

FIGURE 2.2 Life expectancy by per capita GDP (n=143 countries, 2018)

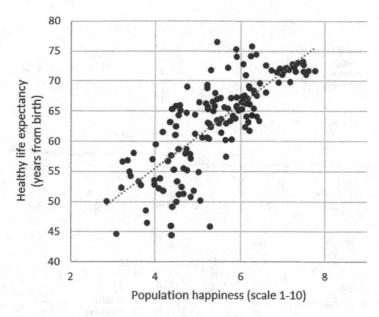

FIGURE 2.3 Impact of happiness on healthy life expectancy (2018, n=146 countries)

We can cross-check the relationship between happiness and longevity. More happiness in countries does lead to more healthy longevity (Figure 2.3: This shows disability-free life expectancy, a different measure to the total life expectancy data in Figure 2.2). The things that are happening in countries to make people happy are also helping them live longer.

On the wind-troubled marsh, you might hear the calls of wild geese and curlew. Look at it this way. Why would you not want to be happy? What good will all those purchased things do you on your deathbed?

The second way to assess the relationship between GDP and happiness is to observe changes over time. Now we can see an even starker picture: every already affluent country has flat happiness over nearly 60 years while GDP per person has grown substantially (Figures 2.4a–d). In Japan, per person GDP has grown by nearly sixfold, in the UK and USA by threefold; in China, recent GDP growth has been extraordinarily fast. Yet, in these countries, people have not on average become happier. These are, of course, the averages for whole populations. We each have our ups and downs, each daily and across the years, yet added together, the patterns are pretty stable.

Each person in Japan, the UK, China and the USA has more stuff than the average person at the beginning of the 1960s, yet has not become happier.

Part of the explanation for this lack of relationship centres on the bad stuff ignored by GDP or measured as good. Averages hide inequalities, which have been growing in most affluent countries. Some people are much richer, but many poorer. These could even up. The relative gaps in income matter for people. GDP also counts creating and cleaning up pollution as good for growth, and ignores the

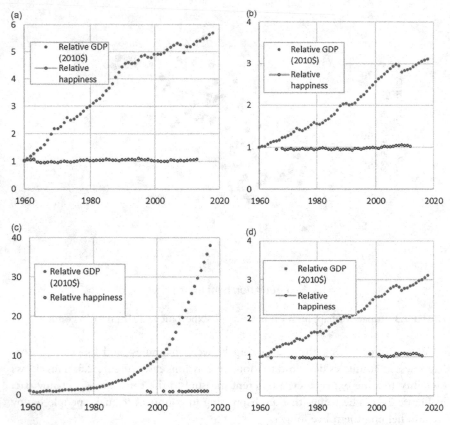

FIGURE 2.4 Relative changes in per capita GDP and happiness: (a) Japan (1958–2018), (b) UK (1960–2018), (c) China (1960–2018) and (d) USA (1958–2018).

value of wage-free labour and volunteering by carers and families. It side-steps the global cost of rainforest destruction.

But we can still conclude: the neon lights of late-stage capitalism have not enhanced happiness. The good life, in short, seems to do with something other than material consumption, bringing us back to the evidence from those ten thousand cultures and their older, better lives.

What the cross-country data also demonstrates is that choices made by governments do matter. They affect the structure of culture and society. Table 2.1 shows data from ten countries with populations of about 10 million. No country is perfect, yet national choices shape life outcomes for their people. People end up living with high or low carbon footprints, with high or low fertility, with high or low health as shown by incidence of non-communicable diseases.

In this batch of ten countries, there are two with very low GDP, high infant mortality, yet very different choices about renewable energy (Burundi and Chad). We have countries at high GDP, low fertility and infant mortality, and again very different investments in renewables (Belgium and Sweden). This is one super-rich country (UAE) with the highest proportion of people with type 2 diabetes,

TABLE 2.1 Comparison of ten countries each with populations of approximately 10 million people (data averages for 2017–2019)

Country	Population (million)	GDP (annual US$ per person)	Under 5 mortality (per 1000 births)	Fertility rate (children per woman)	CO_2 emissions (tonnes per person)	CO_2eq emission (tonnes per person)	Vehicles (per 100 people)	Renewable energy (as % of all electricity)	Proportion of people with Type 2 diabetes (%)
Belarus	9.5	6550	2.0	1.4	6.6	8.5	37	0.7%	5.2
Belgium	11.5	46330	3.0	1.6	8.6	9.5	51	17%	4.3
Bolivia	11.5	2460	21.0	2.7	2.0	13.7	7	28%	6.9
Burundi	11.5	220	40.0	5.3	0.05	0.74	<1	80%	6.1
Chad	15.9	864	69.0	5.6	0.07	5.1	<1	0%	6.1
Cuba	11.3	6640	4.0	1.6	2.3	2.4	4	4%	8.3
Czechia	10.7	22340	3.0	1.7	9.5	10.5	54	11%	6.8
Portugal	10.3	22980	3.0	1.4	4.8	6.7	49	61%	10.0
Serbia	8.8	6500	5.0	1.5	6.2	7.1	29	33%	10.1
Sweden	10.0	57050	2.0	1.7	4.3	4.7	48	56%	4.8
UAE	9.8	40690	6.0	1.4	19.5	27.7	23	0.2%	25.1

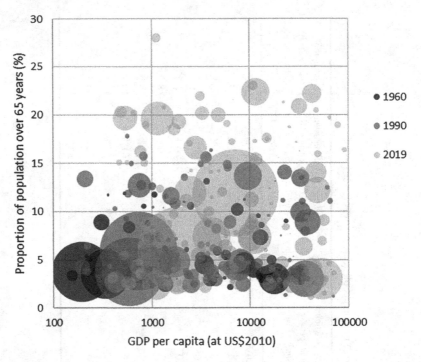

FIGURE 2.5 Changing country population over 65 years of age by per capita GDP
(n=207 countries, 1960–2019)

a pitifully low use of renewables, and the highest carbon footprint. There are two
countries with the same income, yet Portugal produces six times the energy from
renewables than Czechia. There are modest-income countries with very low
infant mortality (Cuba) or high (Bolivia), yet the latter with high use of renewables
(Table 2.1).

At the same time, there are now many more elders. Greater longevity is a major
structural change occurring across all human populations, along with fewer children and greater infant survival (Figure 2.5).

People are living longer, and are more likely to be healthy in later years.

Put another way, an individual is increasingly likely to be able to expect
20 or 30 or more years of active living after stopping the phase in life called
paid work. Older people volunteer more, and volunteers live longer than non-
volunteers. You care for nature, you care for people, you have wisdom accrued
over decades. Now elders have the opportunity to help shape a stable world in
new ways. It is a reasonable expectation to suppose this will lead to divergence,
as choices and values centred on particular places and cultures are expressed
more clearly.

Older people, mostly, also have less need for material goods, and so a shift to
greater longevity will have an effect on consumerist economies. They seem to be
clad in enoughness. Could this, too, affect the journeys towards the good life?

Long conversations
Beside blooming irises –
Joys of life on the road.

[Matsuo Bashō, Edo period, Japan, 1644–1694]

The Rise of Non-Communicable Diseases

There was a mythic dream for affluence, and no one expected it to look like a wreck.

It brought understanding of other cultures: more aviation, more tourism, more cultural mixing. It brought new technologies to make lives easier and reduce drudgery, especially for women in gendered roles within families: the fridge, washing machine, transport, clean stoves. It also brought new lines of food-like substances to restaurants and shelves of supermarkets. It brought cheaper and faster food.

But the side-effects marched in the shadows. Just off the paved road lurked a new monster, invisible at first, then looming large. It seemed so unfair. Just as there was advance and progress, so came the rise of non-communicable diseases (NCDs). So now came, too, the climate and biodiversity crises. As we dreamed, so the ills of affluence grew.

When Odysseus finally came in sight of his home isle of Ithaca, after two decades away, he suddenly fell asleep. He thought he had longed to reach home, to see his wife and son. The crew opened the sack of treasure, let out the wild winds, and they were blown away again. He was asleep just when it mattered most. When he finally made shore, his only skills were in killing and holding an oar, not in making peace and friends for a meaningful life.

At about the start of 1990s, I was part of a project on pesticide transitions, seeking low use and integrated pest management packages that could work for farmers and nature. Every chief executive of the pesticide companies interviewed was asked about their greatest personal challenge. Each of these successful men said something along these lines: it was family life and their children. This was their problem. Their kids wanted them to have a proper job, one that could be spoken of with pride. But those rich executives were stuck. They needed to take up riding a bike, or go fishing. Or volunteer for a food bank. They could buy the rods and flies, resurrect those happy memories when they were ten, running free with friends. But perhaps that would be all too late.

By today, those children will have grown to adults, and the executives will have retired and maybe still not found meaning. It was clear: those younger people seemed to have picked up some understanding of how the world now worked, more than the adults locked into habits of selfishness.

Have we also been asleep this past generation? Change has happened so rapidly, remarkable headway on reducing human fertility and infant mortality, and already the world population is heading for stability. Yet, these past 30 years have brought a disturbing growth in NCDs. Across the world, we solved many infectious diseases, found new ways to fight cancers, transformed medical interventions and treatments, and yet allowed NCDs to stroll on stage. Running alongside the climate crisis, the speed of change has been breathtaking. There are some reasons to be cheerful. There are exemplar countries that have avoided certain NCDs through behaviours defined by cultural norms. There is now the emergence of social and participatory models of health: how best to intervene in the very environments that shape choices.

But the costs are high. For the UK, we calculated that the six NCDs of obesity, type 2 diabetes, cardiovascular disease and strokes, dementias, mental ill-health and loneliness, cost the UK's health systems some £70 billion each year, with a total cost to the economy of some £150 billion. No wonder the UK National Health Service is under stress: its total revenue budget was £120 billion in 2020.

A total of 15 million people in the UK, some 23% of the population, now suffer from long-term health conditions, and these people make up a half of GP (general practitioner) appointments and seven out of ten hospital inpatient days. Long-term health conditions are often termed the diseases of despair. These annual direct costs of NCDs are now a substantial personal and national burden. Poor diet, physical inactivity, fractured social structures, lack of access to green spaces, over-consumption of alcohol and recreational/prescription drugs, combined with some genetic factors, are all causes.

NCDs impose costs on the economy in two ways. Direct costs to health services and the economy, and further side-effects, such as money spent on diets and weight loss, and the emergence of economic sectors that rely on your failure for their success. In no country has obesity or type 2 diabetes fallen once it rose. In some, obesity has stayed low, notably in Japan and South Korea, even with the advent of affluence. Here, strong cultural and spiritual norms have been key.

Recent policy analyses on NCDs by the OECD, World Bank and WHO have this in common: 80% of the recommendations focus on the responsibility of individuals to change. This is convenient and does not challenge underlying economic structures. You respond to economic signals to eat more, you get ill, you are blamed, and then feel guilty. It is notable that no country has yet set binding targets for the elimination of obesity. We can only presume a preference for the food and diet industries. There is welcome talk that smoking may come to be eliminated, in some countries. But there are no NCD targets: not one government has said, we will reduce obesity in our country to less than 5% of adults and children by a defined year. Not one.

This could be why poorer countries may be at an advantage. If they could avoid the rises in NCDs in the first place, they will not have to pay the price of both rises and then the struggle to encourage falls. Creating and sustaining good habits will be key, as will be the structures of whole economies.

According to the WHO, the cost of NCDs worldwide in 2019–2020 was US$10 trillion, about 7% of global GDP. A thought experiment: if this were given to the 800 million people currently hungry and living on less than $1 per day, their annual income would exceed the $10,000 per capita mark, the level at which the worst shortages of poverty and hunger are overcome. One of the reasons for the persistence of poverty and hunger is the economic drag created by the ills of affluence. Real savings would be created by keeping people well and happy, not allowing them to become ill with NCDs and then paying for treatment, and then also measuring this as a contribution to GDP.

Chuang Tzu was one of the founders of Tao, writing 2400 years ago about the time that Lao Tzu was coding and writing the *Tao Te Ching*. He wrote of the importance of enoughness. Perfect happiness came from good times with friends and family, fine foods, beautiful clothes, lovely sights and sweet music, and particularly, "A good position between useful and useless."

Chuang Tzu wrote:

> Great knowledge understands differences between fullness and emptiness,
> Great knowledge knows the straight and quiet road.

Living Well

What does the good life look like today, and how might it help prevent the climate and biodiversity crises. And at the same time make us happier?

The term *the good life* is widely used and has a generally common understanding. It implies contentment and well-being, a life with meaning and a sense of purpose, a life good for us as individuals as well as for others, and suggests doing good is a norm through trust, reciprocity and obligations for people and nature. It is a key component of happiness.

The good life is translated, loosely perhaps, as *buen vivir* in Bolivia, Ecuador and Colombia, *hygge* in Denmark, *ikigai* in Japan, *haoshenghuo* in China, *kalyan-kari jeevan* in Hindi and Sanskrit, *felicidad* in Spain, *bonheur* in France, *het goede leven* in the Netherlands, *koselig* in Norway and *Gemütlichkeit* and *das gute Leben* in Germany. It spans how we might feel at particular moments, say eating together or sitting cosily by the fire (*hygge*), and across a whole lifetime within a particular spiritual framework (*ikigai*). In the UK, *The Good Life* was the title for a popular 1970s TV sitcom, in which good for one couple was associated with not having things, suffering yet being content, and good for their neighbours centred on high material consumption, yet they were constantly discontent. The good life was expressed as dramatic tension: how could any survive with enough and yet be happy?

Guidance on how to live well has a long history. Gilgamesh travelled out into nature, chopped down the sacred cedars, and came to learn that the wilds were key to life. Two and a half thousand years later came the *Tao Te Ching* and *Bible*, Greek and Latin philosophers, the *Koran* and Buddhist sutras, the Norse and Old English sagas and wisdom poems. The worldviews of indigenous groups worldwide matured over thousands of years of continuous settlement into coherent frameworks on the land.

"All my relatives," say the Lakota people, of their old earth culture where everything was connected.

The good life also has Greek and Roman philosophical underpinnings, from Aristotle to the Stoics, where virtue implied good character, an intermediate golden mean point of not too much and not too little, and a contemplative life. Stoics spoke of tranquillity, an acceptance of the transient nature of the world, a gladness in life, serenity and peace, and argued that rich luxuries were counter-productive to the good life. The Stoic Cicero wrote, "If you have a garden and a library, you have everything you need."

Since then the good life has reached national and policy levels at only a few locations. In 1729, the Bhutanese realm observed that the first role of government should be the creation of happiness, leading in the 1970s to proposals about Gross Domestic Happiness as a replacement for GDP. The Buddhist term "right liveli-hood" has come to mean a good life, a living without causing harm to others, and resulted in the foundation of the Right Livelihood Award. China has stated that the good life is now the concern of the whole nation. And the good life known as *buen vivir* has been written into the national constitutions of both Bolivia and Ecuador.

But the more common approach has been to deploy goods and their images to suggest a good life, with explicit national programmes to support material growth. Some governments are tempted each year to encourage more purchases prior to Thanksgiving and Christmas in order to jump-start ailing economies. Others tried to create new aspirations, such as in Turkey the Prime Minister's *two keys to life* for all (a car, a house), and in Sri Lanka the *housemaids to millionaires* programme.

Much environmental literature and recommended changes to behaviours and policy has centred on stopping bad stuff. Modern industrial economies had brought destructive side-effects, and actions were needed to stop or limit pollu-tion, habitat destruction, over-harvesting, human ill-health. The good life remains partially about stopping material consumption, ceasing fossil fuel extraction, but more importantly the focus shifts to creating good activities and sustainable goods.

How can environmental movements spread, it has often been asked, if they ask people to suffer by having to give up something they currently like?

The key to new behaviours will be to create a language about positive choices that address injustice and poverty. This is a journey worth taking, even if it looks scary from the safety of home and existing habits. As it happens, many choices can reduce material consumption and cut carbon footprints, as well as increase well-being. We just need a find a way to help them spread, and develop the social architecture to help take on the harder task of changing habits.

The good life is at the core. All spiritual and religious frameworks contain forms of guidance, both in terms of what we should not do as well as what we should. The cultural and folk concepts of the good life vary by place and over time, yet the central principles are remarkably similar. The tricksters knew. They have been helping for centuries. Respect for nature and respect for people has clearly had evolutionary consequences. It helped individuals, communities and whole cultures survive, it gave them advantage.

The British poet Kathleen Raine wrote in *The Speech of Birds*, "It is not that birds speak, but people learn silence."

The term "good life" retains contemporary salience.

The Domains of a Good Life

In late 2020, I conducted a global survey to seek to understand the elements of the good life. Responses came from 27 countries (see chapter endnotes for some detail). The four most common choices were being in nature, healthy food, togetherness and personal growth/learning (between 12% and 21% of choices). These were followed by physical activity, spiritual and ethical coherence, and sustainable consumption (between 7% and 8%), with the five remaining categories below 3% each. Table 2.2 contains a summary of the rich language and content in each of these 12 domains.

TABLE 2.2 Twelve domains of a good life

Domain	Characteristics
1. Nature	Lying on the grass; big skies; observing and watching nature; beauty; sensate; walking barefoot; open space and long views; sitting in sunshine; walking the dog; pets; healthy ecosystems; the view from home; flowers inside the home; surprise weather; camping; wild swimming; wildlife; chickens, smell of wet soil.
2. Nourishment	Healthy food; food as a gift; preparing, sharing and eating together; sustainable, local and own-grown; tasty; seasonal and fresh; treats; plant-based diets; avoiding highly processed foods; baking and making cake, bread, yoghurt; supporting small farmers; table fellowship.
3. Togetherness	Trust; giving and reciprocity; sharing; gathering of friends and family; intergenerational contacts; volunteering; listening to others; long-term partner, marriage; children; good conversation; circle of friends; the open fire; watching a film together; wearing clothes made by friends; celebration and ceremony; community ritual and festival; visiting and sharing; singing and dancing with friends; attending meetings of protest and prayers; feeling valued; giving gifts.
4. Physical activity and mobility	Regular physical activity; walking, swimming, cycling, boating, gardening, hiking, fishing, skiing, hunting, running, yoga, sport, dance, tai chi; immersion in nature; public transport; electric vehicle and e-bike; no car; reduce and avoid air flights; exploration, discovery, novelty; visiting friends; slow walks; taking time; cleaning the home and do-it-yourself activities.
5. Personal growth	Learning new habits and activities; creativeness; play; active life of mind; making, pottering, tinkering, salvaging, sewing, carpentry; repairing things and goods; craft and art; learning all life; research, data; books; music – playing and listening; gardening – always changing; learning from culture; visiting museums; live theatre and music; charitable work; satisfying and fulfilling work; video games; stimulation; imagination.

(Continued)

TABLE 2.2 (Continued)

Domain	Characteristics
6. Ethical and spiritual coherence	A purpose in life; ideological fulfilment in work; doing good; working for god; optimism; being part of something – social and natural; simplicity; silence, vastness; a path for life; mindfulness, meditation, prayer; spirits in land, water, animals; letting go of things; sharing the good life; relaxing; contentment, happiness; contemplation; accepting things as they are; tranquillity; nature as coherence.
7. Sustainable consumption	Cutting down material consumption; increasing sustainable and green consumption; light footprint; green and ethical choices; buying responsibly; getting rid of stuff; fix and repair; minimise waste, reduce pollution; no air travel; meaningful acquisition; green energy; shop locally; sharing tools and equipment; downsized living; recycling; things made by friends; quality possessions.
8. Enough income	Decent, regular and enough income; job security; financial security; not having to worry; affording comfort.
9. Good health and sleep	Good health; absence of disease; peacefulness at night; comfort; good work; meaningful life; not in pain; slow time; hope for future; refreshing sleep; holidays; inclusive well-being.
10. Home and settlement	Quiet home; solitude; cosiness; sense of place; secure shelter; intergenerational community; comfortable home; living space of home and garden; cooking together; safe and accessible environment; not being too hot or cold; workspace.
11. Supportive public institutions	Health and education accessible for all; public services; local businesses; affordable medical and social care; sense of community; good schools; good health system.
12. Political freedom and trust	Trust in government; freedom of movement and expression; no fear of violence or poverty; free press and freedom of speech; responsible government; human rights; contemplative and caring polity; work that improves the lives of others; capacity to influence.

It is evident there is a link between the detailed components of the good life to health and well-being. It is well-established that being in nature improves mental and physical well-being. Healthy food from sustainable sources improves both personal health and sends market signals to farms about the importance of food production that increases biodiversity and ecosystem services. Greater social capital in the form of togetherness improves health and happiness, and members of social groups are happier than non-members. Regular physical activity increases health and wards off many non-communicable diseases. Personal growth is a key part of engaging with learning and new activities and skills. A coherent spiritual and ethical framework is seen by many as a wrapper for meaning to life, giving further strength to choices and behaviours. Sustainable consumption improves well-being and satisfaction mainly through knowing, as people are acting in ways that do good for the planet.

If we think instrumentally for a moment, we could conclude that the adoption of these components of the good life that lead to greater well-being, and better health would reduce incidence of the ills of affluence.

This would save countries money.

An additional observation from these findings on the good life is that 94.8% of the identified activities and preferences are also low in carbon emissions. Only one response explicitly mentioned material consumption (shopping), and this was framed as an activity for togetherness, namely going shopping with friends. We shall see in the next chapter that the global challenge can be framed around needing to cut per person carbon emissions to one tonne per person. Adopting more components of the good life would help.

A subset of respondents to the survey were residents of 16 low-impact communities and ecovillages in the UK. There are some 300 communities in the UK with an on-site membership of about 10,000 adults and children. Some have religious foundation, others are secular. All share values of self-reliance and common values, living collectively, usually eating together, consideration for others, sustainable consumption and low-impact living. They have come together partly to live lightly on the land, and share concerns about how to limit resource and energy use. Their emphasis in responses was higher on togetherness, healthy food, personal growth and sustainable consumption. There were no responses relating to external and social conditions off the site. Low-impact communities have already taken a deliberate step away from the ways of living common elsewhere in society.

Karen Litfin of the University of Washington has written of the ecovillage movement: the exemplars, experiments, transgressors and inspirations. All over the world are places of "outer simplicity and inner wealth," where people are intentionally trying to create new ways of living. "How then shall we live?" she asks. Some have become famed and iconic: Findhorn, Auroville, Crystal Waters, Sieben Linden, Konohama. A Global Ecovillage Network links these efforts together, helping to share ideas, joys and sorrows. Carbon footprints are lower than average within their countries, as low as 10% in some/this would meet, or come close to, the target of one tonne per person.

Ecovillages seek to promote well-being within limits, creating new norms of positive action and function, values in communal activities of dance, singing and ritual. And yet there remains an unease. For some people, an ecovillage or low-impact community is an escape. There is no desire to influence back the big and ugly system out there. For others, it is an experimental space precisely to develop innovative solutions for wider society, as the Centre for Alternative Energy does at Machynlleth in Wales. Tendai Chitiwere of the San Francisco State University points to another problem. People depart from these projects after dissatisfactions arise, and leave behind only a certain type of person, who ends up in "voluntary quarantine." She is critical of a tendency to exclude social justice from the practice of some ecovillages, and thus they can appear privileged and exclusive. Such "green flight" leaves others to address the wider social and ecological challenges of the world.

A jam tomorrow proposition, in which we suffer now for later benefit, has dogged the environmental movements. People need jam now. We just need a way of framing change about the good things in the good life, not the giving up of things people thought were good but were actually bad. These low-impact communities illustrate what can be achieved with intention. Members have already decided they wish to live in a particular way. They have stepped across a threshold, and have adopted new habits. They are a powerful symbol of the possible and already have low carbon footprints.

Dorothy Schwarz and Walter Schwarz, the then Guardian environment correspondent, travelled to low-impact communities and ecovillages across the world for their book, *Living Lightly*. Such sustainable living, they wrote, meant people could, "Redefine the good life." There are, they also wrote, infinite possibilities. Dorothy lives on the banks of a reservoir and sleeps outside at night by the wildflower meadow. We sat in the garden, and she said, "There are so few insects these days, what can be done?" In the aviaries, the macaws and other parrots chattered.

What more ought she do, especially now?

The way people live in low-impact communities and ecovillages is often called an attentive lifestyle. If we are attentive to nature and community, we may find some good questions pop up, and then possibly some answers too.

The good life can bring happiness and life satisfaction, and at the same time take each of us on a path of positive choices to reduce individual carbon footprints. If this sounds hopeful, then tell a story to someone else, and persuade them to act too. In 2011, Lars Tornstram of Uppsala University wrote of gero-transcedence. Eight out of ten of the 75–100-year-olds he asked agreed, "Today material things mean less than when I was fifty."

They had found a sense of release, took themselves less seriously, felt greater connections with the universe, and found the border between life and death less striking.

One respondent said, "You get a little out of breath, but the view is much better."

These are the kinds of ontological changes that can occur across the individual life course. The global challenge is now how to spread this sense of the good life to everyone at every stage of life, and then create enough collective action to change economies and cultures.

Meister Eckhart, the medieval priest and mystic, wrote:

> Cling if you want, to things,
> But where, then, will your heart be?
> Pay attention to the way you live your calling.

~*~

3

THE CLIMATE CRISIS

The Safety of One Tonne Each

You watch – it's clouded,
You don't watch, and it's clear –
When you view the moon.

[Miura Chora, Edo Period, Japan, 1603–1868]

Abstract: This chapter sets out the details of the crisis of climate change. It has been coming a long time, and then suddenly accelerated. Since the pre-industrial era, world temperature has warmed by +1.2°C (into the early 2020s), over land by 1.5°C, over oceans by +0.75°C. Each season, new records are set for high temperature, deeper drought, rainfall and flood, wildfire, melting permafrost, shrinking glaciers. This is not weather; these are signatures of a greater malice. Our world of 7.8 billion people produces some 53 Gt (billion tonnes) of CO$_2$eq (carbon dioxide equivalent). A safe place is about 1 tonne per person, around 10 Gt. Data on 222 countries is presented: GDP against carbon emissions. A large number of countries and people emit on average less than 1 tonne per person: they need to increase consumption to escape poverty. The majority need to reduce inequality of carbon emissions, which also plays out within countries. The top 1% wealthiest in Europe and North America emit 55 tonnes of CO$_2$eq per person. The world average is 7 tonnes. And a new twist. This transition to 1 tonne may not be hard. "An era is ending," has written Danny Dorling. This chapter concludes by setting out a schedule of 30 personal behaviours and choices that show how to reduce personal emissions. Start somewhere, and the race to net zero is on.

DOI: 10.4324/9781003346944-3

All Things Are Connected

The weavers all agreed, a sacred yarn connects it all.

Looking back, they will say of these days, it seemed the fabric of the world had frayed and faded. Weather was behaving badly, warm days were getting hotter, breeze became storm, there were seasons of drought, then relentless rain fell. Ice was melting at the poles and summits, heat had arrived and fire seasons spread across the year.

It was no longer right to call this climate change. It was a crisis, the sail ripped to shreds and every roof now weighted down with stones.

The Earth's life support systems were in peril. Some were saying, climates had changed before, and would do so once again. Well true, but this time the shift had wholly human causes. And as with many such affairs, the ill effects were distributed unequally.

There were a sizeable number of people who had been drawn to believe the corporates and kings when they said there was much doubt. That we all should wait and see. A fall might await, but you can also fall upwards. I was once a guest of a Louisiana family, out beyond the town by the wild Atchafalaya Basin, the long bottomland swamp at the end of the Mississippi. Katrina and Rita had previously passed by, and I had watched miles of red tail-lights stretching north, the coastward tarmac empty, as the news tracked another inbound hurricane. But words alone could not this time stitch any kind of myth for persuasion. We stood instead in the garden with the kids, by the singing world of frogs in the wetland, counting shooting stars and constellations.

When I had sat with Brenda Dardar-Robichaux, then the principal chief of the Houma Nation at the seaward end of the Atchafalaya, she was vainly fighting for federal recognition of the tribe, and at the same time trying to build back pride for traditions, especially among the children. Basketry, doll making, weaving, community drumming, dance: all these were traditions known well by elders, but less by youngsters. "We are making efforts to revive this knowledge," said Brenda. And yet, they were also losing their land to the sea. Elizabeth Kolbert counted up, in *Under a White Sky*, and found formal records that showed 31 place names had been retired from the coastal parishes of Louisiana. They are already under water.

Yet, even then we knew, this crisis could wreck the economy and social structure for those children. They will need new stories that could lead to hope and transformation.

This had been building for a while. Since the advent of the industrial revolution, from the late 1700s and into the 1800s, total and per capita greenhouse gas emissions, mainly of carbon dioxide, methane and nitrous oxide, into the clean atmosphere have risen steadily then sharply. Figure 3.1 depicts per person carbon emissions for the world and nine selected countries. Carbon-dioxide-equivalent (CO_2eq) is a better measure of total greenhouse impact, as adopted by the United Nations Framework Convention on Climate Change, but the historic data on

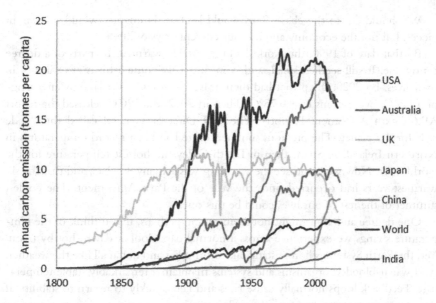

FIGURE 3.1 Carbon emissions per capita by country and world, 1800–2019

methane, nitrous oxide and other minor greenhouse gases is less accurate. Carbon gives us a clear picture of a troubling history.

What we see is this: world population has grown, yet emissions faster. For some affluent countries, there has been a downturn in per capita carbon emitted since the year 2000, but from dizzying heights. Total emissions are too high, if we wish for safety. Rapidly industrializing countries, China for example, have seen large increases since 2000.

But many other countries are still low emitters, and their route to regenerative economies could turn out to be easier.

Back, then, to the year 1990. Since the appearance of modern humans about 200,000 years ago, this climate crisis has crashed ashore during the past one generation of about eight thousand. If this phase of the human world were crammed into a single day, then these 30 years came in the ten seconds before midnight. And in that year of 1990, this was when the carbon content of the atmosphere for the first time exceeded 350 ppm (parts per million).

In the pre-industrial era, the concentration was 280 ppm; by the early 2020s, it had passed 415 ppm and the first monthly averages of 420 ppm were being measured at Mauna Loa Observatory.

Now a clamour has begun: 350 ppm was a safe operating space for humanity; we need to find a way back. It is now a worldwide target, but will take decades to reach, even if emissions stop. The best we can hope for is stabilisation, and then a slow decline. But the path looks perilous. Levels have recently been rising by about 2 ppm per year. By 2050, one more batch of 30 years, concentrations might have exceeded 470 ppm.

We should cut to the chase. This would be bad. Humanity would survive, in pieces, but not the economy and not our current ways of life.

By that date of 1990, the world had measurably warmed, by parts of a degree centigrade. It still seemed small, and who didn't welcome a bit more warmth in their lives. By 2020, the planet had then heated by +1°C, over land by an average of +1.5°C, over oceans by +0.75°C. Then in 2022, the IPCC released the report AR6 (Sixth Assessment Report). The UN Director General called this "code red" for the planet. The previous summer, there had been record temperatures in Northern Ireland, in Sicily, then in Death Valley the hottest temperature for the world, ever. Now we could see the warming was actually +1.2°C, and that the six warmest years had occurred since the year of the Paris Agreement. The coolest summer of the rest of our lives could be this one.

One degree and a bit might not seem much, perhaps, if you think of the temperature swings we experience across a normal day. Cool at dawn, hot by noon. Yet, the Earth System is homeostatic, like our bodies and not at all like the weather. And warm-blooded organisms and systems maintain a remarkably stable temperature. Feedback loops normally cause rises and falls quickly to return to stability. If the temperature stays away from normal, it can be serious.

So think this way: we call a +1°C rise in body temperature a fever. It is a sign of illness. At +2°C, the fever has become serious, the raised heat is fighting the infection. At +3°C, there is organ and structural damage, and we should be on the way to hospital. At +4°C, at this high fever, death could shortly follow.

Some commentators and politicians are still talking calmly of Earth's increases of +2°C to +4°C during this century being survivable.

In some ways, this analogy of the body underplays the danger. Under building pressure, systems can display sudden changes, trip cascade effects, and thresholds can be passed and broken. One set of changes can cause positive feedback, making things turn worse very quickly.

"Tell all the truth," wrote American poet Emily Dickinson, "but tell it slant." She never saw a moor, never saw a sea, and said:

> I gained it so, by climbing slow,
> Between the bliss and me.

Carbon Emissions and Consequences

It is tempting to write: no one could have imagined this.

The International Panel on Climate Change (IPCC) was established in 1988 by the United Nations Environment Programme and the World Meteorological Organisation. It had the support of affluent countries, including the Reagan administration. It was at first a cottage industry, emerging out of wider environmental concerns. The Brundtland Commission on Sustainable Development had published the first worldwide report on sustainable development,

Our Common Future, showing how environments and economies were linked, and set out how we might protect nature and at the same time reduce poverty worldwide to zero. The Rio conference in 1992 then seemed to bring confidence and hope. But there were two snags to both: the speed of climate change was missed, as was the escalation of obesity and other non-communicable diseases of affluence.

The rate of change, since then, has been unprecedented in human history, save for acute natural calamities and volcanic events.

For years, the process was largely called global warming, evoking pleasant summer evenings or lack of winter frost or ice. Now we really should settle for *crisis* or *emergency*.

But to be persuasive to skeptical governments and corporates, the reports of thousands of scientists had to be cautious.

Habits and economic structures do not change easily. Nor did self-interest and fossil fuel extraction. So, scenarios and confidence limits were framed just so, and no one wanted to be called alarmist. And anyway, many economists still believed the economy could be grown out of any trouble. Since 1990, the IPCC produced five assessment reports, then a sixth in 2022. Time passed, there were also 27 conferences-of-the-parties (COPs) by the Egypt event in 2022. Would governments agree on individual responsibilities and collective actions to seal common progress? Or would they speak one way and act in another?

The Kyoto Protocol of 1997 brought the first commitments to reduce greenhouse gas emissions. But by the time of the Paris Agreement of 2015, total carbon emissions had grown over the intervening eighteen years by 45% and per capita by 16%. Aspirations were amended, and now the target was to hold net worldwide temperature increases to +1.5°C or +2°C. Neither is a happy place. Yet, if Paris and later agreements are not successful, the world may be living with a fever of +3°C to +4°C during this century.

Annual total emissions of carbon dioxide alone had grown by 2020 to 40 Gt (gigatonnes or billion tonnes); the total for carbon-dioxide-equivalents (CO_2eq) was 53 Gt. Without effective action, CO_2eq is set to rise to 65 Gt by 2050. The Paris Agreement set new reduction targets for world emissions: to 36 Gt to limit heating by +2°C in total, or to 24 Gt to limit heating to +1.5°C.

But neither will be enough. Net zero has now emerged as a common target – a balance of activities that take up or sequester carbon from the air and those that pollute. Even this will be hard, as total atmospheric reductions depend in part on how long pollutants stay in the air: CO_2 is long-lived, methane short. But if we could get to about 10 Gt, this would have stopped increases, and begin the return towards 350 ppm.

Put another way, 10 Gt for 8–9 billion people, give or take a bit of rounding, and this is about One Tonne each.

Perhaps we might sketch a small portrait of collapse, without being too grim. First a cue, as a complete change in course will be needed, sailing south instead of north, up instead of down. It will happen either by choice, now and over in the next 10–20 years, as green deals and economic packages are agreed upon. Or it will happen after the crashes of the climate crisis, thus a little later, but after ruin of the current economic system reliant on material consumption and fossil fuels.

Sorrow could be heaped on sorrow.

People living close to coasts will be forced to move inland and up as seas rise; some will have to migrate to other countries. Expensive assets in marine-side cities will be stranded. Some will rush to the hills, though wildfires will be more common. Insurance companies will cut cover for the risky businesses, places and assets. Investors will shift away as some cities become unviable. Migration will now comprise the once-rich as well as the currently poor.

Some regions will have become unbearably hot, others torn by wind and storm. International travel will cease, traded food systems collapse.

Some 60%–95% of all the 46,000 glaciers in the Himalaya will have melted, and somehow a new source of drinking water for 1.6 billion people living in the major river basins on the plains of Asia will have to be found. A +1.5°C rise for the world will have translated into +1.8 to +2.2°C for the Himalaya. Melting icecaps in the Andes have already led to severe water rationing for the residents of the Bolivian city of La Paz.

The climate crisis will quickly have become an economic crisis, if it hasn't already.

Some 2500 years ago, Lao Tzu (also known as Laozi) wrote, "The best businesses, serve the common good." But this has not worked this way in the modern era of studied selfishness. The neoliberal economy and material consumption shifts externalities away to others, either in space or over time. A corporation owns ten factories or a thousand trucks: if it does not pay to prevent or clean up pollution of air and water, it looks in better shape.

A food retail company has a thousand restaurants worldwide, a soft drinks company sells a billion servings of its sweet beverages. Neither pays for the obesity or type 2 diabetes that strike many of their own consumers. They are contributing to GDP, the economy is growing.

At the same time, another ingenious play, for the cleaning up of externalities is then counted as a positive addition to GDP. This all works by encouraging more, not less: more consumption, more externalities, more conversion of forests, more pollution, more ill-health, more inequality. Not less.

Externalities are a clever trick born out of the idea that sub-systems of the economy are closed. Yet, nothing in our Earth system is really an externality. The planet has a fairly distinct boundary, letting in and out only light and other forms of radiation (the occasional space vehicle and meteorite too). The way externalities work within the planet system is simply to move negative effects from one place to another, from this business to that person. Externalities are, in short, all about inequality, about who does the shifting and who cannot escape their effects.

Many indicators show what happens to consumption in countries as GDP rises. Just the one fossil fuel, oil, is instructive enough (Figure 3.2). Both scales are log

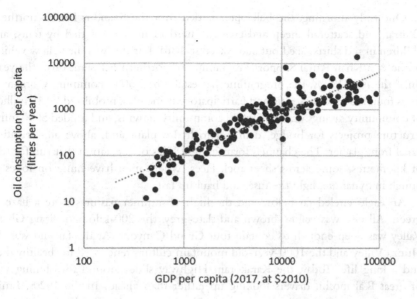

FIGURE 3.2 Per capita oil consumption (2019) by per capita GDP (2017) (n=181 countries)

scales, where each unit increases GDP or oil use by tenfold (an order of magnitude). Each too is the measure per person in each country. As countries have become richer, so oil consumption per person has also leapt. A tenfold increase in GDP brings a further tenfold increase in individual oil consumption.

There is a hint here about which countries and which people have the greater responsibility to change their consumption patterns to meet a One Tonne target. It is also clear: many countries consume less than 100 litres of oil per person per year; many others are approaching or exceeding 10,000 litres per person annually. And so we see, rich people and rich countries consume more and contribute most to the global climate crisis.

You can see how the global COVID pandemic looked like a rehearsal for something much worse. Now the globally connected world could come to an end. And nature return, as the Earth recovers. So, don't worry about the planet. It will be fine. Do worry about many species, many vital habitats: they will not be able to move and will be extinguished.

But nature as a whole? Over time, it survives and regenerates.

Do worry about humanity. We will be in a mess. This could be what the world donated to our children looks like, and to their children. And for you, if you are young now.

Look at it this way. Where today there are glaciers, there is life.

In the mountain desert of the Pamirs and the Hindu Kush at the western Himalaya, there is no gentle transition at the edge of villages. Green one side, a mosaic of barley, maize and orchard trees, on the other dry dust and rock.

One early morning, the helicopter rattled over the Shandur Pass in northern Chitral, and scattered sheep and goats. A herd of horses was tied by tents, and children in red shirts raced out and waved so hard. For this was the yellow vehicle of the Aga Khan Rural Support Programme of northern Pakistan. Over the years since the mid-1980s, the programme has established 5000 community organisations founded on democratic and participatory principles, mobilized US$5 million of community savings, trained 50,000 community activists, and funded 5000 infrastructure projects for bridge, dam, micro-hydro plant and, above all, irrigation canal from glacier. The channels form straight lines of vegetation that run for tens of kilometres, some across sheer rock faces, where men have hung on ropes to punch in dynamite, light the fuse, and haul up fast.

An eagle circled far below and the highest summer pastures were a haze of green. All else was yellow-brown and slate-grey, the 200-kilometre-long Ghizer Valley was deep enough to contain four Grand Canyons. At the far end was the Hunza Valley and the 1000-year-old mountain culture renowned for healthy diets and a long life. Today, the Karakoram Highway slides around the shining wall of great Rakaposhi, drivers looking up rather than ahead. In the 1920s, Emily Lorimer wrote of life in the Karakoram: "The people are hospitable, courteous and polite, and own their soil and live their own lives and can look the whole world in the face."

She recorded a culture centred on sufficiency and ingenuity, and a simple diet of fresh and dried apricot and mulberry, cereal and vegetable, yoghurt, apricot oil and small amounts of lean meat. Several visitors have since then promoted the Hunza diet as having remarkable properties, yet it was just this: healthy, varied and not too much of anything.

Their ways of living were on a land filled with the spring song of lark and cuckoo, and in summer the shadows were cool in poplar plantation and orchard.

All this, and a distinct culture entirely reliant on glacial melt for a way of life that had no negative impacts on others.

> A day will no doubt come,
> When dust flies at the bottom of the sea.
>
> *[Po Chū-i, Tang era, China, 772–846]*

Converging on One Tonne Per Person

We now return to that idea of a One Tonne target, and consider four categories of carbon emitters. The world picture is highly skewed. Not only have affluent countries produced more carbon historically, thus polluting the atmospheric commons first, but their current production of carbon and other greenhouse gases is also still high.

Figures 3.3a and 3.3b show the annual carbon dioxide and CO_2eq emissions per person against GDP per person, with the countries of the world separated into four categories. The world average per capita emissions are 4.6 tonnes of CO_2 and 6.4 tonnes of CO_2eq.

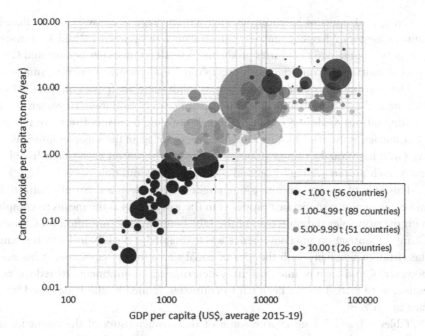

FIGURE 3.3A Per capita carbon dioxide emissions (carbon only) by per capita GDP (countries = 222, populations by bubble size; world average 4.72 t/person)

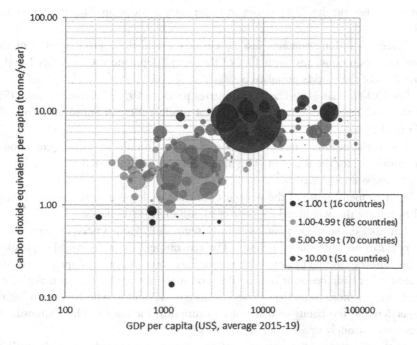

FIGURE 3.3B Per capita carbon dioxide equivalent emissions (all greenhouse gases) by per capita GDP (countries = 189, populations by bubble size; world average 6.89 t/person)

Category 1 contains 56 countries each with average C emission of less than 1 tonne; Category 2 has 89 countries with emissions of between 1.0 and 4.9 tonnes; Category 3 contains 51 countries producing 5.0–10.0 tonnes per person; and Category 4 contains 26 countries, each producing 10–50 tonnes per person annually.

The distribution is similar for CO_2eq, though there are fewer countries below 1 tonne as the CO_2eq data includes the climate costs of land use change, fertilizers emitting nitrous oxide and ruminant animals emitting methane. But there still are 51 countries in the 10–50 tonne category. These are again the super-emitters. Rising GDP has brought rising carbon emissions: both axes on each graph have log scales, each unit increasing by tenfold on the one previous unit.

This idea of a common one tonne target worldwide is a form of contraction and convergence, a model first proposed in the early 1990s as the means to comply with the principles and objectives of the UN Framework Convention on Climate Change. Contraction and convergence was first tabled at the UN in 1996. It retains the virtue of simplicity with the idea of equal allocations of emissions. It has not, however, found its way into nationally determined commitments to reduce net emissions to zero. This has been left to countries, framed by the wider need for a race to net zero.

Tables 3.1 and 3.2 summarise some of the characteristics of the countries in each of these four categories of carbon emissions. Let's called these countries: C1 Poor, C2 Rising, C3 Affluent and C4 Excess. Now inequality becomes apparent.

For CO_2, the 145 countries in Categories 1 and 2 produce 23% of the world's carbon emissions; the 77 countries in Categories 3 and 4 produce the remaining 77%.

Category 4 alone contains just 26 countries, and these produce 30% of the world's carbon emissions, with 10% of the world population. Categories 1 and 2 have 60% of the world's population.

For CO_2eq, Category 1 and 2 countries produce 19% of emissions from 45% of the world population; Category 3 and 4 produce 81% of emissions from 55% of population.

This is the headline: not everyone in the world should be cutting greenhouse gas emissions, for the moment. The moral duty lies with the richest.

We are waiting patiently.

As it happens, some of the poorest countries are leaping to renewables fastest. They can see this would reduce their costs and increase resilience.

The 80% of the world population in Categories 2, 3 and 4 need to reduce their average carbon emissions to 1 tonne. The remaining 20% of the world population in 56 countries should meanwhile be *increasing* consumption to escape poverty and at least meet basic needs. It may be that countries in Category 2 can make structural changes more easily than those in Categories 3 and 4, who have been highly dependent on the habits of material consumption for longer. The responsibility bears most strongly upon Categories 3 and 4.

At the same time, the iron cage of consumption has produced high levels of non-communicable diseases (NCDs) in affluent countries. They have much to

TABLE 3.1 Characteristics of 222 countries in four per capita emissions categories by both carbon dioxide alone and carbon dioxide equivalents (all greenhouse gases), 2019

	Category C1 *<1 tonne per person*	*Category C2* *1.00–4.99 tonne per person*	*Category C3* *5.0–9.99 tonne per person*	*Category C4* *>10 tonne per person*	*Total*
Carbon dioxide only					
Number of countries	56	89	51	26	222
Population (million)	1340.6	3349.1	2320.9	699.0	7709.7
Carbon (Gt)	0.52	7.73	16.74	10.42	35.41
Proportion of CO_2 in each category (%)	1.48%	21.84%	47.27%	29.42%	100%
Carbon dioxide equivalents (all greenhouse gases)					
Number of countries	16	85	70	51	222
Population (million)	144.4	3355.1	3243.3	966.9	7709.7
Carbon dioxide equivalents (Gt)	1.17	8.62	24.65	16.10	49.47
Proportion of CO_2eq in each category (%)	0.22%	17.53%	50.07%	32.18%	100%

gain from changing ways of living. These NCDs are a costly drag on their economies, more so to affected individuals. And all that extra GDP, well it has not made people happier either. And many countries have seen inequality grow within their borders, which makes all people less happy, not just the have-nots.

And costly NCDs are all a bit of a mess.

Not surprisingly, the inequalities of carbon emissions play out sharply within countries too. Diana Ivanova and colleagues at the University of Leeds have published leading work on carbon emissions across Europe, as well as on the best options for individuals to reduce their carbon footprints. Across the continent, the wealthiest 10% of people emit 25% of the carbon. The top 1% have individual carbon footprint emissions of 55 tonnes per year. These top 1% have a greater carbon footprint than the bottom 50% of earners.

Some 5% of people in Europe have footprints already below 2.5 tonnes, and only need to halve their footprints to come close to the One Tonne target. The richest 1% have quite a task, if they feel inclined to be responsible (Table 3.3).

This is why Naomi Klein was saying, all things are connected.

TABLE 3.2 Summary of features of four categories of 222 countries by per capita carbon emissions

Category (number of countries and population, 2019)	Title	Descriptor of current status	Average per capita carbon dioxide emissions within category	GDP range (per capita), US $ (2018)	Ten example countries in each category
Category 1 56 countries 1.34 billion people	Poor	Lacking basics, in poverty, need to increase consumption	Less than 1.0 tonne	Less than $1000	Bangladesh, Burundi, Cambodia, Chad, Haiti, Kenya, Mali, Nepal, Samoa, Tanzania,
Category 2 89 countries 3.35 billion people	Rising	Escaped poverty recently, creative, more aware of enough, need to substitute with green consumption	1.0–4.9 tonnes	$1–10k	Algeria, Botswana, Brazil, Colombia, India, Mexico, Romania, Sweden, Thailand
Category 3 51 countries 2.32 billion people	Affluent	Already locked into material consumption, wanting more, restoration possible with cuts	5.0–10 tonnes	$10–20k	Bahamas, China, Denmark, Germany, Italy, Japan, South Africa, Spain, UK
Category 4 26 countries 0.70 billion people	Excess	Excess consumption, not happier, redesign of economies required	10–50 tonnes	More than $20k	Australia, Bahrain, Canada, Kazakhstan, Kuwait, Qatar, Russia, Saudi Arabia, UAE, USA

TABLE 3.3 Carbon emissions footprints within Europe (former EU-28) according to income category

Income category	Proportion of total carbon footprint for European Union	Average CO_2eq emissions per person per year
Top 10%	27%	23 tonnes per person
Middle 40%	47%	15 tonnes per person
Bottom 50%	26%	5 tonnes per person

The problem is the climate crisis; the source is human emissions of carbon and other greenhouse gases; the largest cause comes from the rich and not the poor. This era has been called not just the Anthropocene, describing how we humans having a geological-scale effect on the planet, but also the Capitalocene and Econocene.

We have also seen that the behaviours and activities that make up the good life tend to be low in carbon emissions. Long and happy lives safely within climate and biodiversity boundaries lie ahead. But first, much will have to change.

This would be a good moment to reflect on the targets set by governments and corporates, and by Article 4 of the Paris Agreement itself.

Many have signed up to net-zero targets by 2050, and some more recently to bring targets towards 2030. The word net implies a balance between taking up of carbon from the atmosphere, and the production of emissions. This is a natural state: plants sequester carbon by photosynthesis on land and in oceans, living things give off carbon dioxide by respiration; methane escapes from swamps and deep oceans, carbon is given off by fires. And now there are human activities that will increase carbon capture and storage. Trees capture carbon, so we should plant more trees. Regenerative agriculture methods increase carbon in soils and above ground, so we should amend agricultural systems. Engineering solutions are being developed to capture carbon in large amounts: forty major projects that each would capture 1 Gt of carbon per year would get us close to the 10 Gt target for emissions.

But some wisdom is needed. A focus on sequestration alone could appear to let the affluent polluters off the hook. If someone can take up carbon, then richer individuals and businesses could pay them through carbon trading schemes for this capture, and then continue with their own emissions. Business as usual, for them, followed by amnesia. The Race to Zero campaign of the UNFCCC does suggest this guideline for signatories: 80% of net emissions change should come from real cuts in emissions, and just a maximum of 20% from carbon capture.

The world needs a halving of emissions by 2030, at a minimum, and then another halving over the next decade. There is an urgent need to act. We are well on the way to 2030, and it would be easy to imagine the talking still happening as each year passes. So, who should be acting? Individuals and governments, businesses and cities.

But we know this too: governments, especially those immersed in neoliberal philosophy, like to put the blame for wrongs on individuals.

You could find them saying: cut your consumption of food to reduce obesity, cut your carbon emissions to reduce climate change – while secretly hoping that people will not consume less. Yet, only national and local governments can change energy systems; only they can put in the infrastructure to encourage walking and cycling; only they can incentivise the price of electric vehicles and solar panels to encourage uptake. Technically, policymakers in national and local governments, in cities and companies, can do much, right away. They would get more support from voters and consumers. Equally, you and I as individuals can change diet, make different travel choices, consume less, choose the good life more.

We could also get caught up in the real difficulties of calculations for countries. A national average for emissions would fall if national population increased, and consumption did not. Should we think of individuals or households as units of assessment and change? What about differences between adults and children? How should we account for national responsibilities: within borders or back to the beginning of supply chains? What about unfairness: there is no easy way to give credit for having taken action already, as there is so far to go collectively. The CO_2eq footprint for meat varies by animal, by system of production for an animal, how they are fed, and whether feed comes from existing farms or from newly cleared forests.

It would be easy to hide behind these complexities, to conclude there is too much doubt to act. This is a well-worn playbook used elsewhere.

In the end, the challenge is simple and moral. We all are crew in the same boat, so the same endpoint applies to everyone. Naomi Klein adds, "There is no grey area when it comes to survival. We are all fighting for our lives." And for our children and grandchildren.

Each person worldwide needs to step over to a path to an average carbon footprint of no more than One Tonne each. This is not about suffering to prove moral superiority. This is about joining a worldwide change in material consumption, a shift to sustainable consumption, and a focus on those activities that make up the good life. For it so happens, many of these are low in carbon too.

And a new twist here. This might not hurt. "An era is ending," says Danny Dorling in *Slowdown*.

> How should I follow a rugged road?
> Go straight, he said.
>
> *[Korean Zen Master Taiwŏn, 1936 –]*

Thirty Options for Carbon Reduction

There are things we can do as individuals and organised in groups, and things we need to encourage governments to do. So, here we go. Let's look at carbon, and focus on affluent countries and their reduction path. We will come back to the need for collective action, the creation of new forms of social capital, and leadership from governments. Meanwhile, the focus here is on what individuals can do to change their own carbon footprints.

Here is an example. We have been using this 30 for 30 campaign in eastern England as part of the Essex Climate Action Commission's work to engage with local communities. If we are going to make the transition to net zero, then saturation across all groups and places is a necessity. This is the key exam question: how to grab attention, and then give people the opportunity to engage in ways they prefer?

The Carbon Schedule has 30 options in the five domains of food, home, mobility–transport, stuff and leisure–work (Figure 3.4). These data are drawn from analyses of affluent countries led by Diana Ivanova and Daniel O'Neill of the University of Leeds. Give or take some differences in baselines, they reasonably

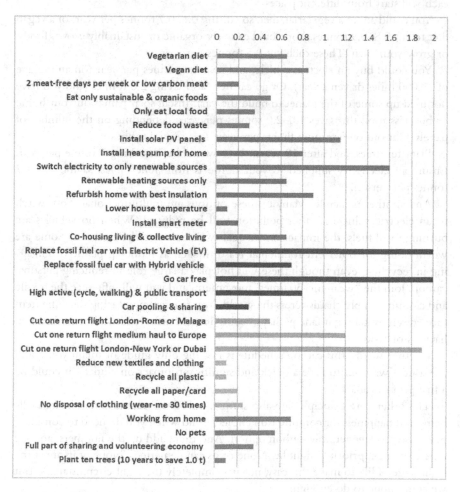

FIGURE 3.4 Thirty for 30: Thirty personal behaviours to reduce your annual carbon footprint for 2030 (tonnes C saved per person per year)

represent those in most affluent C3 and C4 countries, and the wealthier groups within C1 and C2 countries. They point to the priorities. And remember, we're trying to find enough options to reduce our individual carbon footprints by several whole tonnes.

Some of these 30 options require prior resources, for example, to purchase solar panels or an electric car. But they shift investment from running costs to capital: once installed the running costs of daily expenditure come down. Others are choices centred on not doing something so much or at all. Flying is an example: we will have to wait for the development of renewable fuels for planes before they become a low-carbon option.

The point is this. None of us can be perfect. We cannot do everything. We also each will start from different places.

You could adopt a vegetarian diet, so cutting 0.6–0.7 tonnes per year; or a vegan diet to cut 0.8–0.9 tonnes. You could eat only organic or sustainably grown foods, or grow your own. These each cut 0.5 tonnes per year.

You could buy an electric vehicle, and save 2.0 tonnes per year (on an average of 10,000 miles driven a year). Or go car free, also saving 2.0 tonnes (and you have not used up some of the planet to build the car). Install solar panels on your home or business roof: this saves 1.0–2.0 tonnes per year, depending on the number of panels. Cut out one or more flights per year: save several tonnes.

Plant ten trees, and after 10 years, these would be absorbing 1 tonne per year. Ensure all electricity supplied to your home is from renewables, saving 1.5–1.7 tonnes per person.

And another principle. Many of these options are multifunctional. You switch to an electric vehicle, and air pollution will be reduced. When no vehicles are burning fossil fuels, the impact on air quality in cities will be dramatic. Some are worth doing for other environmental reasons, such as reducing plastic or engaging in recycling, even though these will not play a large role in reducing personal carbon. Join the "wear-me-30 times" movement, and you will influence the textile and clothing supply chains across the world. You may enjoy wearing the same item more often, watching it fade perhaps, seeing it grow older with you. And then you have a story too.

In other words, this carbon schedule is not the only guide or menu you could use. But if we want to have a quick and significant impact on climate, it could be a first port of call.

The other core principles are also simple. Substitute activities that bring well-being and happiness for consumption choices. And when you do need to consume, prioritise the greener, low-carbon option. And then, add up the numbers, and see where the next priority might lie. None of us can be perfect. None of us are at the right stage of life to make this easy, nor in completely the right circumstances. But we could hope to do enough.

And enough is all that is needed.

So just this: do your best. And then tell someone else.

Meanwhile, let us have a sip of tea.
The afternoon glow is brightening the bamboos,
The fountains are bubbling with delight,
The soughing of pines is heard in our kettle;
Let us dream of evanescence, and linger
In the beautiful foolishness of things.

[Kakuzo Okakura, The Book of Tea, 1906]

~★~

4

THE INEQUALITY CRISIS

Togetherness Is Better Than Selfishness

There is only one true form of wealth, that of human contact,
When we work merely for material gain, we build our own prison.
[Antoine de Saint-Exupery, *Wind, Sand and Stars,* 1939]

Abstract: This chapter centres on the inequality crises and the importance of togetherness to all humans and cultures. "There is only one true form of wealth," wrote Antoine de St Exupery, "that of human contact." When dusk falls, that is when we gather, to share food and stories. Cooperation stuck people together. And yet, the modern neoliberal project seems to have focused on celebrating selfishness. The advocate Ayn Rand called altruism, "The poison in the blood of western civilisation." This chapter reviews two of the ills of selfishness: rising inequality and loneliness. The evidence is clear. Equal societies with strong connections extend life expectancy for everyone. They keep all people happier. But because social capital has fallen, often dismantled deliberately, a grand project lies ahead. We will want to celebrate community activities and sports, integrated schooling, volunteering, taking collective action in landscapes, the formation of millions of new social groups, especially in emerging economies. Participatory and public engagement help create this groundswell of actions.

DOI: 10.4324/9781003346944-4

True Wealth

Dusk falls. The wanderings of the day are done.

The grain has been pounded, soil washed from roots. A bean stew begins to bubble, bowls are readied. There are shouts from playing children, out beyond the flickering fire. The people gather round, there is light and shadow on their faces.

Now smoke rises, steam from a pot. Soon there will be stories, under the dome of bright stars. Tales from the day itself, sagas of courage and greatness, news and making moral sense. But first, the food is shared, brought from the land and sea to give it to others. In this way, we became mighty. All of us, together on the land.

There was a clearing by the track, by the spruce and birch that will pause each winter in the silence of the snow. It was summer, and swarms of blackfly darkened the evening air. Inside the Innu sweat lodge, the fire stones glowed. On the floor were spruce boughs that filled the dream world with sweet perfume. There was no other light. People huddled before the heat, splashed water on the stones, settled inside the whispering steam. An ancient gate had opened, to this heart of earth, and as the people listened, they grew closer. There was no sense of passing time, and outside the long boreal light would be dimming by degrees.

From time to time, people crawled out, stood and staggered, lay down and stared at the sky. The stories of the Earth continued inside as well as out.

At another place, the wide Siberian steppe was a basin ringed with mountain peaks one hundred leagues away. There were lambs in the byre, beside the single canvas yurt. Over the brow of a low hill white dots appeared, becoming soon a stream of hurrying ewes. The lambs rushed to the fence and started calling. The mother sheep charged on spindly legs. The land was filled with collective joy. The ewes crowded at the fence, the youngsters jostled.

Over the same rise came the shepherd on his brown horse, its head nodding slowly. He tied up the steed, paused, unhooked the spruce gate. And pandemonium. The two worlds of sheep sound clashed, settling soon into murmurs of content. The families walked out in twos and threes. The sun was dropping, and the shepherd stood stiffly in the spring light of the plain. Inside the yurt, through the entrance, a warm world was awaiting. The metal stove sat on a wooden plinth, and the cylindrical chimney led to the still point where light suffused from sky and fell on the pleated silk and wooden lattice walls.

The family gathered before the fire. Steam billowed from a pan, and then a lake pike was cooked in pieces. Wind blew around the canvas. The family drew together, and Altair said, "We are grateful for our isolation."

The old Ohio table was set in the apple orchard. Swallows flew from a hundred nests under the barn eaves. Horses huffed in the stable. The pump was levered to fill a jug with cool water. Dishes were laid on the chequered tablecloth. Elsie Kline said, "This meal is all from our farm."

Her daughter Emily and family arrived on bicycles, over the meadow brow from their farm. Dusk reached the orchard and the heat of the day ebbed. Shall we have ice cream, and the children shouted *yes*, and a tub was fetched from the farm kitchen. How could these Amish of Ohio avoid so many contemporary technologies, how could contentment be so high? Donald Kraybill wrote, "They have dared to snub the tide of progress."

The importance of the good life of these kinds of togetherness was reported by every respondent in our survey.

It was a rich tapestry of value and relation, of celebration and ceremony, of giving and gratefulness. People talked about the importance of their web of friends, their circle, where trust was high and you can rely on others. Such trust implies reciprocity and obligation. You give and you receive. Volunteering was embedded in togetherness, you help your neighbours and others in the community without expecting financial reward or payment. It helps to stick people together.

Togetherness was also love and long-term partnerships, caring for old and young, being cared for. It was good conversation and storytelling, it was singing and dancing with friends. It was making your way in the world with others you care for. It was being curled up on the sofa watching a film. Many of these values are coded into ritual: attendance at temple and church for prayer and silence, at the pub or café with friends, at the celebrations of saints' days and fireworks nights. Food is a stacked part of this togetherness, the collective gathering, preparation and eating.

The significance of social attachment is also brought into sharp relief by losses, the widening gap of inequality within affluent countries, the growth in loneliness, the steady removal of common places and resources during the era of material capitalism. The changes, too, that we might experience across a lifetime. You find you didn't intend to lose contact with old friends, but just did. You didn't mean to stop doing the things that made you happy, you just did.

In the old days, the Timbisha people used to dance much more. Up there on the valley side, tribal elder Pauline Esteves pointed, at the sacred rock. There the women would whistle up a wind for winnowing the mesquite seeds. She told how the Round Dance was led by a singer whose stories were of a world where birds sang in perfumed oasis, and rocks walked on the playa. Rings of people danced around that sacred rock all night, they raised a dust storm with their feet, dancing and singing.

This was the way we spoke to the spirits, she said, and by dawn, the dust would be swirling up the valley sides. We had seen that place, it was now part of the car park of the Death Valley Inn.

There had always been water in the desert, said Pauline. Every place has a story, and the most powerful in the desert are at water sources. Oases have spirits, water babies, little people, and before we consume, we always make a greeting, sprinkle a little water on the land, give thanks. Wherever we go, she said, we are reminded to be respectful: water is not just for cleaning or drinking, it is medicine. But people from distant places just walked in and took it: "That is a bad thing to do, they experienced something bad over there."

Pauline added, "You know the salty playa over by Windrose Canyon, that dry lake, when I was a child we used to play in the water there." The whole valley is a spiritual place to the Timbisha. It has power, it is protective, it is our friend. "Maybe that the reason we're still here," she said.

One of the reasons why togetherness ceremonies fall away is that the world tilts. Often this is deliberate. When Black Elk led the Lakota to their great Ghost Dances of the 1890s, they used the flow of dance to connect together the whole tribe to the sacred hoop of the world. It scared the hell out of the authorities, who rushed to make them illegal, called them war dances.

The Innu were hunter-gatherers until the 1960s in their country home of Nitassinan, an area the size of France spread over Labrador and Quebec. They were forcibly settled in soulless villages, authorities deploying the power of church, children's education and qualification for benefits to stop the old ways of living. Shaman Dominic Pokue had a wooden house in Sheshatshiu, on the banks of Lake Melville, and lived and slept only in the tent in the garden. "All my life I've lived this way," he said.

In *Maps and Dreams*, Hugh Brody described the pathology of first people communities in Canada:

> Many northern reserves appear to be grim and even hateful little places, clusters of houses crowded together by planners in order to achieve economies of administration and services . . . Such compression of a people distinctive for their free roamings through unbounded forest is bizarre and painful.

Out on the land, Dominic said, every day was unique, and he built up stories of a watchful world. They shared with each other, and with the animated world around them.

This profound connectedness to people and nature was not a choice, it was who they were, how they had lived for 10,000 years and more. Yet, instead of active lifestyles eating country food, dense in protein and low in fat, the Innu people have become marooned. Many feel they have lost hope. Mary Adele, a thin elderly woman, said, "It feels lonely here, when you look outside at the sky and the trees. Even though we are here in the village, we feel lonely."

Her husband added, "It was beautiful and happy before. Now I sit in this home, and I feel pain and feel sick. I feel very unhappy in the house".

"In the old days," added grandmother Katnen Pastitchi, "We were always strong and happy."

Brian Hare and Vanessa Woods' book centred on the value of cooperation is called *Survival of the Friendliest*, and it emerged from their research on the domestication of dogs and other animals. Friendliness is powerful, but social stress saps the body energy budget, leaving a weakened immune system. All aggression is costly. Friendliness and trust, by contrast, brings certainty and predictability to relationships, and this helps in the spread of ideas and technologies.

It has long been assumed that domestication makes animals less intelligent, yet this now appears incorrect. It makes them friendlier and more communicative. The half-century of deliberate selection of friendliness in the breeding programme for wild silver foxes in Novosibirsk changed coat colour, brought earlier sexual maturity, lowered corticosteroids and increased serotonin, and produced animals with changed faces and curly tails. Aggression was replaced by friendliness. These were also characteristics found in young foxes, and also in domestic dogs. The bonobo ape is the one ape like humans, more collaborative and unlike the chimpanzee. Female bonobos help other females, prefer males that are non-aggressive, and prefer to eat together. There is some evidence that bonobos and chimpanzees have different structures of amygdala, in the former slowing or preventing aggression.

So perhaps we *Homo sapiens* become self-domesticated? At the age of nine months, a human infant can follow an imaginary line from a pointed finger. By two years, they know what others see and believe. By four years, they can guess the thinking of others, they know too if someone is lying. Children have at an early stage in life *a theory of mind*. And this needs good communication. By about 70,000 years ago, *H. sapiens* had developed face shape changes, shorter and narrower, more juvenile with balloon-like skulls.

Friendliness wins, and over time humans become capable of helping strangers, people whom we did not know.

The Rise of Selfishness

Ontogeny says habits can be changed. This can be for the bad as well as good. There are dark places in the deep woods, but also many more sunny glades.

There seemed a logic for the rise of selfishness, if you want to be generous. Give people the chance to be in control, give them this freedom, and they will struggle harder in competition and eventually do better. They will create more, become wealthier, and then these benefits will trickle to others who had been less successful. Through selfishness, there would be more for all.

And whilst we're at it, public institutions and governments should get out of the game. There was no place for the commons in a world of affluence. In the 1970s, Milton Friedman, soon to be awarded the Nobel Prize for Economics, warned, "We're on the road to serfdom." Grover Norquist created the Taxpayer Pledge Protection in 1985, baking into norms that tax rises of any amount were not only bad for business and the economy, but they were also in some way un-American. Law-makers of a certain hue are still expected still to sign this pledge. Ayn Rand had called altruism, "The poison of death in the blood of Western civilisation."

And yet elsewhere, there was a quite different consensus on the social commons. Take the Nordic countries: higher taxes levied on individuals were supported because they were used to support maternity and paternity leave for all, to pay for care for all the elderly, and provide free education to master's level in universities. The individuals paying the taxes received some of these benefits back by being able to access some of these common public services. These countries are now the happiest in the world.

And now it has been understood, inequality is bad for the rich as well as the poor.

It looks like this era of selfishness is about to come to an end. It was cloaked in purple velvet and fur. It celebrated the breakdown of cultural and social complexities into individual actions. It promised that the wealth of the rich would trickle down and make everyone happy. It believed the pain of environmental harm would later be replaced by worldwide gain. It ignored the evidence from existing good lives, then used the themes of evolution as competitive struggle to suggest other cultures were less developed and less clever.

In the early stages of the first lockdown crisis of 2020, some politicians and commercial leaders seemed mildly astonished that people would help each other for free. They appeared surprised that public institutions could serve public needs, even if they had been dismantling them for years. They joined the clapping by front doors and in the streets. And here perhaps is hidden good news. Despite two generations of accelerated neoliberalism, this core of togetherness still remains. The foundations of social capital, the trust and reciprocity, the mutual obligations and support, these have travelled safely across all of human history.

Two great ills of modern selfishness have been rising inequality and loneliness. Inequality of income can be measured by the Gini coefficient, ranging numerically from 0 to 1, with zero representing perfect equality and 1 complete inequality. It should work like this: as countries become richer and more affluent, as measured by GDP, so income trickles on, and is spread more evenly. Everyone is happier and lives longer. The data is stark. Inequality within countries has grown worse, and the gaps between countries have also widened. It was not supposed to be like this, for the consequences are costly for whole countries and all their people. Non-communicable diseases (NCDs) have risen in incidence, as have the costs of treatment. Life expectancy was also affected. At higher Ginis where equality is worse, healthy life expectancy is lower (Figure 4.1). At higher Ginis, too, happiness within countries is lower.

Twenty years ago, Thomas Insel and Larry Young of Emory University wrote, "It is difficult to think of any behavioural process that is more important to us than attachment." Attachment behaviour in humans has selective advantage and enduring bonds have a neurochemical basis. The neuropeptide, oxytocin, plays a key role, and oxytocin receptors are concentrated in the dopamine-rich regions of the brain. Oxytocin is released by touching, by being in safe environments, and on receiving signals of trust from others. Subjects given oxytocin become more generous, and social attachment increases personal well-being. Mothers with strong attachment to their infants show greater activation of brain reward regions, especially of the oxytocin-associated hypothalamus.

This neurobiology is shared by all humans. We shall come back to this in Chapter 7. It opens up platforms for action that are common to us all.

So how did loneliness come to be permitted to spread so fast and far in affluent countries? High material consumption seems to have brought great fracturing of social bonds within families and communities. There have been changes to family structures, more loneliness, fewer family structures that support children, and a growing number of one-person households.

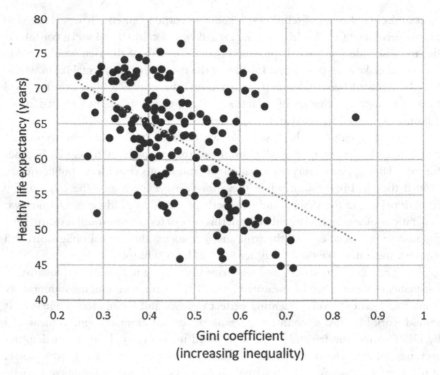

FIGURE 4.1 Inequality of household income and healthy life expectancy (n=154 countries, 2018)

Some of this is by choice, but much is not. In the UK, a tenth of over 65-year-olds are always or very lonely, some 900,000 people, and half of 75-year-olds live alone (though not all are necessarily lonely). In the USA, already a quarter of people live alone, and one in three to one in four people over 60 years of age experience frequent or intense loneliness. Lonely people make more visits to the doctor, and attend the Emergency Department at hospitals more often. Befriending, mentoring and gatekeeper services and group activities, such as walking for health, have been shown to reduce hospital and health care costs, and after such interventions, visits to GPs also fall.

This lack of social attachment brought about by loneliness has a negative effect on health. Lonely adults tend to have higher blood pressure, greater epinephrine secretion at night, higher morning and night-time cortisol levels, and thus poorer sleep patterns. Loneliness increases the gene activity that promotes inflammation, so also produces poorer sleep and reduces night-time repair. Long-term ill-health outcomes increase. Loneliness tracks into adulthood, though not all people alone are lonely, with lonely adults having had a greater number of childhood adversities such as hospitalisation, parental divorce and physical abuse. Shortened visits by carers may not help to remove loneliness: moreover, the chronic stress experienced by overworked carers themselves reduces immune response to vaccines and slows wound healing.

In popular terms, loneliness has the equivalent health effect of consuming 15 cigarettes daily and is twice as harmful as obesity. Loneliness, in short, produces biological effects. In many a culture, the elders would never have been pushed into the darkness beyond the fire circle.

Strong social support keeps elderly people alive: a meta-analysis by Julianne Holt-Lunstad and colleagues at Brigham Young University analysed 150 studies in Denmark, Sweden, Japan and the USA, and found a 50% increased likelihood of survival over seven years for those people with strong social relationships and networks. In the Harvard longitudinal cohort studies led by George Vaillant, the presence of friendships and life-long relationships resulted in ten extra years of healthy life.

Loneliness has become an extreme form of inequality in industrialised countries. And yet this is known, trusting relationships have a positive effect on health and well-being. The value attached to relationships has come to be known as social capital. This covers an individual's contacts and networks, the common rules, norms and sanctions that regulate behaviour, together with the reciprocity and exchanges that build friendships, respect and ultimately trust.

Though we might sing of social capital, the term does have disadvantages. It can seem to suggest the primacy of monetary value, that social bonds and togetherness can be measured by pounds and dollars. Yet, like natural capital and human capital, terms also in common use, all these can be thought of as regenerative capitals that can be built up a particular place, and from which are derived a flow of beneficial services. They have value, but can easily be destroyed.

There are stacking interactions too. One and one make three.

Nature is good for health; green places can be good for social capital: people engage with the outdoors not just for the connection to nature, but to provide the setting for the building of social capital.

Building Social Capital

This is where a check again on the elements of a good life can reveal how important are social structures. Many of these do not feature strongly in GDP measures. Some might be seen as mere leisure, even time-wasting. What is the worth, really, of watching people trying to hit a ball with a bat, kick it, or roll it over grass or down a wooden lane? And why would you want to do this yourself, when you could be spending more?

Once again, we find this: a good number of the components of a good life are low in carbon emissions, and bring relief by quietening the inner mental chatter.

And now we know, social capital has been in decline in the affluent and industrialised countries. One of the most noted examples was researched by Robert Putnam and reported in *Bowling Alone*. The tradition of ten-pin bowling in and around Boston, and elsewhere across the USA, had produced dense networks of local leagues, with teams playing home and away. Matches meant social mixing: one team travelled to another part of the city, and the other acted with the kindness of hosts, even though it was the competition that brought them together. People from different economic and social backgrounds were agreeing to play by common rules. The bowling hall was the fire circle. But as civic disengagement grew, so leagues began to fall apart. Putnam found that just as many people still were bowling, they were just bowling alone. Understanding and friendship across communities had now begun to fall. The bowling hall had been the fire circle.

These changes in civic and social infrastructure are also echoed in the loss of urban roller skating rinks across the USA. Hugely popular amongst black communities, each rink gave rise to a distinctive style of performance, costume and accompanying music. Rink rats travelled to showcase routines and expertise, and national gatherings of skate jams and roller-derbys created intergenerational ties and cross-community social capital. Roller skating came to occupy a central position in many people's lives, producing pride in local identity. But in recent years, 10,000 rinks have been closed, leaving 3000 remaining nationwide. Rinks have large footprints attractive to planners and developers, and costs for maintenance are high. Many city authorities have rezoned, imposed specific taxes and resorted to over-policing. Some say this is a deliberate response to elite panic, analogous to the responses to the indigenous ghost dances of the late 19th century. In the film, *United Skates*, one skater says, "Skating was our hope."

Now it has become harder to express civic pride and identity in physically-active settings that once had brought well-being and connection. When some Delhi schools created an experiment from 2007 to open up a fifth of places to poorer students, it was found that it made the richer students more prosocial and generous, less discriminatory and more willing to socialize. Gautam Rao of Harvard concluded, "Increased interactions across social groups, perhaps especially in childhood, can improve intergroup behaviours."

Community sports and mixed education thus build social capital. The village cricket team travels across rural lanes to play away. The players come from across social groups within communities, and fuse when they play. Some 300,000 football teams in the UK play each weekend, often on flint-strewn or mud-bath pitches, changing in their cars, the wind bitter on winter mornings. Baseball has a particular ethic in the USA and Japan, families and children playing or going to the ball game, spending time watching, often not very much happen. Each team has local identity and meaning. Each gives supporters cause for hope and laughter, despair too, yet all centred on something that not only distracts from the concerns of life, but becomes life as we want to live it.

To some, all this might still seem pointless, but people from different subcultures are coming together in common purpose. They are physically active, so receive a well-being benefit; there is some social mixing, some competition too. Carbon emissions are low for a couple of hours as people play.

Begin an audit in any country, and there very soon becomes too many such activities to list. Climbing walls and bouldering have become popular: you learn a skill, and do it together with others. You join the Park Run on Saturday mornings and run 5 km alongside a thousand others doing the same. You swim at the pool with family and friends, or in the sea, join the yoga or tai chi class, attend the judo *budo*, chat with friends before and after. Sports and activities and their teams are simply a modern manifestation of the value of togetherness and collective action.

This jig is the means to a greater goal: the aspects of the good life that encapsulate improvement in performance and skills, staying healthy, and being with others. It is notable, even in the age of competitive industrialisation whose story is that only winners matter, many activities put more value on the shared food and drink, the baseball hot-dogs and cricket teas, and after the match at the pub or café.

Volunteering is a further form of giving to others and to nature and is good for the givers. Across nearly 40 countries, the International Labour Organisation calculated that one in eight of the adult population is a volunteer, producing the equivalent to 20 million full-time workers. It could be said, these numbers are surprisingly high, especially in the face of modern affluence that seems to give such actions low value and support. Or you could say, people already know that giving is good, for themselves and others. The International Labour Organisation also found that volunteers bring US$400 billion of benefit to world economies each year, a total of US$19 billion alone to the USA.

These are likely to be underestimates. People are not just providing unpaid labour, they are preventing health and social costs. They are spending effort and time, and they receive back meaning, self-worth and pleasure. The data also show: volunteers live two years longer than non-volunteers.

Norman Fischer in *The World Could be Otherwise* tells of care volunteers at a Zen Hospice in San Francisco: "We tell them they will receive more from their service than they give." The patients somehow act unintentionally as guides to the volunteers, who find themselves opening up and feeling whole. This small world of a hospice, on the outside looking like only a place for death, ends up being a gift.

Yet, sports and volunteering are only a subset of other forms of social capital embedded in ritual and ceremony in the annual cycle. These are moments of celebration, relief and togetherness. The harvest festival of rural community, the village fete, the spring carnival and marching bands and dancers. A night for fireworks, the new-year gatherings, independence days. The cheese-rolling down the steepest hill, a mud race across the estuary at lowest tide, the weeks spent together building a birch-bark canoe. Some are ancient rites, some recent innovations.

And all of these activities do this: they stick people together. They send a message to the children, here are many different people, and they are laughing and smiling. And over there are volunteers helping make this happen.

Many of these rituals, once again, are low on material consumption. They emit little carbon, and people who volunteer have higher life satisfaction than those who do not. Trust in others also reduces mental ill-health, and is a key part of well-functioning societies.

It could help the planet too.

> At dusk I came down from the mountain,
> And I rejoiced at a place to rest,
> And good wine, too, to pour out with you,
> Ballads we sang, the wind in the pines,
> We had quite forgotten the world.
> *[Li Po, Coming Down from Chang-Nan Mountain, 701–762]*

Collective Action in Landscapes

The fire circle escorted people across a thousand generations. Togetherness is still a recognised value in the good life, manifest in food, leisure, dance and sport.

Let's go up a few system boundaries.

There is only one common that really matters: it is the Earth itself. It has commons of clean air, thriving biodiversity, land for farm and pasture, river and sea for fish and marine life. Yet, all of these have been in decline, and boundaries and limits have been broken.

A fresh era of collective thinking was launched by economist Elinor Ostrom of Arizona State University, a Nobel Prize winner in 2009. She spoke of this old truth to power: local commons and resources can be managed by people without intervention or regulation by central authorities. The Novel Prize Committee stated, "It was long unanimously held among economists that natural resources used collectively would be over-exploited and destroyed in the long term."

The use of unanimous in the statement was instructive, for this had never been the view of commoners in many water, forest, marine and agricultural systems worldwide. People the world over have been pretty good at building trust, and setting norms and obligations, rules and sanctions, and resources have prospered.

Now a task of devotion lies ahead. The redesign of habits, behaviours and institutions is now vital, and we have to do this in a single human generation. Recent habits and certain technologies are going to have to be dropped.

One way or another, it could get messy.

Tens of thousands of scientists have been contributing to the International Panel on Climate Change, to implementing the Convention on Biodiversity, to seeking foundations of scientific certainty to set out political solutions. There have been urgent international efforts, and in these great dispersed communities, social capital has been held together, at tough times, and so has emerged a strengthened evidence base. And slowly, a tiptoeing towards clearer rejection of nihilistic economics. The hyper-rich do not spread it around, they just hoard more. Some buy their own islands, many put up fences.

Humans have a long history of developing regimes and rules in both hunter-gatherer-forager and agricultural communities to protect and preserve natural resources in a steady state.

These diverse and location-specific rule systems form informal institutional frameworks within communities, legitimated by shared values. These social structures have regulated the use of private and common property throughout history, for instance by defining access rights and appropriate behaviours. Where these systems are robust, they can maintain productivity and diversity without the need for external legal enforcement: compliance derives from shared values and internal rules and obligations. In some agricultural systems, there is evidence that social structures have sustainably governed ecosystem use over millennia, the traditional irrigation groups in South East Asia that manage the flow of water from crater lakes through water temples to the sea, the surface irrigation tanks across India, the institutions that manage the forest, prairie, tundra and steppe homes of indigenous peoples.

In the mid-20th century, Walter Goldschmidt studied the two Californian communities of Arvin and Dinuba in the San Joaquin Valley, similar in all respects except for farm size.

Dinuba was characterised by small family farms, and Arvin by large corporate enterprises. The impact of these differing structures was striking. In Dinuba, he found a better quality of life, superior public services and facilities, more parks, shops and retail trade, twice the number of organisations for civic and social improvement, and better participation by the public. The small farm community was a better place to live because, as Michael Perelman later put it, "The small farm offered the opportunity for attachment to local culture and care for the surrounding land".

A study 30 years later by Linda Lobao confirmed these findings: social connectedness, trust and participation in community life had remained greater where farm scale was smaller.

In the UK, there is much to answer for when it comes to land structure, the commons and the emergent social structures. The enclosure of the commons began in our own country and then was spread in the empire. It began in the vast Fens, once a common wetland, then drained. Eight rivers flowed in, and out northward to the sea. On islands were established monasteries and the tallest cathedral in the land. At that time, the commoners of the Fens travelled by punt, walked on stilts, lived by fishing, cut willows, raised geese and hunted wildfowl. It was a world of neon kingfisher, flashing over streams, blue dragonfly and rust-red hawker, rising mayflies sparkling in the sun, the sound of an otter splash, a beaver slap.

But there came a national plan to drain the wetlands dry. Landowners learned drainage skills from visits to the Flanders' lowlands, soon grasped that unproductive wetlands could be converted into farms for crops and cattle. In the 1620s, Dutch engineers led by Cornelius Vermuyden were commissioned. Two memes were deployed to speed this drainage plan: call the land a worthless wilderness, call the people backward.

So the Fens were called, "A wilderness of bog, pool, and reed shoal, a vast morass, with only a few islands of solid earth," and people, "A barbarous sort of lazy and beggarly people." The commons, wrote Adam Moore in his 1653 *Bread for the Poor* treatise, were "Pest-houses of disease . . . hither come the poor, the blinde, lame, tired and scabbed." Knighted Vermuyden called his new land *summer ground*, as he dug and drained across the reigns of many a king and past a Civil War.

This was a low point, but not the lowest. Works were spoiled by the Fen people, and commissioners reported, "Outbursts of indignation and disturbances of lewd persons." Wetland people burned rick, opened floodgate, drowned again the land. Riot and discontent flowed back and forth for 60 years. Dark as this was, now were launched 4500 Enclosure Acts signed by Parliament, and the one Waltham Black Act of 1723. E P Thomson wrote how this act created 50 capital offences for those who took from forest, waste, marsh or clearing, for anyone found with their face *black* or who might, "Appear in any forest, close, park or in any warren, or on any high road, heath, common or down."

Calling these habitats *wastelands* was a term to last for centuries, for the empire then exported measures to more than 60 countries overseas.

You can see the link; biodiversity equaled wasted land. This helped boost growth in the industrial revolution, moving displaced people to city factories for waged labour. The backdrop was again active destruction of public institutions and local groups, especially those based on and around common property resources. In the UK, the Enclosure Acts privatized common field systems and the wild commons. Now came an agricultural revolution.

Over time, inherited and legacy institutions have continued to be undermined by choices made by the modern agricultural political economy: social institutions have been ignored, co-opted, undermined and deliberately broken. Often, state institutions were imposed on farmers as the price for obtaining modern varieties, fertilizers and pesticides, such as in Malaysia and the Philippines. Elsewhere, local institutions withered, the *kokwet* water systems in Kenya, *warabandi* in Pakistan, forest and water communities in India.

The loss of the rules of institutions then allowed over-extraction by individuals. In Gujarat, for example, groundwater was depleted so fast that salty seawater rushed in to fill its place. Empty and paper institutions were also formed by some states without local participation, such as for grazing in China or irrigation in Thailand. At the same time, much forestry management had also become centralised into state and private enterprises that took little account of existing cultural institutions and norms of co-management. This all happened so fast it became an unchallenged norm.

Further changes to social structures were promoted by the conditional policies of structural adjustment adopted by international finance institutions from the 1970s and 1980s. These resulted in the dismantling of public institutions, and the adoption of the Training and Visit (T&V) system of agricultural extension. The T&V system had a linear diffusion model, also called transfer of technology, first implemented on recommendation of the World Bank in 1967, and resulting in the disbursement of US$3 billion to 512 projects over 1977–1992. Structural adjustment brought free-market policies to 135 countries between 1980 and 2014.

In return for loans, governments played their part by deregulating, shrinking public institutions, and opening up their markets to imports. Many researchers now conclude: this also brought greater inequality.

Michael Cernea at the World Bank already had concerns in the late 1980s over the cost of ignoring local institutions. He found that the creation of farmer and rural institutions led both to sustained performance after project completion and to more efficient and fair use of natural resources. New forms of participatory inquiry and systems of collective learning and action were field-tested, putting farmer knowledge and capacity to experiment at the centre of practices for improvement. By the mid-1990s, the diffusion of knowledge model had become ineffective: non-adopters were termed laggards, extension staff had lost motivation, and research systems had been prevented from becoming learning systems.

Out of this wasteland, though, emerged something bright: one of the most important stories about contemporary social capital and the subsequent rise of social groups on the land. It happened like this.

Participatory approaches with farmers and rural groups began to be deployed in the 1980s and expanded through the 1990s. If you listen to local people, you can build on their knowledge. If you hear their priorities and reasons for living, you can find a common platform to suggest improvements. If people experiment and evaluate, they themselves may become convinced of the outcomes. It all works much better than walking in and telling people what to do. Whilst at the International Institute for Environment and Development, and working with Robert Chambers and colleagues at the Institute for Development Studies, I ran with colleagues training courses for scientists, extension workers and policymakers in more than 40 countries.

At every single course, someone with crossed arms would frown and say: well, that might have worked over there, with those people, but it won't work here. We were asking, after all, for trained professionals to give up their pre-existing paradigms, and think about social and ecological systems in new ways. We were asking them to drop old habits.

This was nothing more, it should be said, than the approaches to pedagogy set out so vividly by Rolf Lynton and Paolo Freire in the 1960s and 1970s in India and Brazil, and then by Richard Bawden at Hawkesbury College in New South Wales in the 1980s. At the centre of these participatory approaches was the need to form social capital in village and sub-village groups, particularly for the collective and joint management of irrigation water, forests, community finance and watersheds.

In 2020, a group of 29 of us completed a global assessment of this emergence of social capital and togetherness. We found that more than 8 million groups with a membership of some 250 million people had been formed to focus on integrated pest management, forests and soil management, irrigation water, pastures and ranges, support and finance services, innovation platforms and small-scale systems. Across 122 initiatives in 55 countries, the number of groups had grown from 0.5 million in the year 2000 to 8.54 million by 2020 (Figure 4.2).

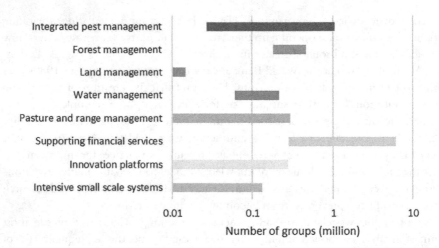

FIGURE 4.2 Increase in numbers of groups in eight categories of sustainable agricultural and land management (2000 to 2020)

The area of land already transformed by these members was three hundred million hectares, mostly in the less-affluent C1 and C2 countries. Farmers and land managers working with scientists and extensionists in these groups have improved environmental outcomes and agricultural productivity. In a few cases, changes to national or regional policy supported this growth in groups, though this remains a barrier.

As yet, not many policymakers have got the message.

Other forms of social capital have also emerged in support of transitions towards greater sustainability and equity. These include transnational farmer movements, the La Vía Campesina with two hundred million families represented worldwide, and national land rights and anti-land-grab movements, the Movimento dos Trabalhadores Rurais Sen Terra of Brazil, who have supported the resettlement of nearly 400,000 families on 8 million hectares over a decade.

In Australia, the 30-year-old Landcare movement that was formed to bring farmers and community members together for joint action in landscape now has 6000 groups and 140,000 members. This has led to environmental volunteering in coastcare, bushcare, dunecare, rivercare and many other types of groups. Nearly half of members have increased well-being, a greater sense of belonging and social connectedness, and members are saving the country more than US$140 million in health care costs annually.

At the same time, organisation around food has advanced in the form of food sovereignty and justice movements, and alternative food networks and movements, particularly from urban food production landscapes and many involving consumers as well as growers and farmers. In these ways, small patches are linked to consumers through community-supported agriculture (CSA), box schemes, tekei women's groups in Japan who promote healthy food, simple living, self-help and fair trade, the food guilds in France and Switzerland, and widespread farmers' markets. In the USA, CSAs have grown in number from just two in the 1980s to more than 7000, with 50,000 consumers as members; the number of farmers' markets is nearly 9000.

Food systems are being changed by togetherness.

> How can I possibly sleep, this moonlit evening?
> Come my friends,
> Let's sing and dance, all night long.
>
> *[Ryokan, 1758–1831]*

The Value of Social Groups

The value of these social groups can be seen in several ways.

There is evidence of personal transformation, particularly in the emergence of new leaders of groups, especially by women. There have been changes in the relationships between women and men, new group effectiveness and conflict resolution over common resources, changes in the worldviews of farmers and scientists and extensionists, increases in the savings and repayment rates of members of microfinance groups.

At the agricultural system level, there have been increases in crop productivity, such as in farmer field schools on all crops, and in grazing and pasture productivity. There have been increases in tree and agroforestry cover on farms, reductions in the use of pesticides by integrated pest management, and the adoption of organic and natural and zero-budget systems.

For natural capital and ecosystem services, there is evidence of increased biodiversity and reductions in the use of harmful or potentially-harmful compounds, increases in irrigation water availability and efficiency of use, improvements in forest productivity of wood, forage and secondary products, and increases in carbon sequestration in soils by conservation agriculture.

This platform of more than 8 million social groups could comprise an opportunity to consider greater challenges, such as advances towards globally meeting the Sustainable Development Goals and addressing climate change. As new worlds are being brought forth as a way to transform the way we live, this could also change modern consumption cultures. Some social capital is already influencing global systems, resulting for example, in mitigations of climate change, biodiversity loss and air pollution, as well as increases in net food production.

We could imagine, for example, platforms of groups engaging in co-production of new patterns of material consumption and ways of a good life that keep within global boundaries and limits. It is clear that considerable changes will be required worldwide to limit the advance of the climate crisis, both in individual choices and behaviours and in the policies developed by all countries. An era of stable economies and degrowth is needed, certainly the green restructuring of economies directly to reduce material consumption and substitute with sustainable or green alternatives.

Clearly, individuals will have many reasons for organising and taking collective action, and given the context for these changes, it is likely that many individuals will support sustainability and equity outcomes. But there is no guarantee that such values will remain unchanged. Can these relationships and groups survive and flourish? And is something universal happening?

Oliver Scott Curry and colleagues at the University of Oxford Kindlab have studied behaviours in 60 countries and found many forms of cooperation centred on the allocation of resources, coordination of mutual advantage, social exchange, conflict resolution, fairness and loyalty. Of 962 observations, all but one had positive moral outcomes, leading Curry to suggest these looked like laws or *human universals*. In *Human Kind*, Rutger Bregman added, "People, deep down, are pretty decent."

Kindness is both our common state and response to threat. It is selfishness that is the outlier.

But threats will come. External sources could include major social and economic disruption, such as following the COVID pandemic; climate-driven forced abandonment of farms and territory; policy changes in support of land grabs and large commercial monoculture operations, such as for oil palm plantations; and state support for only empty or non-credible groups. Internal disruptors might include stresses arising from benefit capture by individuals; gender imbalances in the flow of benefits; and farm abandonment in favour of employment in urban areas.

Nonetheless, many advantages have already been found in the sharing economies of connected food systems where goods and services are pooled, through better distribution of power, increased collaborative consumption, higher trust, and more efficient use of resources. Agricultural transformations will be critical in the coming years both for helping to reduce climate forcing and mitigate negative effects.

It would appear that these social groups, structures and movements have already created opportunities for individual and collective transformations. In *Blessed Unrest*, Paul Hawken called such organisations an immunity system for the world with their millions of members.

At the heart of this togetherness and personal growth is not some wide-eyed form of optimism in which everyone has to think and act the same. A stable economy creates the space for many forms of divergent personal growth. We can grow and change as individuals, as can our groups and institutions. And it is worth saying again, togetherness and social relationships can result in greater well-being and happiness, lower costs imposed on health systems, and help people live longer and create a thriving planet.

Even in the most extreme of circumstances, imprisoned on an island or in a concentration camp, it has been possible to use the idea of togetherness to keep open avenues for life.

Nelson Mandela spent 18 years on Robben Island and another 9 on the mainland, yet forgave and befriended his captor guards. He went on to invite one,

Christo Brand, to his presidential inauguration. Viktor Frankl and his family were deported to a concentration camp in 1942, where his wife and parents were killed. There he was, he wrote, with no hope, the complete loss of his former life, the mental agony of injustice and unreasonableness of it all.

And yet, he also wrote,

> We do have a choice. We prisoners can remember the men who walked through huts comforting others, giving away their last piece of bread. Everything can be taken from a person but one thing: the last of basic freedoms, to choose your attitude in any given set of circumstances.

> Just off to hoe my melons one day,
> Shouts and call along the narrow lane,
> And suddenly my old friends were with me,
> I led them where a cool breeze would blow,
> To relax and look upon the heaven and earth.

[Wang Wei, 699–761]

~★~

5

THE NATURE CRISIS

Regaining Earthsong and Attentiveness

Attention is the beginning of devotion.

[Mary Oliver, *Upstream*, 2016]

Abstract: This chapter tunes to Earthsong and the nature crises. What shall remain if we carry on this way? Nature is our home and more than that. "Attention is the beginning of devotion," wrote poet Mary Oliver. Look closely, and you have to sing. And be humble, for nature knows way more than we do. Here nature is discussed as sacred place, alongside the values and ethics of care, the grim norm of exploitation. And yet, a pilgrimage walk, a swim in the river, tending the garden, these are low-carbon activities. The term nature-based solution is becoming a common currency. These "expand the solution space," said the International Panel on Climate Change in 2022. They also offer opportunities for attentiveness and immersion, and these, as we shall see in Chapter 7, unlock the quiet and contentment all people seek in one way or another. Nature is essential infrastructure, environments structure behaviours, and we receive health services from nature. The chapter closes with nine trickster questions: what are your best moments in life? If you listen carefully and are attentive for long enough, you will hear.

DOI: 10.4324/9781003346944-5

What Shall Remain

Nature knows way more than we do.

It is watchful. It has intent. It requires of us the highest responsibility and behaviour. Every plant and animal. Each cloud and mountain. You could teach a stone to stand, carve intaglio and petroglyph, and it will sing long after we have gone.

Some of the world's oldest individuals live on slopes high above the south-western American deserts; they germinated from cones 5000 years ago. Since our invention of writing, these weather-beaten bristlecone pines have seen flow and ebb. The trunks today are striated red and white, scarred by lightning, 10,000 feet above the hottest place in the world. Each individual is home to bird and insect, and still, green needles grow under dry cornflower skies.

Nature is our home too and more than that. It is shared food and a cool breeze on a summer night. It is an urban park sweet with cherry blossoms; it is a glacier supplying drinking water. It is where we set the tent and yurt, it is mystery, it is happiness and danger. It was where, in this age, we came to clear the trees and lay down concrete, blast mountain-top for coal and ore, fix silver pipelines end-to-end.

You could ask a fish, what do you make of this water in which you swim? And say to us, how do you find the usefulness of oxygen? Nature is so much home we do not see it. And yet, every culture developed rules and guidance. You gather those fruits, but not all. You divert this irrigation water, but not all. You will be invited to catch this antelope, that caribou, if your mind is clear and you act with respect. And all cultures have asked, what should remain, for each one of us will die and we must think of the children too.

When we act, we should ask again, what shall remain?

If you sit and listen to birdsong on a spring dawn, when you have finished will the birds still be there? If you watch the sky flame at sunset, reflected in the rivers of the marsh, will the sky and sun still be there? If you walk a pilgrim route, ten steps or ten thousand, at the close of play will it still be there?

This has been the way of Earthsong: to live our good lives and not use up the Earth. Still today, we cannot explain how the cuckoo and nightingale navigate from the tropics each spring to the same wood for nesting. The Lakota shaman and leader, Black Elk, said to watch the black-horned ground beetle in spring, for they turn to walk in lines towards the far-distant unseen herds of buffalo.

I once saw a rook's parliament on a cold clear winter morning. Thirty gloss-black birds were perched on the leafless branches of a grey poplar. They faced inwards at a silent pair. Incoming birds received an update, the talks continued, and then long silence. They rose suddenly, chattering upward to the cloud, smoke swirling and were gone. Jack Kornfield once came upon a group of eight robed lamas pouring offerings on a fire, high in the Himalaya, ringing bell and beating hand drum, and sitting around them was a ring of 60 blackbirds. He felt he was, "Witnessing something ancient, before humans and animals were separated."

"This tiny bird," wrote singer and folklorist Sam Lee of *The Nightingale*, "led me across a threshold."

Each year, 5500 future pairs fly from Gambia, Guinea and Senegal to return to the very same singing grounds as the previous year. "I thought I was, an angel bright," sang Sam Lee. And everywhere they go, the nightingale is in narrative song and ballad, part of cuckoo fair and Sufi dhikr song, part of Persian conference of the birds and Greek shepherd tale, there to give wonder to the walkers in Berlin's city Tiergarten. Beatrice Harrison played her cello to a nightingale singing in an Oxted garden in Surrey in May 1927, the BBC's first live outside broadcast. And this too entered myth, to remind us of these angels bright.

The author and aviator, Guy Murchie, wrote in *Song of the Sky* of birds flying with gliders, copying every manoeuvre; also of a friendly meadowlark who sang from the tails of C54s at Los Angeles International in the 1950s, the bird splattered with oil from too much time hanging around planes: "How he could sing!" At Travis Air Base, he watched many a time gulls flying deliberately behind ten-engine B-36 bombers as their motors were being fired up. The birds dropped into the blast, into the soul-shaking roar, and would simply disappear. And then, seconds later appeared a quarter-mile downwind. They turned, flew back, and dived in for more.

The poet Mary Oliver asked:

> Tell me what it is you plan to do,
> With your wild and precious life.

Yet, everywhere we look today, there are planet-busting losses.

Whole rainforests levelled and burned, the tropical habitat that contains half the world's species. The greatest beasts for 70 million years hunted out of oceans for oil to light the city street, their blubber for soap, and only later we understood whales could sing and had distinctive culture. Herds of 50 million buffalo mown down on prairies for their fur and leather. And today, tar sands squeezed for one last drop of oil, sacred land and lake harmed by explosive fracturing of the ground beneath our feet. In 2016, the Dakota Access Pipeline crews used a fleet of bulldozers to clear a two-mile swathe through 80 cultural sites and 27 burials of Lakota Sioux, even though the frontline village had flags and representatives of 350 Native American Nations.

In 2020, another mining company stepped up to blast yet more sacred sites in north-west Australia, blowing many thousand years of culture to pieces. They wanted to ensure they could harvest all the diamonds from the rock. They named their new line of jewelry *Dreamtime*.

At the time that *The Limits to Growth* was being published in 1972, the Norwegian Arne Naess was at the World Future Research Conference in Bucharest, and he presented for the first time his ideas for a Deep Ecology Movement. He was a mountaineer and Gandhian boxer, academic and activist, and worked for social change. Deep Ecology recognises the intrinsic value of all living things, that richness and diversity had essential value too, and that ideological change was required so that we could seek a quality of life better than one centred on rising material consumption.

"Earth was home," Arne Naess said, so Deep Ecology was all about, "How to live well."

He wrote:

> Mountains are big, very big,
> But they are also great. Very great.
> They have dignity and other aspects of greatness.

The world is never knowable, he said, and deserves respect. His friend Alan Drengson wrote of this new-old sacred wisdom of Naess, listing his favoured phrases: "Everything hangs together," we should "Act beautifully" and "Find joy in simple things," "The more diversity the better," "Trust don't doubt," and "All things are open to enquiry."

When I had lunch with the President, we talked of the soil, the very earth beneath his Georgian farm of groundnut and cotton. I had been at Emory University, talking to students in groups small and large, and anthropologist Peggy Barlett took me on a tour of Atlanta's urban food systems. Here was nature in the city, in corner plot and community garden, here was empty lot turned green, here was café and children's garden, box scheme and all manner of people linked through food. The former President, Jimmy Carter, was keen to hear this detail of city movements on the doorstep, holding a copy of my book called *Agri-Culture*. He still had great plans for all kinds of policy reform, how to make improvements good for nature and for food. Look, he said, at the land.

So Peggy and I left to climb a sacred Indian hill, the dome of granite monzonite where 60,000 soldiers died in the Civil War.

Kennesaw Mountain was a steady climb to the 500-metre summit. At the top, the pink stone had been turned by querns into hollow circles. Some were filled with water, one with bright red moss. No one else alive was there, but the land did not feel empty. We could see the circle of the world, the sky touching the Earth on every side. We had to catch our breath. This hub had been home to mound builders for a thousand years. Down below, people would be consuming much, arguing, raising hopes. Once the smoke from fires would have ascended slowly. This day, a falcon circled, halfway between the stone and sky.

Nature always knows much more than we do. And this is what we keep finding: the interconnectedness of natural systems.

A change in one place tugs at the threads of a larger whole. Suzanne Simard in her ground-breaking book, *The Mother Tree*, showed that trees in forests are linked to one another via fungal hyphae in the soil. They share carbohydrate products of photosynthesis, they transfer nutrients, they send signals about pest and disease attack. Foresters long had a monoculture model: for best results, take out the weedy species and keep only the target trees. Yet, Simard found the so-called weedy birch was helping the Douglas Fir in certain seasons by transferring food, and the fir helping the birch at other times. In the middle, of course, were the dense networks of soil fungi, capturing carbon and keeping the soil healthy.

A dozen years after Armenian independence, people sat in the shade of street trees that filled the green squares of Yerevan. Breezes were drawn down from the hills, and by evening string quartets were playing and people laughed. Beyond the 3000-year-old city, we stopped at roadside vendors to fill bags with watermelon and oranges and drove through the salt-burned wasteland. The Aral Sea had been drained, the lakes of the Caucasus also shrunk by over-irrigation. And now the Armenian soil was salt and all the trees were gone. When you see like a state, as James Scott observed, grand plans emerge. They are often disastrous.

And yet at Meera eco-farm, there was restoration. An orchard of apricot, tamarisk and rustling poplar, cattle grazing beneath. There were thirty types of vegetables. Two hoopoes, messenger birds of Solomon, flashed across the broken blocks of state-farm structures. Snow-capped Mount Ararat was tall in the brilliant sky. There was the chatter of songbirds, and two concrete tanks had been sealed and swirled with the shadows of 3000 rainbow trout.

As we sat on old chairs at dusk, the world began to tilt. We looked into the shadows. A pair of red foxes strolled from the trees. They walked to a pond and paused, one reversed, dipped its tail in the water. Snap. A trout bit. The fisher fox flipped the fish high and round upon the shore. They ate, we smiled.

The 7th-century Armenian mathematician, Anania Shirakatsi, concluded that the earth travelled around the sun, the night sky really was dense stars, and the moon was lit by reflected sunlight. Nine centuries would pass before other European cultures stopped believing the world was flat.

In *Pilgrim at Tinker Creek*, Pulitzer Prize winner Annie Dillard wrote of her life in the cottage by the creek: "It's a great life, luxurious really . . . spring comes up in yellow daffodils, all the way up to the peach orchard."

All Is Sacred Nature

The better course might have been, keep all of nature sacred.

Under modern designations, 15% of Earth's land has been protected, and 7% of sea and ocean. Under the older ways, there are 100,000 wooded hectares of Shinto shrine and Buddhist temple in Japan, 200 sacred forest sites in Ghana, 400 in China, 900 in Tanzania and a 100,000 in India. Almost every churchyard in Britain has sacred veteran trees, the oldest yew was young at Fortingall when Pontius Pilate was a boy and lived nearby. In Aotearoa/New Zealand, the River Wanganui was declared sacred as a person in 2017: say Māori people, "I am the river, the river is me." In Bangladesh, all rivers were granted the same legal rights as persons in 2019. There are good reasons why in the desert, every oasis is sacred.

Occasionally, the land slips beyond imagination, and a door opens.

I once climbed sacred Keshege Mountain in Tuva, up ice-slick steps set inside a rock chimney, and at the summit, the land just dropped gently away, and all the grasshoppers and bees fell completely silent. There was not a sound as I sat beneath the dome of bright Siberian sky. The world came to a halt. I have been leached of every ounce of energy in a ruined church set on top of a Neolithic burial site in the North Sea, frozen to the ground. I have inched to the top of the highest Mayan Jaguar temple at Tikal, and far away a dark storm of ghosts dashed across the rainforest, the monkeys howling down below. I have gone in search of an Innu birch-bark canoe with an elder and his grandson, and in the clearing, there was a host of small mammal skulls hung on branches and a canoe-sized hole left by thieves.

In each old place, a song deeper than silence still was coming radiant from the land. In these ways, a sense of eternity in the landscape can emerge.

Time passes in different ways. If we wish to hurry, to insert hope and expectation, to leave and arrive at a precise time, then this could be in error. The cow needs milking, for as long as it takes; the boat will leave the shore, when the gale dies down; the dance will begin at sunrise, but maybe on another day. In a stable economy, the future matters, but you do not need to be rushing at it.

This kind of interconnectedness leads us to think less of boundaries between things (people, species, systems) and more of coexistence. This is partly why many indigenous and traditional cultures see people and animals as interchangeable. An animal is a person, as is a plant, and a person can become an animal, also a rock, the moon, a star, and all with the capacity for intention. In Blackfoot culture, Rock is a trickster, rolling down the hill to catch people by surprise.

The Māori whakapapa is the family line of both identity and genealogy back to the first arrivers in Aoteoroa. In many a parish church in the UK, you will find a list of names of vicars and priests stretching back to the building of the church, often near rafters and porch containing images of green men and women. The Mayan people's calendar expected that one cycle would take 63 million years to complete. In Tibet, the origination of Earth is said to be 100 kalpas ago, taken to amount to more than four billion years. The Guich'in people-of-the-caribou have had a 20,000-year model of pattern and restraint, details laid down for each generation. Then along came opencast mining and the oil drill and pipeline. And away went the caribou.

Time anyway has different properties on the land. The Inuit word *uvatiarru* means long ago and in the future. Wade Davis recounts the story of a band of Greenland Inuit who walked a huge distance to gather wild grasses from a verdant valley. When they arrived, the grass had not sprouted. So, they camped and waited, watching it grow, and flower, and set seed.

Human cultures evolve, and timescales for change vary. Cities rise and die, hunter-gatherer-foragers and small-farmer cultures live stably over thousands of years. The island of Bali has had irrigated rice managed by egalitarian *subak* social groups for 5000 years. Over time, ways of living became fitted to ecological circumstances, choices that bring divergence from other places. In the central African rainforest, Aba pygmy people see a different perspective in views, as they mostly lived inside tightly enclosed trees. In the Amazon, some peoples have no verbs for past and future, all their time existing in the here and now.

Just as today, many cultures lived with knowledge of others, yet did not converge or adopt. The Nahuatl hunter-gatherer nomads lived with knowledge of the settled city-states of Toltec and Aztec, yet kept to their own way of living. The Amish today talk explicitly of the importance of nature and togetherness, farm organically with horses, and do not find themselves tempted by the highly visible material consumption economy nearby.

When we humans live inside the rhythm of an animated world, we seem to be content not to assume a supreme position. In the stable economy-environment, each day is always different.

For foragers and gardeners, being out on the land is intense and vivid, unique and unpredictable. Events form the content for stories that are accepted as highly specific. In the boreal north of Canada, Innu, Koyukon and other Peoples will say, nodding, it could have happened that way. Outsiders thought the Apache were slow and lazy, as if, of course, these were both a bad thing. Yet, the Apache felt they were always doing something, going somewhere, preparing food and baskets, telling stories. Their world was socially and personally dynamic within stability.

Today we find some have dared row out, beyond the pounding waves. Subcultures within affluence have taken deliberate decisions to diverge, seeking also to show too that modern selfishness was the wreck on the shore. In *The Abundance of Less*, Andy Couturier wrote of people in Japan who have taken different paths to find a good life. They seemed to have a lot of time, he wrote, did not use money to entertain, and they were not overwhelmed or rushed. A woodblock craftsman, Osamu Nakamura, said his work was half meditation, and so he poured his energy into making things.

A female farmer also on Shikoku Island, Asha Amemiya said, "Us lazy people just ruin capitalist society," and former engineer and now children's book artist, Akira Ito, said he liked to doze, "Under the wide mosquito net, listening to cicadas in the early afternoon."

Those drowsy days, he said, were worth to him, "A thousand pieces of gold."

In the morning calm, in the evening calm: a stable and regenerative economy-environment is not going to see progress cease. It was how people lived in ten thousand cultures and places for one hundred thousand years. True farmers, said Ohio's Gene Logsdon in *The Contrary Farmer*, see their small farms as a source of never-ending discovery. They are small, biodiverse, community-oriented, unstressed, and above all successful. "We are pioneers," he wrote, "seeking a new kind of religious and economic freedom."

Farmers who pay attention are, in short, likely to be successful and happy.

In early 2021, Partha Dasgupta's comprehensive review on *The Economics of Biodiversity* was published for the UK government. The key headline: nature had never really been part of economic models, even though wherever we look the economy is no more than a sub-system of the Earth. Dasgupta's choice of language was meant for economists and policymakers. Biodiversity, he wrote, delivers provisioning, regulatory and cultural services. We live in a bounded global economy, and so it is logically impossible to conceive of any negative impacts of the economy as externalities. You can calculate the damage, as we have done for a number of agricultural sub-systems, but nothing is external. And also, nature is mobile, invisible and silent, and thus we need to talk more of sacred and spiritual values.

The grim norm, also observed Partha Dasgupta, was that we had become accustomed to seeing exploitation masquerading as cooperation. Put another way, there can be no such thing as an externality. Above all, it was time for a restatement of the values of the commons, and what have been called Common Pool Resources. In 1968, Garret Hardin thought he was being helpful in pointing out that unregulated or unmanaged commons are always heading for disaster, for he said individuals would always over-use and under-invest if they felt they could get away with it. But most commentators seemed to get no further than the title, and so the phrase, *The Tragedy of the Commons*, came to suggest that everything held in common was heading for a fall.

Dasgupta reminded us, the Earth system is a common, the finite biosphere, each habitat and every species. Throughout all of human history we have been commoners, setting rules based on trust, reciprocity and mutual obligation, and then building institutions that survived for hundreds of generations. The leading thinker and writer on community-based conservation, Fikret Berkes, has written that sacredness of land and commons is, "Truly a global heritage."

It is universal: all cultures have nature stewardship traditions. It is just that these ways of living have been recently subverted and undermined. Valuing the commons helps make place and thus increases care.

As soon as you start to document the values of common green space in cities, you find they are creating huge individual and national value precisely because many people use them regularly. A study in England and Wales for the UK government department Defra in 2022 found that urban green space was worth £25 billion to local people each year. Hyde Park in London was valued at £24 million alone, a great green and open space used by 200,000 people on a typical summer's evening. The values are partly from well-being created, and partly from illness and hospital visits avoided.

Well, perhaps bitter winds weathered good sense. As we saw earlier, enclosure of the wilds and wastes along with breaking up of the common fields was pioneered in England in the 1600s and 1700s, and then rapidly expanded to aid the industrial revolution. Now we hear welcome talk of commoning and reclaiming, deliberate efforts to bring people together on the land. And what happens with carbon footprints when we think of the commons?

Public transport produces less carbon per kilometre travelled than the private car. Public health systems are lower carbon than private, tap water is lower carbon than bottled, urban parks provide access to green space at no cost, and are lower carbon emitters than gyms.

Swimming in a river or the sea uses less carbon than a private pool, as does food from your garden or allotment than a supermarket. In many cities have emerged reclaim-the-streets campaigns and guerilla-gardening to create new green and quiet space to encourage place-pride and cooperation. In Toronto has come the YIMBY movement, yes in my backyard, in which gardens and communities are linked up with backyard owners who are not using their own garden. In Detroit, heart once of the auto-industry and now a by-word for post-industrial decline, has emerged green belt from rust belt, with 1400 urban farms and community gardens established on industrial plots.

If you grow your own food, clean your house, care for relatives, go for a walk, then GDP largely does not notice. Put on a wolf skin, and you will understand the language of nature and animals. Writing about 300 BCE, Zhuangzi wrote that nature offered ways to discover inner tranquillity. In the Song Dynasty, 1300 years later, Guo Xi indicated that landscapes should be "Marchable, visible, accessible and liveable."

And so emerged the *shanshui* mountain-water tradition, in which landscapes had nature, and road and track; stream and river, and fishing boat; trees and woods, and woodpile; limpid lake, and arched bridge to pavilion. I walked up the sacred Huangshan Mountains in Anhui, a land that had been kept vacant for more than a thousand years. Only selected artists had been permitted by emperors; then only party leaders came, keeping the mountains to themselves. But these Yellow Mountains of Anhui were opened up in the 1990s, and have since been declared a World Heritage Site.

They have been reclaimed. On a typical day, there may be 10,000 pilgrim walkers, during the holiday weeks in October and May up to 30,000. The people came back, they walked out of mist, through bamboo grove, up three thousand granite steps and into the heavens. They looked down on seas of clouds where ancient pines grew from rock, and waterfalls poured past pagodas offering green tea and noodles.

At the Welcoming Pine, *Ying Ke Song*, pilgrims jostled for photographs in front of the thousand-year-old pine. Like a cascade or slant-style bonsai, its purpose was contemplative. On his narrow road to the deep north, Matsuo Bashō wrote, "Go to the pine if you want to learn about the pine." This one, and all the others, surviving the punishment of the axe for centuries.

Near Cloud Dispelling Pavilion, where the clouds did not dispel, locks had been attached to chain-link fences by the precipice. There were 60 sections with some 40 locks on each, each with names engraved. We were here, they said. Through such gestures we believe we can influence the world. Somewhere in the abyss were the keys, thrown over symbolically. This public repossessing is a celebration. The Huangshan Mountains had become China's Kinder Scout, the mountain claimed in England in 1932 as a common. People once could only know these places through art, not personal experience. Now they could immerse themselves.

And we might observe: all this walking was zero-carbon. It slows the climate crisis.

This is Li Po's most famous poem:

> Before my bed, there is bright moonlight,
> So that is seems, like frost on the ground,
> Lifting my head, I watch the bright moon,
> Lowering my head, I dream that I'm home.

Li Po (Quiet Night Thoughts, 701–762)

Nature-Based Solutions for Good Health

The 13th-century monk and writer, Dogen Zenji, wrote, "A bird flies in the sky, and no matter how far it flies there is always air."

We could ask the bird about the value of air, as we did the fish about its water, and we will meet with puzzlement. And we might ask, too, what is the value of this nature-home in which we live?

Shall we deploy the term green to demonstrate the instrumental values of nature? It seems we still do need the evidence that Earthsong is good for health, that nature must be redesigned into our lived environments, that health systems would advance if green social prescribing were adopted to prevent ill-health arising from non-communicable diseases. Evidence on the value of greening the economy is, in short, one way to remind us that there are workable ways to create stable material economies that will be filled with life and opportunity for personal and social growth.

Let's start with the health benefits of nature. We have been researching the impact of natural places on mental and physical health for twenty years.

We came up with the term *green exercise* to suggest that activities in natural places could improve health and well-being. Thousands of people have been studied, and there are no counterfactuals. All people seem to benefit. All green environments work, whether urban or rural, high or low biodiversity, protected or farmed, garden or park. Every activity worked, whether walking, angling, cycling, gardening, and hundred others. This is not to say we each respond in the same way in every green environment. An urban park with a monoculture of grass offers different messages about the changing seasons than a wildlife reserve. It has better options for walking your dog. A garden changes week by week across the seasons, the tundra and the desert in different ways.

We also deliberately medicalised terminology by suggesting there could be appropriate *doses of nature* that also would improve well-being. These echoed the way dose–response curves are used to calibrate medical drugs and environmental toxins.

A danger centres on suggesting an instrumentalism in our complex and fundamental relationships with nature. The upside, though, was to indicate the wider values of natural capital and ecosystem services through the application of health concepts. Nature could be a place for therapy and getting better, for living well and thus increasing happiness and life satisfaction, for living long and thus for healthy longevity. A great deal of research has now shown that the deliberate therapeutic use of natural environments, in garden, allotment, care farm and wild place, has short-term and long-term positive effects on people under mental stress, including at-risk children and young people, refugees, probationers, dementia sufferers, office workers, and mental health patients.

The natural environment delivers vital health services as well as other ecosystem services. It contains the thrilling song of the skylark. Here the story changes up, for there has been the further adoption of intervention acronyms: Nature-Based Intervention (NBI) and Nature-Based Solution (NBS).

The IPCC Sixth Assessment Report, published in 2022, says NBSs provide an example of how innovative ideas, "can expand the solution space." The IUCN-World Conservation Union defines NBSs as "Actions to protect, sustainably manage, and restore natural or modified ecosystems, that address societal challenges effectively and adaptively, simultaneously providing human well-being and biodiversity benefits." Previously the Paris Agreement on climate change called on all parties to acknowledge, "The importance of ensuring the integrity of all ecosystems, including oceans, and the protection of biodiversity, recognized by some cultures as Mother Earth."

Well-functioning natural systems do, in short, play a significant role in buffering social and physical infrastructure from climate hazards and the ills of affluence. The modern evidence base began in the 1970s, when Rachel and Stephen Kaplan at the University of Michigan worked with the US Forest Service to reveal human preferences for green environments. Their Attention Restoration Theory would go on to articulate how attention was a limited resource that could be fatigued, and how nature had restorative effects. We'd put it this way today: turning our attention towards nature quietens the inner mental chatter (see Chapter 7 for discussion on the Default Mode Network).

Two key studies then followed in the early 1980s from Ernest Moore also at the University of Michigan and then Roger Ulrich at the University of Delaware. Moore worked in the state prison, and showed prisoners with green views of farmland were less often ill. Ulrich worked in a suburban Pennsylvania hospital with post-operative patients, and those on a ward with a green view were found to spend less time in hospital and need fewer pain-control analgesics.

An appealing theory followed, built around the term *biophilia*. In those days, the notion of an innate, that is genetically determined, preference for natural places seemed to fit. After all, we humans had evolved for millions of years in such places. Were we simply responding to deeper evolutionary echoes? In truth, such explanations were unnecessary. It is not so much the characteristics of the environment or green place that matters; it is how we choose to behave in and around nature, or indeed are made to behave, at any time or stage of life.

Think again of that sunset, and what we do. We stop and stare, we take pictures, so often that perhaps every phone or camera in the world has at least one image of a spectacular sky, the silhouette of a tree, the rising moon and the appearance of the first stars. We have simply learned to stop, to be attentive and become immersed, even if only for a few moments. Sunsets also start stories, as next might follow the lighting of the fire, the gathering round the hearth for food, the opportunity for further storytelling. Sunsets provoke attentiveness, and so they make us feel good about the world. In this way, the liking of a sunset becomes a habit, perhaps without us realising why.

Something similar happens at the beach. Millions today find the beach a place where sitting and doing nothing is permitted. You watch the waves, the sky and birds, the children running round. Times passes slowly. You lie in the sun, pretend to read and your mind is quiet. You might then sleep a while.

Yet, not so long ago beaches were places for work, locations for fishing village and boats pulled up from the high tide, places where people looked out for danger. There were fish, but also storm and wave and silent deadly current. On the north Norfolk coast in the UK, there are fisher cottages with no seaward windows. They did not wish to look upon the old enemy. Yet, if we were to say, we are moving to the seaside, you might imagine a large picture window, the glittering blue light, a deck perhaps. Not a blank wall.

The habits of attentiveness and immersion are again mediated by the changing norms of culture and economy.

In our good life survey, almost all respondents mentioned the personal value of regular contact with nature, whether plants, animals and pets or whole habitats of meaning. There was lying on the grass, looking at the clouds, watching and being attentive, walking barefoot; there was sitting in the sunshine, the open space and long views and sense of freedom, the smell of wet soil. Some spoke of daily walking the dog, looking after the chickens, flowers inside the home. Others valued gardening and wild swimming, and the surprise of weather: what could happen next, what will be the rhythm of the day or season? Joseph Campbell said that myth offers a framework for personal growth, a way to lead a life in tune with nature.

We already knew, it seems, that nature in all these forms was good for us.

And over time, we each can become experts in our places. Our ecological literacy grows, we can read the land. And the more we understand, the more we care.

Further research has filled gaps, explored inner mechanisms and external social processes, documented interactions with place and behaviour changes across the life course. It has been shown that greener environments reduce social inequality and have particularly positive impacts on mental well-being, that physical labour in the home is important for health and longevity, and that blue space is as important as green: it is not the colour that matters but the opportunity to behave in a way that improves well-being. Meta-analyses of nature connectivity and well-being have shown the more connected we are to nature, the greater the life satisfaction. Life course and longitudinal studies, especially of Caerphilly men, Dunedin children to adulthood, Maudsley and Cambridge cohorts, Milwaukee nuns, and Harvard alumni and local cohorts taken to old age, have shown how choices on behaviours, consumption and mental states directly affect health and well-being over many decades.

All demonstrated the value of early intervention amongst children whose cognitive outcomes are improved when regularly exposed to activity in nature settings in playground, garden and woodland.

This research also shows the value of stacking, a term used by Stephen Kotler in his book *The Art of the Impossible* to illustrate the value of several synchronous choices of behaviour. Many of these engagements with nature involve physical activity, which alone is good for health. Many occur in groups, and togetherness and social capital are also good for health. Many might involve healthy food or lead to changed diets, again good on their own for health. In many ways, we do not need to press on with reductionism, in creating models to show the relative importance of each factor.

What we know about the good life is that it is the combination of all these that can lead, step by step, to low-carbon living that also brings happiness and well-being.

Here is one example from history. Asclepius was the Greek god of healing and medicine. From about 500 BCE, some 800 Asclepian healing temples were raised across the eastern and central Mediterranean. Typically, these were situated on promontories overlooking the sea, at Epidaurus, Pergamon and Kos, where light was multidirectional from sky and water, the breeze cool, and aromatics from pine, lemon grove and thyme-rich meadows filled the air. These temples healed by using medicinal plants, quiet dream rooms, hydrotherapy, exercise and healthy foods.

Today, the wooden staff of Asclepius, with the coiled temple serpent, is still used as a symbol of medicine, the snake revered for rejuvenation and ability to form the circle of eternity. It came to represent this pluralistic approach to medical healing and treatment. Theocritus of Syracuse wrote in *Idyll I* in the 200s BCE, "There is sweet music in that pine tree's whisper, there by the spring."

Five centuries earlier, the poet Hesiod, had written in *Works and Days*,

> At the end of the day, you went to drink,
> Sitting in the shade, have the heart's fill of food,
> Facing into a fresh westerly breeze.
>
> *[Hesiod,* Works and Days, *700s BCE]*

Nature Is Essential Infrastructure

We might now ask: if nature is good for health, how then might we think about the deliberate design of places where we live and work?

And how, too, might we guarantee provision and access for every group of people? On many an affluent country's shore, the beach has become available to only paying customers. In many a city, urban parks have gates with locks. Neighbourhood design matters, and so do policies. We should be looking after natural assets, but also reclaim them, as there are nature-based solutions to many of the contemporary ills of affluence. Closing off nature reduces the options for all to be healthier.

Let's look again at the evidence. The design of settlements and buildings shapes health. The view from the window matters, dementia onset is delayed by walking, and mental health and happiness are improved by being in nature. Green space close to home reduces mortality, cuts levels of stress hormones, increases levels of physical activity, changes dietary decisions and habits, and affects longevity of the elderly. Walkers live longer and healthy plant-based diets play a key role in reducing mortality: one study of elders showed regular outdoor exercise and rich social networks added five years to life.

It is also clear that life satisfaction and happiness play a direct role in longevity. The happiest third of the population aged over fifty have a 10% chance of dying over the next nine years, the middle third a 20% chance, and the least happy a 30% chance.

John Ji and colleagues, then at the Duke Kunshan University, showed how the design of neighbourhoods matters amongst the oldest old in China. Physical activity was higher when green space was accessible within 500 metre radius of home. Social and leisure activity was also more common when green space was accessible. As a result, mortality was lower: people lived longer where their home was near green space. In the UK, the national Monitor of Engagement with the Natural Environment survey has shown that people who visit green space on average 50 times per year are twice more likely to have good health than those who make no visits. Catherine Ward-Thompson of the University of Edinburgh found that those living more than 500 metres from green space had higher levels of stress hormones. David Rojas-Rueda of Colorado State University and colleagues in Spain assessed more than 9000 papers and undertook a meta-analysis on studies with more than 8 million participants. They found an inverse link between surrounding greenness and all-cause mortality. They concluded, "Interventions to increase and manage green spaces should be a strategic public health intervention."

Accessible nature provides a health service. Richard Mitchell of the University of Glasgow has shown that green urban space can also be equigenic. It has greater positive impact on poorer social and economic groups. Green space closes equality gaps.

Epidemiological studies have shown the importance of green space and green neighbourhoods to health indicators. You have to have the green space, and it must also be accessible. This is why religious and spiritual locations play such an important role in Japan. Every village, every urban neighbourhood, has a Shinto shrine or Zen temple, and through cultural norms, they encourage regular access. The oldest old in Japan thus have easy access to natural places as part of their daily routines. It improves their well-being and longevity. Japan has the highest national proportion of centenarians in the world.

Once we know that hospital patients in east-facing rooms, where the light is bright early in the day, spend less time on the wards with green views, there are implications for redesign. This has been hard for hospitals, where the pressing focus is always going to be on illness and acute treatment. There is, though, a welcome movement in the UK now to establish a garden in every hospital for patients and visitors. It is, though, hospices and cancer treatment centres, especially the twenty-six Maggie's Centres in the UK and three in Barcelona, Hong Kong and Singapore, that have incorporated nature into building design as a norm.

Each has light and air, plants and large windows, access to the outdoors, a sense of space and slow time. All these encourage attentiveness and immersion among patients, carers and visitors.

Yet, some still are deaf to people's cries. Nature does not have an advertising budget. It does not spend to create wants. The world advertising budget is US$500 billion annually. It creates dissatisfaction, it fashions suffering, it hopes you will buy more stuff to overcome the very anxieties it creates. Nature cannot compete on the same terms. This is another reason why the target for stories and evidence about nature benefits is the public, its policymakers and health practitioners.

And so, look at what we already do. Every major city in the world has public parks and green space, commons created over the past hundred and fifty years. Singapore has put them in the air, New York has greened an elevated railway line and made the High Line park, which now attracts 5 million visitors annually. Across the year there can be an average of more than 100,000 people daily in New York's Central Park, and 50,000 daily in Beijing's Forbidden Palace and gardens. Each shrine and temple in Japan has garden, grounds, forest, waterfall or sacred mountain: in Tokyo there are 1400, in Kyoto 2000. Nature is already an essential health infrastructure in cities.

And so environments structure behaviours. In London, the average person walks 490 kilometres per year. People can walk to work, to the theatre or restaurant, they walk to and from tube station and bus stop. In rural England, the average person walks only 190 kilometres annually. The roads are dangerous as cars now travel fast, local shops have closed, and the bus service is irregular or has disappeared. You need a car for the basics.

In New York City, the obesity rate amongst adults is 22%, in Manhattan 19%, and in the state 26%: all much lower that national averages. The city has sidewalks, people can walk, and there is a public metro system. Many other American cities have no such common infrastructure. These New York citizens are more physically-active during the course of their day, just moving in the spaces between work and leisure. Now cities across the world are signing up as part of a biophilic cities movement, and Stephanie Panlasigui of the San Francisco Estuary Institute has observed, "This has the potential to enrich our experience in daily life."

Our feet were made for walking, and this helped make stories. The land revealed itself as we walked forwards. There were signs, an edible plant, a hoof print, a large paw, a broken branch, a bush with red berries, a spring bubbling from the slope. We returned to camp with food and water, and a story about the journey, what was food, what was learned, so others could walk in our footsteps. For walking laid down memory. Walking in this way is attentive. You look for signs, you are ready to run. But you are calm too.

At any time in life, walking is known to increase the size of the hippocampus, the centre for memory. Walking also sets back the onset of dementias and is effective therapy for some people already suffering symptoms of degenerative brain conditions. Walking changes the brain as well as the body. Once upon a time, we all walked. Every day, and eventually over every continent. From shore to blue shore, over land bridge and plain, across ridge and marsh. Then came domestic beast and the wheel, the sail and engine. Today, much has been ceded: too little daily physical activity has become another ill of affluence.

It came to pass that physical activity came to be seen a drudgery, a deployment of labour that was wasteful. Was it because when we walk, we do not spend? King Henry VIII banned pilgrimages in the 1530s, he and advisors felt people were not contributing enough to the economy. In hunter–gatherer–forager cultures, women and men are active every day. They also seem to have more time to rest, to lounge around the hearth and talk. They work enough, but not more. There was evidence from hundreds of studies of hunter-gatherer communities: people work fewer hours daily than do working modern people. There was personal growth, yet stability within their territories.

And anyway, as Jay Griffiths in *Why Rebel* reminds, "If I sleep out on the earth, I dream differently there."

Today, in the USA, there are 80 cars/autos for every 100 people, and this includes in the denominator all babies, children and the elderly. As countries become richer by GDP, so the motor car seemed to bring freedom and status. The relationship across countries between GDP and car numbers is instructive (Figure 5.1). There

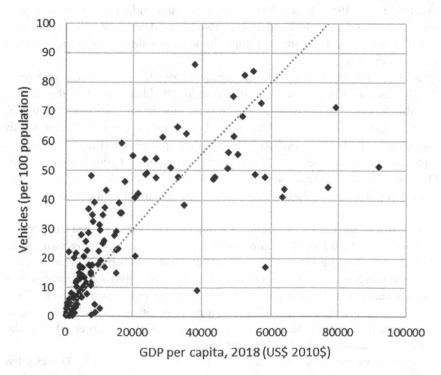

FIGURE 5.1 Car/auto vehicles (per 100 people) by per capita GDP, 2018 (n=164 countries)

is no tailing off, though some affluent countries are below average for numbers. As countries get richer, so the number of cars per 100 people rises. And so must the public expenditure on roads, the tarmac to cover nature, the replacement house and community torn down for roads.

And when we have this freedom, what next? To sit in cars, in endless jams. For if you build a road, cars will fill it. The average speed of a person travelling in UK cities is 8 mph, no different from 100 years ago when horses where the main source of transport power. This, some say, is progress. There are now one billion cars in the world. And then there is the air pollution. And road traffic collisions causing 1.4 million deaths a year.

Physical activity improves both mental and physical health. Inactivity causes 1.9 million deaths worldwide each year, roughly 1 in 25 of all deaths. Inactivity increases the likelihood of obesity and is a key risk factor in many chronic diseases of later life. We should not forget that physical activity also occurs at home. The term *non-exercise physical activity* was used by Elin Ekblom-Bak and colleagues at the Swedish School of Sport and Human Sciences to focus on the health benefits of home repairs and cleaning, mowing the lawn, car maintenance, and errand-based bicycle rides. They found that sixty-year-old Swedish women and men with high NEPA reduced risk of first-time cardiovascular disease by a quarter and all-cause mortality by a third over a 12-year period.

You are active at home, you live longer. You are active at work, you live longer too. In the early 1950s, Jeremy Morris and colleagues studied 25,000 bus and tram drivers and conductors in London over two years. There was a significant and much higher rate of heart attacks among the sedentary drivers compared with the active conductors.

So this: individuals who do not engage in regular physical activity have a 20–30% increased risk of cardiovascular disease. Worldwide eight of ten cardiovascular disease deaths are preventable. The cost of sedentary behaviour in the UK is £8.2 billion per year, in the USA some US$90 billion per year. Annual health care costs in the USA are on average twice those per person than in the Nordic high-tax countries, roughly US$9900 per person compared with US$4–6000 for Denmark, Finland, Iceland, Norway and Sweden. The Nordic model is available to each population, and cheaper at the national level partly because it keeps people out of illness.

In Norway, the outdoor tradition of *friluftsliv* will see roughly half the population outside at the weekend: running, walking, cycling, skiing, sailing, swimming, sunbathing, eating together. In Finland, where equity and happiness seem normal now, and everyone values time away in nature, it was not always so: education was very unequal until the end of the 1960s, healthcare systems in poor shape until the early 2000s; a century ago, it had the lowest measures of welfare and well-being in all Europe. In *Finntopia*, Danny Dorling and Annika Koljonen observed that "the story of Finland shows that anything is possible."

It was fiery, the hottest spell for Copenhagen since the 1920s. The city has become in recent years geared to the outdoors, to activity and well-being. More people cycle than in other cities, it is easy to walk, there is clean water in the harbour. Along each dockside were crowds of people. A single white cloud was in the sky. There was laughter. Art galleries had emptied, the indoor zones of cafés too. There were outdoor yoga groups, dancers round a music box, groups on the water with sail and kayak, and others leaping from the dockside. No one had felt the like of such city heat before.

There was clamour, yet people did not have to choke on dust. Copenhagen calls itself a green tiger. More than half of commuters bike to work. This did not happen by accident. White crosses were painted on streets where cyclists had been killed by cars, then cycle routes were constructed and more curbs placed on motor vehicles. The rapid growth in cycling happened fast, just in twenty years. Other cities still consider the car as far too important. Yet take away the noise and air pollution, keep the spaces green, ensure it is accessible, and cities soon change their character.

The past five years have seen a 50% growth in the green sector in Copenhagen, improvements in health and quality of life, US$43 million in avoided health costs, and increased tourism and associated jobs. And these have fallen: air pollution, noise, traffic accidents and congestion. Every 10 kilometres of bike travel has been calculated to save 4.5 krone ($0.7) for the Danish economy. Café numbers in the harbour have risen fivefold. A district cooling system uses water from the sea, and overall the city has reduced water consumption by a quarter, saving millions for public authorities.

At the Institute of Happiness in Copenhagen, Meik Wiking has researched and written of *hygge*. He's not saying the country is perfect but is sure they have got more things right than not. *Hygge* is happiness, giving value to coziness, friends and togetherness, equality and gratitude. It is sharing food, baking cakes and lighting candles. It is also co-housing, it is volunteering and looking after neighbours, it is being in nature, and feeling you have a meaning in life. In the northern latitudes, it is always waiting for the sun. And all these activities, they are low in material consumption and low in carbon emissions.

Meik Wiking writes *hygge* may be bad for market capitalism, yet is already good for personal happiness. Which actually is good for the economy too.

The aim, he says, is to change the story, "How to make cities great, how to make them livable again. By working on happiness, we are doing more now to change the city."

The headline: welcome to our backyard.

> In my heart, I listen to evening rain,
> And stretch my legs without a care in the world.
>
> *[Ryōkan Taigu, Edo poet, 1758–1831]*

Health Services From Nature

Even as the winds sweep in, we know the natural environment provides vital health services as well as other ecosystem services.

In 2019, the UK government's *Twenty-Five Year Environment Plan* sets out six priorities, one of which was, "To connect people with environments to increase health and well-being," and substantial focus has been put on making use of green spaces and natural habitats to improve the health of whole populations. A wide range of nature-based interventions (NBIs) has been deployed by conservation, wildlife and community organisations and charities. These natural and social therapies have been offered to groups under mental stress, to at-risk children and young people, to military veterans, refugees, probationers, and dementia sufferers. Fields in Trust has recently calculated that parks and green spaces already provide some £34 billion of health and well-being benefits annually, saving the UK's National Health Service more than £100 million per year.

The UK Chief Medical Officer recommended that the costs arising from ways of living now require *a new canon for prevention*. With an ageing population, growing numbers of people with long-term health conditions, cost inflation, and pressures on revenue funding, the health systems in affluent countries need to find ways to invest in prevention to slow the pipeline of people requiring primary and secondary treatment. This concern for prevention is not new: in 1736, Benjamin Franklin advised, "An ounce of prevention is worth a pound of cure," referring specifically to fire safety.

What has changed recently in affluent countries has been the scale of costs of non-communicable diseases and the pressing need for changes to policy and investment priorities. The treatment costs of obesity, type 2 diabetes, loneliness, cardiovascular diseases, mental ill-health, dementias and physical inactivity run well into the hundreds of millions of pounds and dollars for affluent countries. In the UK, the prevention-pays policy was expanded by the Chief Medical Officer with a raised focus on health as an asset that needed protecting: "Prevention creates the right conditions for good health and well-being – helping everyone to live well for longer."

This raises the prospect of a great financial unlocking. We already spend to treat the ill-health arising from NCDs. If some of these were prevented, then lives and economies would benefit.

But there is a small problem with this prevent-to-save agenda. When health systems are entirely privatised, health providers are paid by the volume of work, that is by the number of ill people. They have no incentive to reduce supply. When health systems are in the public sphere and supported by the tax base, then there is a real economic benefit to prevention. In private health systems, medical insurance companies may provide incentives to reduce demand, but this also presupposes that all people can afford insurance. It was estimated that 44 million Americans had no health insurance in 2020, and another 38 million had only partial coverage.

The UK health system has extended prevention-pays into a new intervention model called Social Prescribing (SP). This was launched by General Practitioner surgeries and in primary health care as an alternative to medical intervention: non-clinical community interventions (NCCIs) are particularly focused on people and patients with NCDs. And now interest is expanding into green social prescribing

(GSP). The National Academy for Social Prescribing was established by the NHS in 2019, and is seeking to expand SP as a new social movement, defined as "supporting people via social prescribing link workers to make community connections and discover new opportunities, building on individual strengths and preferences, to improve health and well-being."

The most effective Social Prescribing programmes employ Link Advisers as guides to patients, interviewing and assessing needs, and then ensuring each person can take advantage of the community-based options available with good signposting. These fall into distinct categories: for advice and knowledge, such as on benefits or housing; for skills development, such as computing, food and cooking; for activities in social groups, such as befriending and self-help groups, dance, art or crafts; and for activities with therapeutic design, especially nature-based such as walking for health or woodland therapies, and formal counselling.

Most Social Prescribing operations in the UK have more than 50 social options for patient referral; some end up with more than 200. This is beginning to look like the rebuilding of social institutions shattered by decades of selfishness and inequality.

Evaluations of SP and GSP programmes have found benefits for health providers, particularly in cuts in average numbers of appointments at primary care providers, falls in numbers of Emergency Department visits, and reductions in the need for secondary care appointments. Health providers have seen declines at all three stages of 12–35%. The Healthy London partnership for Social Prescribing calculated that London could have saved £110 million over three years if social prescribing had been widely available and used.

At the University of Essex, we recently conducted an evaluation of four programmes of nature-based and mind-and-body interventions in the UK, using the key measures of impacts on life satisfaction and happiness together with demand for public services. We assessed the impact on 642 people of woodland therapy, ecotherapy, community horticulture and outdoor tai chi. These were programmes based in natural places, usually stacked with other potential benefits arising from togetherness, personal growth, physical activity, and food/diet, and lasted for long enough potentially to cause long-term changes to habits and thus neural structure.

The primary measure was the new standard of impact on life satisfaction/happiness. These have proven to be robust and meaningful in other contexts. The UK government's Office for National Statistics has been measuring well-being since 2013, and the World Happiness Reports measure annual changes in happiness across more than 150 countries and another 150 cities.

Figure 5.2 shows the improvements in life satisfaction/happiness, benchmarked against the UK average of 7.69. All the participants began with levels below this national average. All made improvements as a result of being on these programmes.

The literature on life satisfaction and happiness makes a clear observation: it is difficult to achieve movements of +1 point on the 1–10 point scale. In the World Happiness Reports, the maximum change across whole populations in life satisfaction/happiness over a decade is +1.0–1.3 for five countries, then +0.5–1.0 for

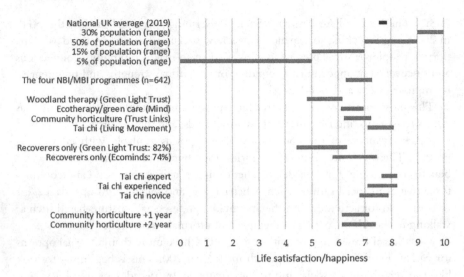

FIGURE 5.2 Improvements in life satisfaction/happiness following four nature-based and mind–body interventions (the national data are ranges for each category; the lines for the NBI/MBI programmes indicate start and end points due to each intervention)

thirty countries, and +0.0–0.5 for 43 countries. Existing levels are stubborn: the positive and negative changes as a result of significant life events have been shown by Andrew Clark and colleagues in *The Origins of Happiness* in drawing upon 100,000 observations of life events to be at most +1 or −1 for marriage and birth of a child, for divorce and loss of job. The What Matters Course run by Action for Happiness has trained 1500 people and produces a +1 point average increase after two months on the course.

Here we found the woodland therapy of the Green Light Trust increased life satisfaction and happiness by +1.36; the ecotherapy and green care of Ecominds by +0.87; the therapeutic horticulture of Trust Links by +1.03; and the tai chi of Living Movement by +0.97.

There were three further findings. There was a positive effect on inequality, with those starting lowest having the greatest benefit. The effects lasted up to two years. The economic benefits were significant: we used the UK government's evaluation methods from the Treasury's *Green Book* to show that each individual provided net present economic benefits to society of £6500 to £12,500. Such impacts justify early spending on prevention, and this raises an important question.

This raises a big question: how could such nature interventions be scaled up so that thousands and then millions of people would change habits and behaviours permanently?

This will require bold local and national government action, backed up by serious pushing through social movements. Institutional inertia and self-interest will be hard to overcome. After all, if populations were to walk more, eat healthier food, be more active and spend time with others, perhaps they would spend less on material consumption. Healthy nature produces healthy people, and nationally we save the economy money.

A proper system of accounting would recognize this, but GDP fails again. It still counts expenditure on treating illness and disease as a contribution to economic growth.

Dogen Zenji wrote in the 13th century:

> Although mountains belong to the nation,
> Mountains belong to people who love them.

Trickster Questions

Nature and physical activity are good for health. Earthsong helps prevent displaced costs, assisting the transition to a stable economy. It helps us live longer.

Leslie Marmon Silko describes her homeland of north New Mexico, where there have been some 18,000 years of continuous living: it is an animated landscape, full of intent. She walked how a hunter walks, "Not too fast, stopping frequently to listen, listening the way a deer listens," and one day found a large piece of turquoise on the trail. How odd, she said, it was not on the path yesterday. "Let it be," she concluded, "Leave things as they are, all things in the world."

When Antoine de Saint-Exupery was learning the route for the air postal services in the 1920s and 1930s, flying in planes with open-air cockpits, he wrote, "The earth teaches me more about ourselves than all the books in the world."

The night routes were filled with trap and snare, cliffs that loomed and whirling currents that could uproot cedars. On being talked through a new route by an experienced pilot, noting each obstacle and sign, little by little, he wrote, "In the lamplight, the Spain of my map became a fairy tale landscape. Guillaumet didn't teach me about Spain, he made Spain my friend."

Dorothy and Walter Schwarz asked in their book, *Breaking Through*, well, how much do you feel connected to your place? Pause a moment, they said, and try answering these seven trickster questions:

> Question 1: What is the route of rainfall to your tap?
> Question 2: How many days till the moon is full?
> Question 3: What were the subsistence methods of the cultures that dwelled
>   at your home place before you?
> Question 4: What are the names of five edible wild plants in your bioregion?
> Question 5: What are the names of five trees nearest your home?
> Question 6: From where you are reading this, point north.
> Question 7: How many people live next door, and what are their names?

And here is another, asked by George Vaillant in his long-term prospective studies of ageing in the USA:

Question 8: What have you learned from your children?

Vaillant found some respondents thought this was a ridiculous question. They were more likely to be unhappy, inactive and disconnected to nature. They tended to die early.

And thus you could also ask:

Question 9: What were your best moments in life?

Our best moments in life: few people say it was having or buying things. Mostly it was events and experiences, friends and relationships, stories about making sense. Earth is our home. We should leave things as they are, and live as well as we can. The Koyukon of the boreal forests say to animals and plants, "I hope you will return again, and that we will be here to see you." Elsewhere, there is a job of recovery and restoration awaiting.

"That's why we're here," said Tyson Yunkaporta in *Sand Talk* of the First Peoples of Australia. "We look after things on the Earth and in the sky and the places in-between."

That's why we're all here. To look after stuff and each other. Nature will survive, even though it has been diminished. The question remains whether we can act soon enough to prevent collapse from the ills of affluence. Engagement with nature reduces material consumption. It can make us happier with what we have got, live longer, appreciate nature and seek to care for it and be more humble.

If you listen carefully and are attentive for long enough, all around you will find some inspiration in the Earthsong.

> Wild peonies, now at their peak,
> In glorious full bloom,
> Too precious to pick, too precious not to pick.
>
> *[Ryōkan Taigu, Edo period, 1758–1831]*

~★~

6

THE FOOD AND AGRICULTURE CRISIS

It's Nourishment, Not Calories

In the tea room, unlikely bowls are prized,
Yet it takes years of hard work, to create these vessels,
They are perfect, in imperfection,
Homely and friendly to the touch.
[William Scott Wilson, *The One Taste of Truth*, 2012]

Abstract: This chapter lays out on a long table the food and agriculture crises. How did the world turn from nourishment through apparent plenty to the new silent costs imposed on public health? It was, it seemed, a wondrous revolution, more food from the land, more food per person worldwide. Then in supermarkets, thousands of new foodstuffs each year, all remarkable invention and packaging of desire. Yet, we all know, the right amount is best, as a Swedish proverb puts it. Food that brought people together became a world of sorrow. It imposed huge and hidden costs on nature and environments, it provoked the obesity and type 2 diabetes crises. The world agriculture and food systems contribute 25% of global greenhouse gases. We could do good by eating well, but this is a big burden of responsibility. Help could come with policies and practices for regenerative agriculture, slow food, adoption of new diets, celebration of local and distinctive foods. No country worldwide has been able to reduce obesity rates once they rose. Two countries managed to keep levels very low: Japan and South Korea. Put a rainbow on your plate, say people in Okinawa. Eating less meat will favourably change food systems worldwide. "Every meal should be one of the best meals of your life," writes Claire Greenwood. The chapter closes with a famed story of giving: three grapefruits.

DOI: 10.4324/9781003346944-6

Food Is More Than Calories

Sure enough, it had become a time of plenty.

And this turned out to create a different sort of shame. Agricultural yields have increased steadily for 60 years, producing more food per person worldwide. Poverty and hunger fell. And each year several thousand new food-like items came to be launched by food companies.

Yet, as happened in almost all economic sectors during this late stage capitalism, the benefits were spread unevenly, within and across countries. Many agricultural and food systems looked successful because they were able to shift costs outwards, imposing silent harm on biodiversity and natural capital. They produced the calories, but diets and dishes started making people ill.

Food had imposed another form of silent cost, now on people's health.

The UK is one of the richest countries in the world, if we revert to GDP as the measure, yet it had more than 2000 foodbanks in 2022, staffed by volunteers and delivering emergency food to more than two million people. The USA is a food superpower, and hunger affected one in ten of the national population daily in the early 2020s, more than 30 million people. The pandemic made things worse, but these have been problems growing over many years.

On the global scale, affluent countries and international agencies have repeatedly stated that the elimination of hunger was one of the most pressing priorities of their age. There has been progress: the proportion of the world population hungry has fallen. But there remained 800 million people malnourished and hungry every day by the early 2020s. None of us can call this success.

A sixth of the responses in the survey on the components of a good life centred on food. No surprise here. We need to find or purchase food and drink every day, unlike clothing, transport and housing. And because of its regularity as a consumption activity, it can produce ceremonies and rituals, bring people together, heal rather than harm. The good life respondents said this about the good aspects of their relationships with food: it is about health, it is a gift; it is about preparing, sharing and eating together, it is table fellowship. It is about tasty, seasonal and fresh dishes; for many, it is plant-based diets. It is about avoiding highly-processed foods; it is about baking and making cake, bread, yoghurt; it is about supporting small farmers and eating sustainable, local and own-grown foods.

You can see: many of these are about attentiveness to food and people, and immersion in the processes of growing, preparation, cooking and eating. Nourishment is necessary for growth and health. There are suggestions here of the inherent qualities of diversity, taste and goodness, as well as elements of togetherness in the devotion of preparation, and time taken eating with others. And as food comes from land and sea systems shaped by specific ecological conditions, it has differed from place to place across the Earth. We come to like and value difference, partly because of habit, partly because it affirms who we are.

Food is more than calories, it has sub-story. You might also say, in many places it does not look like this anymore.

Here is a short story from years gone by. It was the 1970s, and the North Sea oil rush had just begun. To our home territory in the east of England came oil families from the plains and nodding rigs of Texas. It was soon after the three-day week and dark malaise, only 20 years after post-war food rationing had ended. The oilers imported their home culture, and hatched a plan to set up a football team for their children. The colonel of the four local US airbases agreed, the civilian Rough-necks could join their league, and so the new team turned to local schools to fill the roster: we upset the military, won the league. But those young cargo-culters experienced something more than sport, for on each match day, as the afternoon sun shone, great fires on charcoal were lit.

Barbecue smoke drifted across the grassy field, and there were burgers as never smelled before, also bins of coke and root beer bobbing in ice. All was intoxicating to the natives, these tastes and all the social wrapping.

So let's start with wondrous revolution. Humans to go out into natural places for food and then bring it home for other adults. By sharing and giving, we created trust and reciprocal bonds. Friendliness rewarded everyone. Along came agriculture, and all seemed well. Agriculture was marked by periods of sharp innovation and discontinuity, the rapid adoption of new practices and technologies by millions of farmers. From the domestication of wild grasses and animals to the privatisation of the commons, from the breeding of modern cereal seeds to the manufacture of fertilizer and pesticide, from draught animals to computerised machinery. This all looked like progress.

Since about 1960, two generations ago, with post-war expansion of farm size in affluent countries and the advent of the green revolution in developing countries, world food production from agriculture more than tripled; fish caught from open oceans rose by sixfold. Food production grew faster than population: today's Earth population of nearly 8 billion has a half more food per person than the 3 billion people had in 1960.

Now the production of food shapes every corner of the planet. It is a major driver of the climate crisis, it causes ill-health. Growth brought cost, and a clever twist, for as with other economic sub-systems these side-effects and externalities were counted as contributors to economic growth. Clean up water pollution, restore a wetland, pay to go on a diet, pay to treat type 2 diabetes: GDP likes all these just as much as making cars or TVs.

Let us pause a moment to consider the record of indigenous, tribal and world peoples who have lived their good lives in jungle, forest, prairie, marsh, desert and tundra. For tens of thousands of years, modern humans managed the wilds, lived inside nature, built sophisticated systems of management, and did not resort to destruction of their own home. We know what it is to persist in particular places, and how to use hundreds of species of wild plants and animals. Wild edible species still form part of the diets of a billion people, and it is estimated that 7000 species have been used for human food. Even today individual agricultural communities commonly use 100 wild species per location, aggregating up to 300–800 for whole countries.

With the need for food and medicine, so also came knowledge, developed and transmitted across generations about what was good and bad for diet and health. And as each place developed its own ecological characteristics, so too did human diets.

Across every habitation, too, emerged ethics of care, respect and responsibility. Food was ceremony at the threshold of natural and human worlds. There is only one Earth system, yet many sub-systems of food and agriculture. We eat, as Partha Dasgupta has recently written, so "We are all asset managers." The trouble is, we as commoners have become less able to calibrate enoughness. Why would we, when everyone else is saying, eat and drink more?

In the 1950s, Alan Watts said in *The Way of Zen*, "In walking, just walk. In sitting, just sit. Above all, don't wobble." We could say the same of food.

"Lagom är bäst: the right amount is best"

[Swedish proverb]

How Food Became a World of Sorrow

The sun went down, and many ways became dark. How did food systems go so wrong, and why do they seem to stay so broken?

It is a tragedy. Just as many agricultural systems have made progress towards regenerative and prosocial methods of production over 30 years, so diets and dishes have grown worse. Just as sustainability began to spread, so food systems were actively redesigned. Or perhaps broken.

There were *evil geniuses* at work, as Kurt Andersen has written. And almost no one predicted the consequences. The influential reports of the Club of Rome and Brundtland Commission in the 1970s and 1980s both missed obesity and other non-communicable diseases. And almost no one saw the coming pace of change. The key inflection year was again about 1990, when atmospheric carbon first passed 350 ppm, the last safe space for humanity.

Since that time, the incidence of obesity has risen in affluent countries and communities. As GDP rose, so did the proportion of people whose weight exceeded a body-mass index of 30 (BMI is weight in kilogrammes divided by height in metres squared). In some cultures, being fat had been a sign of prosperity and power, and so was rare. It was caused as often by genetics as culture.

And this too: agricultural and land-use systems now produce more than a fifth of greenhouse gases, through releases of carbon from cutting trees and degradation of soils, through methane from ruminants, through nitrous oxide released by fertilizers. Forests still are being cleared and burned, mainly for soybean to feed the meat demand of industrialised diets and fast-food systems, and for oil palm to bulk up foods and make soap and cosmetics.

This is why diets matter for the climate crisis. Food has become a world of sorrow. The Rockefeller Foundation recently analysed the US food system: it supports 22 million full-time and part-time jobs, some 11% of national employment, and is worth US$1.1 trillion per year. And yet, the full and true costs are $2.1 trillion more. A big economy looks good in one way, but not in another. These external costs include $360 million of costs for obesity, and another $600 million for other diet-related NCDs. Says the report, "Our food system has deep impacts that reach far beyond our plates."

They also noted that these additional costs were disproportionately borne by marginalised and underserved communities. I have written in the past that consumers pay three times for food. You pay the purchase price, you pay by taxes for governments to subsidise certain systems of agriculture, and you pay a third time to clean up the environmental and health costs.

The direct subsidies paid by governments to the food and agriculture systems amounted to US$700 billion worldwide in 2020, plus another $35 billion to fisheries. This is enough to give every current hungry person in the world an annual basic income of about $1000. This is not going to happen, but choices still matter. You select a food or one whole diet, and you shape farms and nature at the front end of the food chain. You influence how subsidies are distributed by both voting and eating. Everything is connected.

And yet this type of responsibility can feel like a burden. We can do good by eating, we can do bad. Some governments prefer this, for they wish to suggest it is not economic and social conditions shaped by them that matter, nor is it weak regulation of food content and labelling. It is only down to you. So if you fail to eat well, the fault is yours alone. The British Prime Minister of the early 2020s, Boris Johnson, once wrote about obesity, speaking directly at food consumers in a national newspaper article. He said, "Face it, it's all your own fat fault."

Mostly we do not have the time to make informed choices each time we shop. Poverty limits choices even more. We also don't know many of the consequences of purchases; we are being deliberately misled by food labelling. We just wish all could be well.

The Small Bits in Ecosystems

Let's digress to farms and the small bits that make up food systems. Take insect life, a major source of economic damage. They eat the stuff farmers and gardeners want to grow, they harm animals. For thousands of years, farmers have had their own methods to control damage, often encouraging good predators to consume bad pests. In ancient China, farmers strung bamboo between fruit trees to encourage ants to travel between trees to control pests; in medieval Britain, farmers grew mixtures of cereals and legumes together, thus fixing nitrogen and helping to push away pests.

In Kenya, a remarkable system has been adopted by 250,000 small farmers. Researchers found that maize mixed with grasses around the field edge with legumes intercropped pushed away the corn stalkborer and pulled it into the resinous grasses that were toxic to the stalkborer's eggs and larvae. The mix of plants emitted semiochemical hormones that pulled in parasitic wasps and predators, and then also suppressed weeds within the fields. The modern simplified system of monocropped maize had required costly pesticides and fertilizers to do this work and was now replaced by this diverse one called *vutu sukumu* (push-pull) that needed no external and costly inputs.

If you had the choice of maize cob at the shop, which might you choose? The one from the monoculture, or the one from the push-pull farmer? The choices made for pest control over the past century tell us a great deal about food systems and their side-effects. Most consumers seem to have little role to play beyond lobbying or selecting certain foods.

By 2020, agriculture worldwide was using 3.5 billion kilogrammes of pesticides each year in the hope of controlling insects, weeds, fungi and other pests. Farmers spent US$45 billion annually to buy the 17,000 types of compounds. We recently calculated the environmental costs of these pesticides to be US$10–60 billion each year. These are costs not paid by the manufacturers. In addition, we now know that pesticides are annually causing 380 million cases of acute poisoning of farmers, affecting more than 40% of all farmers worldwide. Chronic effects are harder to trace, yet we can generalize like this: all are biocides. They are intended to kill other organisms.

Back to insects. Perhaps we need not worry. There was an old woman who swallowed a fly, you might recall, who swallowed a spider to catch the fly. You may also remember rumours of the night-time moth blizzard, insects caught in headlamps when driving, smeared on windscreen, on bonnet and headlamps.

By the 1990s and early 2000s, the insects had largely gone. No more moth blizzard. Truck drivers in the USA reported they could drive across whole states and yet their windshields remained clear. Research in Germany and the UK has shown that insect numbers fell by 75% over 30 years. These were the insects once eaten by other organisms higher up the food chain, by birds and by ground animals. This collapse could have been down to insecticides killing insects and herbicides killing host plants, it could have been a specific compound. Yet, over time, we find almost every class of pesticide has fallen from claimed grace. A compound was a marvel at release, then the gathering of information on harm, the growing concern, the years of partial policy intervention and business lobbying for delay. For years, the deliberate creation of doubt: let us wait a little longer. Then, very occasionally, a ban.

Neonicotinoids were first used in 1990s and seemed wondrous for their targeted effects and low toxicity to animals. Then they were found to kill bees and other pollinators. They were banned from use on crops in flower, and then remarkable research tracked these neonics moving from fields through soil to be taken up by wild flowers, and then expressed in their pollen. The burden on bees was found to be greater from the wild flowers than from the crops.

For several years, there was not a single honeybee in my garden or on nearby wildlands, even though we live in a formally protected area in the UK. When neonics were banned, the bees came back. Now a reversal of the ban has been permitted. Growth at any cost could return again.

"I wish that everyone who said they believed in angels would actually believe in insects," wrote Jay Griffiths in *Why Rebel*.

Wait, I see something. Here's a short story.

Gather round, for there is a large passenger jet parked on a concrete runway. We could investigate how this tightly-engineered system works by taking it apart, and so will lay all 6 million pieces across the ground. Then we could experiment and remove all the smallest pieces, each rivet, screw and tiny spring. And throw them away.

Now we could rebuild the aircraft without them. It will look from a distance much like the original aircraft. All the big bits are in place, the wings and windows, the engines and wheels, and there looks to be the promise of comfort. But if you saw what happened to all the small bits, would you join the queue to fly?

The researcher and observer of food systems, Tim Lang of City University, wrote in *Feeding Britain*:

> There are now vulnerabilities with the intense, just-in-time, efficient, hyper-active food system created in our name and with our tacit acceptance.

Regenerative Agriculture

And yet, we know how to do this.

Agriculture is rather unique as an economic sector as it directly affects many of the very natural and social assets on which it relies for success. These influences can be both good and bad. Industrialised and high-input agricultural systems rely for their productivity on simplifying agroecosystems, bringing in external inputs to augment or substitute for natural ecosystem functions, and externalising costs and impacts. Pests tend to be dealt with by the application of synthetic and fossil-fuel-derived compounds, wastes flow out of farms to water supplies, and nutrients leach to the soil and groundwater.

By contrast, sustainable approaches to agriculture seek to use ecosystem services without significantly trading off desired productivity. When successful, the resulting agroecosystems have a positive impact on natural, social and human capital. A wide range of terms for more sustainable agriculture has been developed and deployed, all part of the same broad church of common principles. These include sustainable agriculture itself, regenerative agriculture, sustainable intensification, agroeco-logical intensification, organic farming, natural farming, conservation agriculture, agroecology, biodynamic farming, restoration agriculture, attentive agriculture, permaculture, nature-friendly farming, evergreen agriculture, green food systems, pasture-fed livestock, rotational and mob grazing, and save-and-grow agriculture.

It is a long list of successes. Sustainable farming is now central to the UN's Sustainable Development Goals and wider efforts to improve global food and nutritional security, and the FAO has defined regenerative agriculture in 2021 as "an inclusive agroecosystem approach for conserving land, soil and biodiversity and improve ecosystem services within farming systems."

These concepts tend to be open, emphasising outcomes rather than means, applying to any size of enterprise, and not predetermining technologies, production type, or particular design components. They seek synergies between agricultural and landscape-wide system components and can be distinguished from earlier manifestations of intensification because of the explicit emphasis on a wider set of environmental and socially-progressive outcomes. Central to the ideas of sustainable and regenerative agriculture is an acceptance that there will be no perfect endpoint due to the multi-objective nature of sustainability. Thus, no designed system is expected to succeed forever, with no package of practices fitting the shifting dynamics at every location.

The central understanding from these developments is this: modern, industrialised agricultural systems have depleted, eroded and lost carbon. In order to turn agriculture and land use systems into carbon sinks again, new methods of agriculture will have to be deployed and adopted widely on hundreds of millions of hectares.

These are the core components of regenerative agricultural systems: ploughing is replaced by no-tillage cultivation; monocultures of crops are replaced by complex crop rotations and patterns; and organic matter is deliberately added to soil in the form of plants (cover crops, green manures, legumes, agroforestry and residues) and additives (biochar, fermented mixtures, compost).

And these are some of the soil-based blunders of industrialised agriculture: ploughing destroys soil microbes (bacteria and fungi) and releases bound carbon, large machinery damages soil structure and increases soil erosion, pesticide compounds are widely toxic to microbes and beneficial insects, and parasitic-control compounds and antibiotics routinely fed to livestock kill microbes and beneficial insects, such as dung beetles. The result: agricultural soils have lost 25–75% of their carbon stocks compared with natural or undisturbed neighbouring soils.

The secret to regenerative systems is to focus on the microbial world of bacteria and fungi. At any location you choose, it is possible to find 25,000 species of bacteria, 10,000 species of fungi, and tens of kilometres of fungal hyphae. These microbes are fed by plant roots through exudates of complex carbohydrates, and so a greater diversity of plants creates a greater diversity of below-ground ecosystems. Diverse microbial communities help to control pest outbreaks, provide nutrients for plants, transfer messages between crops and trees, and drive the food chain for larger soil animals. At the same time, they are harmed or killed by residues of insecticide, fungicide and herbicide, and by inversion ploughing. Soil life makes soils fertile, and captures carbon for the long term.

David Montgomery of the University of Washington has called for a soil health revolution, noting that soils are neither dirt nor reservoir of chemicals. We need to think of healthy soils as a "sea of biology."

And a hint, again, that modest rethinking can lead to a dramatic change in landscapes. It was a simple idea, a revelation that came to Tony Rinaudo after years of futile attempts to work with people in dryland Niger and elsewhere in West Africa to plant, raise and protect trees. The forest, he suddenly realised, was already underground. Trees provide fodder, wood and medicine, they support soil structure, influence the microclimate, shade animals and people, yet they rarely reached maturity. Once cut, people had been taught and told: remove the stumps as crops need unobstructed soil and fields. But protect the stumps, and the trees will quickly regenerate. They are connected to sources of water and nutrients underground, they are connected to each other. Through this, Farmer Managed Natural Regeneration has developed a thousand faces, spreading to 10 million hectares of dryland, producing landscapes of tree, crop, livestock and healthier people. All without planting a single tree.

This landscape has been called the Great Green Wall of the Sahel, such has been the surprise and transformation. There is clear evidence from fields that crops under trees are greener and healthier. The trees draw up deep water and make it available to the crops. Through fungal links, they are probably sharing photosynthates too. The forest of trees and soil microbes underground are also a rich stock of captured carbon. It has become another nature-based solution to address the climate crisis.

Ben Okri wrote of the famished road, and how an iroko tree had been felled, how the tree was mighty and its trunk gnarled and rough: "It looked like a great dead soul dead at the road's end." An above-ground iroko can live for 500 years.

Regenerative economic systems are multifunctional; they tend to build natural capital and create new sources and flows of ecosystem services. In agriculture, these benefit farmers and the planet. The side-effects are beneficial rather than negative. In many systems, trees are thought of having primary value as shade trees. Yet, trees can also fix nitrogen, be a source of livestock feed, and as perennials support soil fungal and bacterial communities in the soil. Regenerative systems tend to be more biodiverse at field and landscape level, and so provide other valued ecosystem services, such as insect pollinators and water-holding capacity.

And now, carbon too.

Put simply, it has become clear that a soil with organic matter supports rich communities of micro-organisms and macro-organisms, and these interact to support crop and pasture plants, and carbon storage grows over time. Regenerative agriculture systems capture carbon and can hold it for many years, so long as the agricultural management stays favourable.

The amount of carbon that can be captured in soils and above-ground biomass in agricultural and land use systems varies according to management practices, existing soil types and structure, and climate/weather variables. At any given place, the baseline conditions for soil and climate/weather are a given; it is the agricultural and forest management that can be varied. Soil carbon is thus plastic: it varies according to the direct decisions and practices of farmers and land managers. Figure 6.1 contains a summary of the scientific consensus on annual carbon capture per hectare according to the range of management options available to farmers and land managers at any given location.

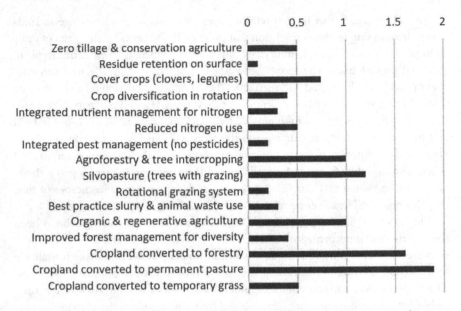

FIGURE 6.1 Carbon capture by temperate agriculture and land use (tonnes carbon per hectare per year)

Applying a cautious deflator of 25%, this would indicate regenerative approaches can increase annual carbon stocks by 0.2–1.0 tonnes per hectare. There are nearly 5 billion hectares of farmland worldwide. If these could be capturing carbon rather than releasing it, here could be a significant contribution to the race to net zero. But it needs adoption and uptake by millions of farmers, supported by dietary choices by many more millions of consumers.

So how are we doing on adoption? Are there signs of saturation in landscapes?

We recently conducted a global assessment on the adoption of sustainable and regenerative approaches, detailing the adoption of seven system sub-types: integrated pest management, conservation agriculture, integrated crop and biodiversity, pasture and forage, trees, irrigation management, and small/patch systems. This found that 163 million farms (29% of all those worldwide) had been redesigned for sustainability on 453 million hectares of agricultural land (9% of worldwide total). There was good evidence of increased system outputs, more crop and pasture productivity, more tree and agroforestry cover on farms, less use of pesticides, fertilizers and other inputs. For natural capital and key ecosystem services, there was more water availability and efficiency of use, improvements in wood, forage and secondary products, increases in carbon sequestration in soils and above-ground, and reductions in surface water flows and soil erosion. All examples involved building of social capital to create changes across whole landscapes.

Here are two tasters of what can be done with this kind of attention to building natural and social capital.

Sawaeng Ruaysoongnern helped to establish Farmer Wisdom Networks in north-east Thailand. This was after the widespread let-down of modern systems. He observed, "Extreme failure caused people to think differently." Farmers recreated polycultures, diversified vegetable, herb, animal, fruit and multipurpose tree mixes. They organised into groups and connected into regional networks. Each member recruits two more farmers each year, and each group of ten farmers seeks to establish another group. Farmers have become more self-sufficient, growing more of their own food, but also earning more from better links to local consumers. These networks use smartphones to share knowledge, ideas and innovations. They create video stories to share across the region. These also speak to consumers and policymakers. Even though these farmers' networks are influencing local and national policies, many farmers themselves say happiness is the most important thing for them.

One of the finest innovations in public engagement has come from the remarkable social innovation of farmer field schools. These were launched in the 1980s when Peter Kenmore, Kevin Gallagher, Russ Dilts, and Dada Morula Abubakr and others incorporated adult education methods with agroecology, drawing on Paolo Freire, the civil rights movement, the approaches of the Danish Folk High Schools. Research had shown that in irrigated rice systems the more pesticide used, the greater the pest damage: insecticides were killing beneficial insects and arthropods, which has been exerting pest control for free. It was also known that most farmers would not know this: detailed entomological knowledge is rarely a feature of local knowledge systems. They asked: could rice farming be amended to reduce insecticide use, could the beneficial insects do enough, and could a system of learning be created to allow farmers to demonstrate to themselves that they would not lose their crops? The first farmer field schools were established in south-east Asia, and over a 30-year period have spread to 90 countries and nearly 20 million farmers.

More than 1.1 million farmer field schools have been run, and the resulting groups cover 25 million hectares of farmland. Notable country leads include Indonesia, Burkina Faso, China, Kenya, Philippines, Sri Lanka and Vietnam. Farmer-field schools are called "schools without walls," and groups of about 25 farmers meet weekly during the entire 100–120 day crop season to engage in experiential learning. The aim is to develop human capital in the form of field observation, analytical skills, and understanding of agro-ecological principles. Farmer cooperation also increases, and over the years, these schools have evolved to include other crops, livestock, agroforestry and fisheries.

The Deliberate Unravelling of Food

Yet, the jumbo jets of food economies still are vast and faulty.

The great shift began in the 1980s. Food still mostly looked like food, fast food outlets were small in number, sugar-sweetened beverages were seen as luxuries, obesity was rare, and the diet-industry was small.

In the UK, the food system has become the largest national employer and manufacturing sector. It has an annual turnover of £121 billion. At the same time, the treatment costs of obesity and type 2 diabetes are £27 billion per year, greater than the annual spend on police and fire services. In the USA, the food system is US$1770 billion in size, and obesity affects some 40% of adults, costing the economy US$500 billion per year.

In these two affluent countries, obesity externalities comprise 23%–28% of the size of the food system. In China, health costs have risen to be about the same as in the USA, some US$500 billion per year. These are costs not borne by manufacturer or retailer. This is hard to tolerate, for these are not abstract inefficiencies; they are harmful to individual lives of nearly a quarter of the world. Worldwide by 2020, there were over 2 billion people overweight.

So, no dancing, no gathering for applause.

And anyway, as Karen Lykke Syse of the University of Oslo exclaimed: "You aren't what you eat." There is a danger of emphasising individual choice in the face of a whole complex food system of good, bad and ugly practice. You can make the right kinds of choices, but cannot alone change the food system.

This is how it all went sideways. Food manufacture, retail and fast food restaurant companies created a desire for modern foods high in sugar and salt. And these were convenient and accessible, prices kept low. In the 1960s, a chicken for a meal was a luxury, and children jumped up and down when there was a rabbit in the pot, shouting, "We're having chicken!"

Now you can purchase a dozen fried pieces for two or three dollars or pounds. We could be perverse, and say the chicken has become one of the Earth's most successful animals, having increased in number to more than 25 billion across the year. But jump in here, for there is no such thing as a cheap chicken. Too many suffer short, cramped and debeaked lives, then the human consumer suffers too.

There then came a moment of genius for food growthers in the 1980s. A whole new class of foods was invented in the era of sudden expansion in fast food outlets and growth in supermarkets. Medical doctor Carlos Monteiro of the University of São Pãolo has classified food into four categories. The first is wholefoods, things recognisable as foods. The second is processed ingredients that help to turn food into dishes: butter, salt, oil and sugar. The third is simple processed foods, such as cheese, canned and smoked fish, canned vegetables.

And the fourth came along in the 1980s and 1990s, the ultra-processed foods (UPFs) centred on radical reinvention to make food manufacture cheap.

These UPFs tend to be low in nutrients, yet high in sugar and fat. They cannot be altered by cooks, so obliterate old rituals. They have high profit margins and high externalities and now account for 50% of calorie consumption in Canada, the UK and USA, and 20–35% in Brazil, Chile and Mexico. They are guilty of much of our diet-related ill-health. Kevin Hall and colleagues at the US National Institutes of Diabetes at Bethesda conducted a clinical trial on UPF consumption in 2018. Those participants who ate the ultra-processed diet consumed an extra

500 calories a day, and yet their blood hormones responsible for hunger remained elevated compared with the unprocessed diet.

The people ate more calories, and still were hungry.

Bee Wilson observed in *The Way We Eat Now* that these foodstuffs have silently fledged in food systems, pushing out traditions. The food writer, Michael Pollan, calls them "edible food-like substances." Food, he said, should be about pleasure, community, family, identity, relationships with nature, and spirituality. Carlos Monteiro concluded: a ban on these category four UPFs would improve the health of nations. UPFs are, he said, "Not real food, they are formulations of food substances."

The second piece of genius played out with the sales of the simple soda or soft drink, also called the sugar-sweetened beverage (SSB). Delicious, refreshing, better than water, and full of empty calories: they do not fill us up, but do make us put on weight. And then, again, still leave us hungry.

In the UK, USA and Mexico, the average person consumes more than 200 litres of soda/soft drinks each year, more than half a litre daily. In France, annual consumption is 37 litres per person. At an average of 400–500 calories per litre, this comes to be a substantial intake of empty calories. Many brands have zero-calorie lines, using sweetener substitutes for sugar. But recent evidence suggests the sweet taste still disrupts the body's appetite and insulin systems. And it is not just soft drinks. Sugar is smuggled into other beverages: a single branded coffee-milk-ice-caramel drink from a large well-known coffee retailer contains 20 teaspoons of sugar, as much as a litre of SSB.

You eat or drink for pleasure, to overcome or avoid hunger, yet modern foods and drink are sneaking in calories you do not need. Tim Lang has further witnessed, "Food can make or break economies." It breaks us too.

The third play of the 1980s centred on changes in access to these new foods and drinks. More fast food outlets, more supermarkets, more ill-health.

The top 100 fast food brands worldwide had 376,000 outlets in 2019, and the top 20 brands (who hold 276,000 outlets) had a turnover of $280 billion worldwide. In the USA, spending on food away from the home grew from 25% of food spend in 1970 to 50% today. In the UK, it was 45% of total spend before the pandemic. There is one fast food and takeaway outlet for every 1500 people in the UK. The highest densities are in urban areas, where there are strong associations between indices of multiple deprivation and incidence of fast food outlets. There is also some evidence from the USA that higher rates of obesity occur in communities with high concentrations of fast food outlets.

In the USA, life expectancy has begun to fall, in counties where there is both high income inequality and high levels of obesity. Thus, the consequences. The ultra-processed foods, SSBs and fast food outlets have brought the dispiriting story of the rise of obesity. And we should state again: this is not, "The people's own fault."

The architecture of choice has changed, the spend on advertising and marketing is relentless. The growing incidence of poverty limits the capacity for many

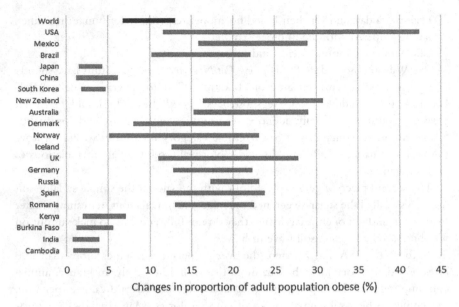

FIGURE 6.2 Changes in obesity incidence by country (% adult population, 1990–2018)
Source: [left end of bar is 1990; right end is 2018 value]

to choose health over calories. In the USA, the proportion of people obese has grown from 10% in the 1990s to 40% today; in the UK, up from 10% to nearly 30% (Figure 6.2). In most affluent countries, the incidence of obesity now lies in the range of 20–30%.

Worldwide, obesity now affects 15% of all the adult population. Let's put this more bluntly: despite fine words and warm policy aims, no country has intervened successfully to reverse the incidence of food-caused ill-health. Not a single one.

A number of countries have kept rates very low. You might think this is no surprise for Kenya, Burkina Faso and Cambodia, perhaps modern food systems and diets are yet to strike, maybe opportunities have not yet come for people to eat differently. Or perhaps traditional food cultures are still strong. For affluent Japan and South Korea, rates have stayed below 4%, largely by keeping the cultural value of food and health at the core of policy and national discussion. For China and India, obesity rates have increased amongst the wealthy, but are still below 6% for the whole nations.

We will come back to Japan and South Korea shortly, for this success did not happen by accident. In the 1940s, Ruth Benedict in her anthropological observations on Japan in *The Chrysanthemum and The Sword* wrote what matters is the amount people feel they give to the world. There was an assumption in Japan that to be short of food was to create a stronger spirit. It was also felt that physical activity makes hungry people strong and vigorous. More food was thus not automatically a good thing.

A reminder, too, of why change can be hard. The UK and USA have weak food policies, but an economically successful diet industry. The diet industry is a side-effect of a broken food system and appears as a positive in GDP. If you eat less food, you subtract from GDP and make the economy look worse.

When people are hungry, you don't need a diet industry; when people have enough, you still don't need one. When we have succumbed to advertising, taste and ready access, we need to find a way to become thinner again. We have to work on creating new food-related habits. But it is very hard, for those external signals and temptations are relentless.

Our lived-niche is full of ultra-processed and fast foods. The diet industry has an income of £2 billion in the UK and US$72 billion in the USA. This sector needs repeat customers, with a few high-profile winners to offer hope. Wendy Wood in *Good Habits Bad Habits* talked to a former executive of a large diet and weight loss company. He admitted, five years after a weight-loss programme, 85% of their participants were back at their same weight or were heavier. Some had lost 100 pounds (45 kg) and had put it all back on again. One company, Weight Watchers, has 3.5 million active users.

Other estimates suggest that 80% of diet clients end up unhappy, as diets fail and people are taught to blame themselves. The greatest barrier to permanent weight loss is that dieting is built around the idea of willpower rather than seeking to change ingrained habits repeatedly strengthened by food systems. Wendy Wood suggests that 75% of people think this willpower problem is a personal weakness, even though some had tried permanently to lose weight more than 20 times, and failed on every occasion.

At the same time, the stigmas of weight grow more common. As people eat more food and worry about it, so stress increases, cortisol levels rise and immune repair declines, all leading to wider ill-health. Not much changes in the food system, though.

Data on the global burden of disease shows food itself is a major cause of death. Each year, there are now about 55 million deaths worldwide, up from 48 million in 1990 (Figure 6.3). About 4 million of these are from injuries. Over 30 years, communicable disease deaths fell from 15 million to 10 million, resulting in a fall in the proportion of deaths from 32% to 19% (these data are all pre-pandemic). Deaths from NCDs, by contrast, rose from 27 million to 41 million, accounting for 73% of all deaths worldwide. In the category of NCDs, tobacco causes 8 million deaths, alcohol 2.9 million, drug use 0.6 million, and malnutrition 3.2 million. Within food consumption, a diet with too little fruit and vegetable causes 3.9 million deaths, with high meat (especially processed) causes 0.2 million, with low grain and legumes 3.6 million, with SSBs 0.14 million, and with high sodium 3.2 million.

The OECD and World Bank in recent updates on obesity are calling the obesity crisis "a ticking time bomb." It is true it could get worse. But the problem itself long ago exploded. No country has set binding targets for reduction. Most say they wish to stop the bad stuff, a constant reminder to people about what they are missing, and turn responsibility back on individual choice and moral obligation.

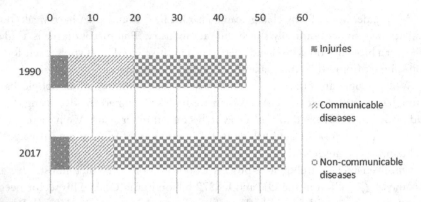

FIGURE 6.3 Global deaths per year (million)

There is a vast industry of labelling, created to confuse consumers. There is mass communication and health promotion. And, a little, here and there, of advertising regulation.

A reminder about what happened to us as individuals in affluent countries and communities. We stopped walking, it was too good to have a car. We began to eat new foods, how tasty, how easy. We let social connections lapse, why bother as only individuals count now. We spent less time in nature. We sought satisfaction and happiness from things and stuff, how easy to buy, how easy to replace with another.

The second tranche of bad news is that both overweight and obese body status bring co-morbidities, in particular cardiovascular disease and type 2 diabetes. There has been a sharp increase in the incidence of type 2 diabetes in 205 countries over the past ten years (Figure 6.4). Over time, the drift is towards the north-east of the diagram: GDP growth has brought a higher incidence of type 2 diabetes, in some countries now affecting 15–20% of the population. In 2019, diabetes affected 9.3% of people worldwide, some 460 million people. Levels are higher in urban areas than rural (11% compared with 7%), and higher in high-income communities than lower (11% compared with 4%).

If you are urban and wealthy, type 2 diabetes is more likely to strike. It then reduces life expectancy by 10 years. These are some of the not-very-hidden ills of affluence.

And not surprisingly, there is a link to carbon emissions too. Figure 6.5 shows the increase in incidence of obesity by per capita emissions of carbon. Almost all countries emitting less than 1 tonne of carbon (CO_2eq) per person have obesity incidences below 10%; almost all those with carbon emissions greater than 10 tonnes per person have incidence above 20%, many above 30%, and one, the USA, now above 40% of the adult population.

There is one further link to the carbon costs of food: meat.

Ruminant livestock emit methane, which has a direct effect on the climate crisis. Intensively reared livestock need animal feed rather than grass, and so introduce energetic inefficiencies into the food system.

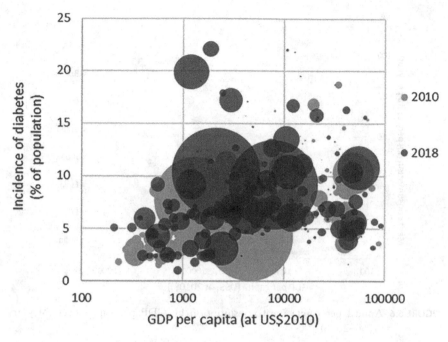

FIGURE 6.4 Incidence of diabetes by GDP per capita (n=205 countries, 2010–2018)

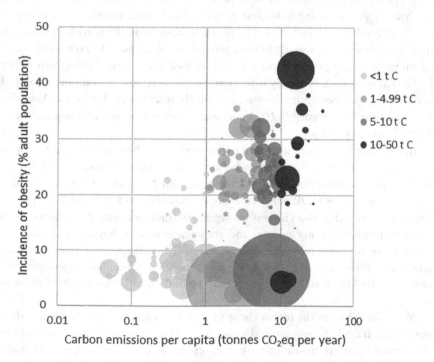

FIGURE 6.5 Incidence of obesity by carbon emissions category (n=183 countries, 2018)

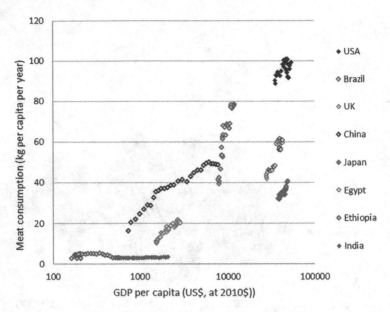

FIGURE 6.6 Annual per capita meat consumption by GDP per capita over 30 years (1990–2018)

Much of this feed, especially soya, now come from countries where rainforests are being felled to create fields and pastures. High meat consumption is now a feature of food systems and diets. The trends indicate again the remarkable changes that have occurred in a single human generation. With notable exceptions, once again, meat consumption per person has increased as countries have grown richer by GDP (Figure 6.6). Not only have more people been able to consume meat, but consumption per person has increased, annually reaching in Argentina, Australia, Brazil and the USA some 80–100 kilogrammes per person per year (equivalent to 220–270 grams per day).

Many countries remain low meat consumers, partly because modern food systems have not yet arrived, and partly because of existing social and cultural norms. Average meat consumption in India has remained stable at 3.0–3.5 kg per person per year even as GDP per person grew from $580 to $2100 (at 2010 prices) between 1990 and 2018. The reason: vegetarian diets are central to religious and cultural norms for many people, and these determine behavioural choices. In India, there are some 500 million vegetarians. This wrapper of culture has been effective in keeping meat consumption low. There are 1.2 million vegetarians and vegans in the UK, 8 million in the USA, 11 million in Japan and 30 million in China.

You might also say, the reason chosen for cutting meat consumption really does not matter. It could be religion and legacy, it could be intentional and new. To limit the climate crisis, we now might ask, how could these cultural wrappers be made more popular?

And yet, now something new comes along. Recent years have seen an expansion of commercially-oriented innovations in meat alternatives. These focus on meat analogues derived from two sources: plant or fungal feedstock, and cultures using the cells of livestock. Both routes offer products that look and taste like livestock meat. These are variously termed analogue, alternative, plant-based, cultured cell and fake meat. It had been assumed that these would represent a tiny part of growing plant-based markets for food and dishes: not all vegetarian or vegan consumers wish to consume products intended to appear and taste like meat. But perhaps now a triple tipping point approaches: i) the technologies could be scaled to high levels of production; ii) the quality has increased to the point products can be indistinguishable from livestock meat; and iii) unit costs are falling rapidly, with some products likely to retail at much less than the cost of meat.

Low cost and high quality could mean analogue meat undermines some livestock production to the point it is no longer viable. Most likely to fail will be intensive systems with high negative environmental externalities, and low animal welfare; most likely to prevail will be grass-fed sustainable systems for which positive marketing propositions can be made. The impact on global animal feed, especially soya, sourced from agricultural systems that have caused (and continue to cause) tropical deforestation and biodiversity loss could be substantial. There would be an additional benefit to the race to net zero from reduced methane emissions from ruminants, increased carbon capture, and more energetically efficient food systems.

In Japan, *oryoki* is the term for the right amount of the right food, just enough. The *oryoki* spirit says eat slowly to appreciate the colour and tastes, and think of what is hidden beneath the floorboards: what is the sub-story of your dish, how did it come to your plate? Norman Fischer says of food, "Every gift extends the heart and opens the spirit." Perhaps each meal should be like a bento box: prepared with care, love and deliberate design, a work of art.

Some of these options seem within grasp. A significant question could now be: how to create new norms for healthy diets and dishes that also deliver low carbon emissions? "Close the gap, make the leap," urged Naomi Klein.

Food is one way to say farewell to some ills of affluence. It could lead to behaving in ways to save the Earth. And create something of a good life.

Are you showing me the answer now?
Have some tea.

[Korean Zen master Taewŏn, 1963]

Put a Rainbow on Your Plate

In many a culture, people will be thinking, there must be evil spirits out there. How are we going to get this right?

Once there were ten thousand ways of making meals. Food was local, fitted to the foodshed. Now has come convergence, across cultures and within countries.

And why resist this creation of desire, these cheap calories wrapped in taste, the deep advertising budgets. Some 90% of politicians are said to believe that a failure in willpower is the main cause of obesity. Individual choice is celebrated, eat what you want, but be ready to be blamed if the consequences are bad.

At the same time, we know what works. Most affluent countries promote variations of healthy food messages based on about five portions of fruit and vegetable per day. This was initiated in Californian by Director of Health Services, Kenneth Kizer in 1988. The primary aim was to sell more fruit and vegetables produced by Californian farmers, but the campaign was quickly adopted by other countries and by the World Health Organisation.

It was an easy message to remember. It would track through to improved health and longevity, and also lead to reduced health system costs. But in the first six years to 1994, the proportion of Americans eating five fruit and vegetable remained unchanged at 11%. The Centers for Disease Control are still reporting that only 10% of adults eat this target amount of fruit and vegetable, with some states as low as 7%. The national average is 2.7 servings per day. In the UK, the average daily consumption is slightly higher at 3 portions of fruit and vegetable per person, but 20% of people consume none at all.

Michael Pollan has written of a simple heuristic: have 5 colours on your plate or dish, and only eat things that look like food. In the islands of Okinawa in Japan, some 200 food items are eaten regularly, and wrote Héctor Garcia and Francesc Miralles in *Ikigai*, this is called, "Eating the rainbow."

So, put a rainbow on your plate.

These kinds of advice are evidence-based, and they work for health when adopted. But in affluent systems, such knowledge of what is good might not be enough. Food is more than items of fruit and vegetable. We need to pay attention to the architecture of choice, the social, cultural and economic conditions that wrap up food. Unless you have made the choice to eat vegetarian or vegan diets, it often seems that fruit and vegetables are hardships to be avoided or at best tolerated.

The contrast with South Korea is striking, where 300 types of vegetables are regularly eaten. In her 50th anniversary edition of *Diet for a Small Planet*, Frances Moore Lappé observed, "I began this journey with the realization that growing and eating plant-centered diets was a great choice. Today, it is a no-contest necessity."

In the late 1960s and 1970s, Barry Popkin identified how nutrition transitions had begun to occur, and how values and preferences for food in some countries and regions had shifted away from traditional diets and dishes. Popkin first pointed to the Mediterranean diet and its favourable health outcomes. High in fruit and vegetable, low in meat, high in anti-oxidants and vitamins, a little tannin-rich wine, and dishes distinctive to particular places, all evoking pride and regularity of consumption.

What later became clear was that people in these countries also took time to eat together, to consume slowly, to use food to cement social bonds. The ceremonies of cooking and eating were important, as was the role of food in collective identity and unique links to place. It is not just that the Mediterranean diet was protective.

It is that the relationship with food was more about togetherness than calories and convenience.

"We shall prepare food," wrote Dogen Zenji 800 years ago, "with close attention, vigorous exertion and a sincere mind."

Partha Dasgupta also added, "Eating food communally is a salient feature of societies everywhere." Except, of course, when it is no more.

This could be widened now: a healthy diet, low in meat and high in vegetable and fruit, from local producers and sustainable systems, will increase personal longevity and other indicators of health, reduce the costs imposed on public health and care systems, improve natural capital and biodiversity, and reduce carbon emissions.

You might well ask: is this not enough to change the food systems across the world? How have countries intervened in food systems, and did anything work? The examples of success are few, and policies with sharp teeth are rare.

We have to conclude: either we do not know how to do this, or we are simply content to continue to blame individuals for their food choices. Not a single country has seen falls in obesity or type 2 diabetes once they had increased. The EAT-Lancet Commission in 2019 called for a great food transformation, nothing less. But Tim Lang comments that comprehensive and integrated food policies remain far from the priorities of governments wishing to emphasise the economic success of the sector, regardless of the health costs that have to be paid for later.

Nonetheless, there have been advances. In Mexico, sugar taxes have led to a reduction in SSB consumption. Finland has banned trans-fats, and ensured public institutions purchase 60% of their food from organic and local sources. Finland has launched a *sapere* food education programme in schools, focusing on the noise and taste of foods to increase awareness and attentiveness amongst children. Increasing sensory perception of foods changes the relationship children have with food and eating. Chile changed food policy in 2016, adopting laws against unhealthy foods, and imposed labelling on sugary drinks to demonstrate the negative effects, resulting in a 24% reduction in consumption.

By contrast, the government of India created the post of Minister for Processed Foods, with a budget of US$200 million. An incumbent in the role recently lamented, "We only process 10% of the food we produce in India."

At the same time in many affluent countries, there has been a growth in interest and adoption vegetarian and vegan diets. In the UK, Veganary platform had a record half-million sign-ups in 2021 for a month of changed diet. Could governments do this sort of thing too?

Well, yes of course. But they seem scared of existing economic interests, worry about the consequences for the economy, quite forgetting that healthy food would help reduce national health costs. Tao Huang and colleagues at Zhejiang University studied 125,000 vegetarians in comparison with non-vegetarians, and found all-cause mortality was lowered by 9%, cancer incidence reduced by 18% and mortality from cardiovascular diseases lowered by 29%. Yet, as we have seen with the climate crisis, once the tide turns, then policy targets can follow, especially where there are also economic benefits in innovation and change.

Here it is very clear. Healthier diets and dishes would reduce health costs, save governments money, and improve well-being. And help meet climate targets. As indicated earlier, things have got better as well as worse.

Governments set conditions for food and agricultural systems. They influence the social fabric. I was once in the rural communities around Anuradhapura, for 1200 years an ancient capital in the centre of Sri Lanka, and at the time staggering to recovery after 20 years of civil war. "The land is improved by having farmers on it," said one Singhalese woman. "My farm is constantly changing now," she said, and there was agroforestry, perennials and 16 types of vegetable.

"We feel we can do anything," she also said, but when a storm raced across the fields and rain beat on the tin roofs, and lightning struck, the children were crying from the echoes of the war. During that time, said a neighbouring Tamil farmer, "We didn't know if we could start life again." Yet, he had just hosted a visit of 15 farmers from both south and north to his organic farm of banana and 40 other crops. "I felt very proud," he said. Things can get better.

Some compelling research on life choices comes from tracking populations over time. Life course and longitudinal studies have shown how behaviours and consumption choices directly affect health and well-being over many decades. The Harvard Alumni and New England centenarian studies showed quite clearly what kept people alive and happy into later years: it was diet, physical activity and social togetherness.

A sizeable proportion of people did none of these, and died eight to ten years sooner.

In 2020, Yanping Li and her colleagues at the Harvard Chan School of Public Health studied 40,000 nurses and health professionals and their habits for five behaviours: diet, smoking, physical activity, moderate alcohol, and healthy body weight. At the age of 50, those with four of the five behaviours could expect 34 more healthy years of life. Those with none of the five could expect only 24 more years. John Ji of Duke Kunshan University has worked with the oldest old in China using a remarkable panel of some 6000 people. He assessed the role of fruit and vegetables combined with physical activity, such as jogging, playing ball, tai chi and gardening, amongst people in their late 90s. Those with one of fruit, vegetable or physical activity daily could expect an extra half-year on life, two brought one and a half years more, and all three increased longevity by two years.

There is a saying about the oldest-old in China: "Vegetables and tofu keep you healthy."

In *Just Enough*, Gesshin Claire Greenwood wrote: "Every meal should be one of the best meals of your life."

Over a 20-year period, Cuba became part of an instructive but unintentional experiment that revealed much about food and health. It showed how national economic and political conditions can induce positive health consequences, and then how fast such advances can unravel.

Economic sanctions were imposed on Cuba in 1990, resulting in a dramatic fall in daily per capita energy intake from 2900 kcal per day to 1850 kcal. This was concurrent with increases in physical activity from walking and use of public transport, up from 30% to 70% of adults, and a reduction in smoking. Over the decade to the year 2000, obesity incidence fell from 14% to 7% of adults, producing a 50% fall in diabetes deaths, a 35% fall in cardiovascular deaths, and a 20% fall in stroke deaths. This showed that positive public health outcomes can happen over short periods, even though this was as a result of an involuntary economic crisis causing poverty and shortages of food.

Perhaps the rebound was inevitable, especially as this was not by national choice. After sanctions ended, over the next ten years to 2010, both obesity and diabetes rates increased, and health indicators worsened.

The importance of the social wrapper to food and health was also demonstrated by remarkable research led by Michael Marmot and colleagues in the 1970s. He studied acculturation by several thousand Japanese migrants who had moved to mainland USA, later supplemented by data from further migrant waves from Japan to Hawai'i and Canada. The migrants brought hard-wired habits and their existing health status and would have wished to fit into the new norms of the host country. When migrants moved to their new home communities, rates of diabetes and heart disease soon doubled and increased by fourfold for hypertension. Rates also increased over time: after four years in Canada, hypertension affected 3% of migrants, over years 5–9 it had increased to 7%, and then for over 10 years it had grown again to 13% of the population.

The Japanese lifestyle had been protective: high levels of physical activity and low consumption of fat, sugar and animal protein in the diet, together with an emphasis on group cohesion and social interaction, both of which reduce stress. The new ways of living were less protective. In *Doughnut Economics*, Kate Raworth says the top priority for human and planetary health is to "change the game" to a far bigger one.

Here then is a new goal for food policy: think of food as having social, health and environmental value, not solely as a vehicle for economic growth, and thus encourage plant-based diets, dishes that value identity, and food systems that eliminate inequality. Healthy diets save money for primary and secondary healthcare systems and thus create economic benefits for countries.

They allow people to live longer and happier lives. Then connect to regenerative and diverse agriculture too.

A food version of the Hippocratic Oath might look like this: do no harm, do only good.

> Eat and drink with your whole heart,
> As you enjoy a dish,
> Think about the people who cooked it,
> Visualise the field where the ingredients were grown.
> [*Shunmyō Masuno,* Zen: The Simple Art of Living, *2019*]

Three Grapefruits

There is evidence we are eating more healthy foods, yet at the same time are becoming more unhealthy. The Global Burden on Disease shows that with economic development and increases in GDP, so came falls in unsafe water, increases in household air pollution and child wasting, and rises in smoking, alcohol consumption, drug use and dietary ill-health. You also know you can eat well to help stop the climate crisis and to reduce the costs of public health services. This might too be more fun and friendly and help us live longer. And being social is not an ideology. It is a way of living together and not using up the planet.

In this, each person has a role as an individual consumer, and so do governments in helping create healthier food systems valued by citizens.

These are some of the routes towards a good life. The greatest danger might be this. In a single generation, diet-related ill-health has become completely normalized. Habits are hard to form and harder to break. The brain, once it has learned a habit, does not want to devote energy to breaking it and learning another. Fast and unhealthy food habits, once formed, are tough to change. Even harder when a whole dish is cheap and full of the sugar that we once so craved. But it can be done.

Roy Taylor of the University of Newcastle has shown how type 2 diabetes can be reversed through dramatic diet restriction to 700–800 calories per day. The habit change is needed for 56 days, and nine out of ten people lost 15 kilogrammes of weight, enough to cure them of the life-threatening condition.

Habit change needs three things. First, a change in the cues around us, which used to be natural and social but now are commercial. Second, a change in our responses to these cues. And third, a change in our readiness to adopt different and healthier alternatives.

A transgressive trickster might also be needed, to get you in the game. And then you need to walk across the threshold to take on the heroic journey to create a new personal food culture.

In the late 1960s, after a six-year stay in Japan, the author Taitetsu Unno and family were returning to California. A friend had come and given them three grapefruits. At the time, overseas fruit was rare and expensive because of import restrictions. Unno and his wife decided to give away the three grapefruits. His wife gave them to her *ikebana* flower-arranging teacher.

A few days later, they received a letter written with brush and ink on traditional paper. The teacher thanked them after commenting on the weather. She wrote that she shared the first grapefruit with her grandchildren, who were thrilled with the fragrance and taste of a fruit they had never seen before. The second she ate with a friend whom she'd not seen for twenty years, making the reunion more powerful.

The third she took to a hospital where a best friend was dying. The friend hadn't eaten for a week, but tried a segment. She asked for another, then another. The family was in tears. The gifts of grapefruit, the sweet aroma and taste, the memories created. One small act of giving caused many ripples. True gifts involve giving up of something we could cherish, and in doing so Unno and his family created something softer and kinder.

The grapefruit story had grown greater with the giving. It had become a new currency.

In the continuance of the stories and song,
The Earth shall continue.

[Simon Ortiz, Taos Pueblo poet, from Joan Halifax,
The Fruitful Darkness*]*

~★~

7

THE BEST THINGS IN LIFE

How Immersion and Flow Make the World a Better Place

Without a peaceful mind,
Elephants, horses and the seven treasures,
Are worthless things; palaces and
Fine towers mean nothing.

[Kamono Chōmei, *Hōjōki*, 1155–1216)

Abstract: This chapter now comes inside to explore the mechanisms and structures of brain and mind that are common to every human. This legacy made us in a certain way, ready to respond to natural and social environments around us. We find: the best things in life are not things. We all start life in the same place, with a beginner's mind. And then it all starts to develop. The chapter commences with an analysis of the goods and bads of habits. Habits mark us as human, they automate behaviours leaving space and time for cognition. Yet, they can be hard to change. It takes about 50 hours of hard effort to break or make a habit. Flow activities help, and are central to the contemplative and immersive states that are common across all cultures. Flow is the state when we feel and perform our best, when the self and ego disappears. Problems vanish, the inner chatter is quietened. Flow activities tend, too, to be low carbon. They create the state of transient hypofrontality: the mental chatter that is a significant source of ill-health is switched off, for a while. The chapter discusses the structures of blue and red brains, and how attentiveness and flow can help create calm. Evolution gave us no off-switch for the mind. We have to choose body behaviours to press the switch. Be attentive, become immersed, be focused, become in flow. The green mind created in this way is pro-social and more connected to nature.

DOI: 10.4324/9781003346944-7

The Best Things in Life Are Not Things

We have been cycling back and forth between modes of change that centre on the individual and the collective. Much needs to change, much will change if the climate wheels come off the wagon very fast. Yet, there are many ways of living that bring contentment. We humans have been good at finding a good life, despite it all.

And always looming, the big bear of a question: is there enough to go round?

To address the four crises of climate, inequality, nature and food, there will have to be structural change in the way modern economies and societies are organised. There will have to be social movements to bring people together, to propose and push and build political pressure. There will have to be change in us too, you and me as individuals.

And this is the slope. Each one of us starts at the same place. With a beginner's mind. Empty of knowledge and habits, empty of experience. Ready to respond to the social and natural environment, ready to learn and grow. Inside, we are a match. And this means there might be universal truths available to aid these transformations.

In those days when so many thousand cultures came to settle deep, closely into ecosystems, meandering apart over time, the trickster was host and architect of many a saga. They made us laugh, they taught us how to behave. There was Loki and Eshu, Coyote and Raven, Monkey King and Mink and Krishna. And many more. The hero who crossed the boundaries had at least a thousand faces.

They knitted together wounds, those wise fools getting life going, as one hundred billion humans lived and died. Every one of us, it is clear, inherited highly similar internal structures of body and brain. Cultures differed, yet biology largely did not. In these ways of living, many found what felt like types of the good life. We knew there would be blossoms next spring, even if the times were troubling.

If we are to chart a route away from the crises brought about by consumption and the ills of affluence, we are going to want to be able to choose habits that are low in carbon emissions. This is currently hard to do. Yet, there are hints. Some are very old, dating back at least to the earliest surviving writings. Some are very recent, coming from modern neuroscience and the study of how we behave and respond within our social and ecological homes.

The headline is this: the things and activities that make us happy, that help us live long, are very often low in carbon emissions and adverse biodiversity impact. And many do not need to be bought.

Well, over there is a rabbit hole, there the back of the wardrobe, there the path winding into the dark forest. To find the elixir, the boon that the world does not know it needs, we will need to take one step, and then another.

Beginner's Mind: A Trickster Story

The Great Hare was a clever fellow who was always playing tricks on animals and people.

He was always haring-away, was hare-brained and the leaper, the one who leaps to a truth. In English folklore, he has eighty names, the frisky one, the lurker in the ditch, the one who doesn't go straight, the skulker and the springer. His mind is always active. He is the trickster called Glooskap in the American north, and is so clever he always defeats his opponents by talk. Now it came to pass, when Glooskap had defeated every giant and sorcerer, magician and all manner of evil spirits, and he thought his work was done, a certain woman said, not so fast Mister, for there remains One who will remain unconquered for all time.

"And who is this One?", inquired the Hare.

"It is the mighty Wasis, and there she sits."

Now Wasis was the Baby, and she sat on the floor sucking a piece of maple sugar, greatly contented, troubling no one. As the Lord of the Beasts had never had a child, he knew nothing of managing children. He turned to baby with a bewitching smile and told her to come to him, his voice sweet like the summer bird.

Baby smiled, and did not budge.

Glooskap spoke terribly, and Baby burst out crying, but did not move for all that.

He used his most awful spells and sang songs which raise the dead and scared devils. And still, she sat and looked on admiringly and seemed to find it interesting, but all the same, never moved an inch. Glooskap gave up in despair, and Wasis sitting on the floor in the sunshine went goo, goo! And cooed yet more.

And to this day, you can see all babes are well-contented and going goo, goo! For of all the beings that have ever been since the beginning, Baby is alone the only invincible one.

In the legendary 1930 classic, *Zen and Japanese Culture*, Daisetz Suzuki said, go back to your infancy and observe. The great earth may crack, and the child remains unconcerned. A burglar may break into the home, and he smiles at him. "Would a baby be overjoyed if the empire was given to him, or he was decorated with a medal," asked Suzuki. What concerns the child is the absolute present.

The infant as yet knows no fear, no insecurity, does not need to be something nor live up to some expectation. There is anxiety about food and comfort, but not much more. Quite soon, though, this begins to change.

Ruth Benedict wrote of a saying in Japan: the young child is born happy, it is the task of parents to keep them happy.

Christina Feldman and Willem Kuyken added in their recent book, *Mindfulness*, what does it mean to be happy and to live the good life? They observed that mental ill-health was now a one billion person worldwide problem and that through a combination of ancient wisdom and modern science we might just, quite literally, "Come to our senses."

> Always keep your mind as bright and clear,
> As the vast sky, the great ocean,
> And the highest peak, empty of all thoughts.
> *[Morikei Ueshiba, founder of Aikido, 1899–1969]*

So, let's begin with habits.

The Goods and Bads of Habits

Habits are a wonderful thing.

They are both learned behaviours and fixed action patterns of thought and language. Habits are hard to form, needing high cognitive input, and are harder still to change or break. You learn to walk and ride a bike, then no longer need to think. But equally, you learn to like the taste and convenience of fast foods or the culture of smoking, and the brain does not want to assign resources to unlearn.

Can you imagine unlearning how to ride a bike? It's as tough as unlearning how to smoke.

Habits made us human, by freeing up space for other thought and learning. Social psychologists and habit experts, Bas Verplanken and Sheina Orbell, called habits "Durable and resistant to persuasion." They are not the ills of society, they are how we as humans work. They are, they also said, "Structures for the long haul."

Habits are one of the brain's cleverest tricks. You expend a large amount of energy learning something, and when the routine is perfected, it is shunted away from the cortex to the ancient brain and basal ganglia, the primal parts of the brain. Now only small amounts of energy are required to initiate and implement.

You spend months as an infant learning to crawl, and then to walk. Our developed gait becomes unique, having emerged from this process of trial and error. For the rest of our life, we do not need to think about how to walk. Consider too of learning to drive a car, so much concentration at first, on multiple and simultaneous tasks, and then about 50 hours later you are driving without thought.

Habits thus create the cognitive space to focus on something different, on thinking, planning and reflecting. Humanoid robots have not solved this problem: they can move as if walking, but require huge amounts of processing power for every decision and physical move, taking in and processing signals from the outside world. Such robots use about 16 times as much energy to transport a kilogramme when walking compared with a real human.

Good and bad habits have the same origins. They save us energy, but can also become restrictions. They often result in us missing the richness of the world.

Life begins with a beginner's mind. The mind is clear, and the days are wondrous as the infant takes in signals from the world. At this start was the source, this empty stillness at the core of human existence, and now attentiveness brings in external signals. A layer of habits, memory and thoughts now begins to build up, created by experiences in the world, and this layer expands as life unfolds. The hippocampus forms memories and increases in size with cognitive engagement with the world.

When the sub-routine of a learned habit shifts from cortex to the basal ganglia, you do not need to think about walking, tying your shoelaces, picking up a pen to write. The activity now seems effortless. You no longer have to think about knit-one, pearl-one, no longer have to wonder how best to hold a chisel or how to ride that bike.

All habits require hours of practice, then suddenly no longer require active attention.

Many non-communicable diseases arise from behaviours we have gradually adopted over time. You slip into patterns of eating and drinking, smoking, a sedentary lifestyle, reduced contact with family and community, less active transport, reduced engagement with nature, more use of pharmaceutical interventions to ill-health. Once you have learned to smoke or drink alcohol, both unpleasant at first, or learned to like the convenience and taste of fast food, once these have become habits, you have walked willingly into a trap.

Commerce understands this well. They want you to throw away old things and buy new. They want regular consuming to become a habit in itself. They want you not to think, and certainly not to change. Habits are palaces of the familiar, but also enclosures. "Habits are frighteningly subtle," said mindfulness experts Mark Williams and Danny Penman, "Yet can be incredibly powerful."

If you want to break a habit, it is hard. Your brain shouts: you don't want to do that; we up here spent all that effort and time learning this habit, and now you are trying to change it.

Change uses up resources, we have to burn energy and effort. These are the whispers, the very words of many devils, the dust and noise that threatened the desert monks and nuns and the island saints as they sought peace. You long ago laid down the pathways in the brain, and now you want to break them. This is going to hurt, says the brain again. Really, don't do it.

Joan Borysenko calls this the dirty tricks department of the mind.

It can't let go. It resolves the world into simple binaries, it keeps asking *what if* and *if only*. She gives the example of hunters of monkeys in south-east Asian forests, who use a gourd with a small hole, placing a banana inside. When the monkey puts its hand inside, it cannot remove both hand and banana. But it cannot let go. It is trapped by its own mind and is eventually captured by the hunters. The hugely popular *Seven Habits of Highly Effective People* by Stephen Covey was one book among many seeking to reveal a secret ingredient for business success. His view was to look inside, seek right living, and think of excellence as a habit: "Habits can be learned and unlearned, but I know it isn't a quick fix. It involves tremendous commitment." He had an analogy. Habits are gravity. If you want to fly to space, you will have to burn a lot of fuel to get off the planet.

The journey of a thousand miles starts with a single step, acknowledged Stephen Kotler of the Flow Research Collective. "But it is still a thousand miles. Uphill, in the dark."

Habits are the way we work. And some are destructive. When the old habit recedes, there can be unexpected joy. Catherine Gray has written with wit of her dispatch of alcohol after years under water. The connected addictions: alcohol to escape, the debt and illness, the friendship loss, the normalizing behaviours, the belief that alcohol is necessary for a good life yet never seeing the hook and trap. Those who don't drink alcohol are called stone cold sober by those who do; she inverts the term with "sunshine warm sober." The good life is sunshine warm, but it is hell to escape the habit, even though she says you had been living constantly with shame and existential dread. Alcohol does not release your inner self, it is just a harmful habit.

Rainer Maria Rilke wrote, "You are not dead yet, it's not too late."

Here's another example. Norman Doidge in *The Brain's Way of Healing* described how a Parkinson's disease sufferer, John Pepper, helped another patient create new motor habits. Damage to the brain had impaired her motor functions for picking up a cup. She had developed hand-and-arm tremors whenever she tried to bring a cup to her mouth. Pepper himself had learned to hold a fork and spoon in completely new ways, bringing food smoothly to his mouth by a new circuitous path. The therapy for her was to practice picking up a cup in a completely new way: arm outstretched, circling round the back, grasping the cup anew, and bringing it to the mouth in a curve. Pre-existing brain circuits that had been damaged by the disease were thus being side-stepped by creating entirely new ones. Now her hand did not shake. It was hard, though, to devote the cognitive effort required for long enough.

Neuroplasticity and the 50-Hour Rule

Habits are all about plasticity. This is a term that means the brain changes by doing something with our body. Plastic means malleable and pliable.

The brain is not fixed at birth, it responds to changing conditions outside the body. Braille readers use the visual cortex when reading, in what is called cross-modal plasticity: their visual centre is converted for use by the sense of touch instead. Amputees have to find ways to put an unneeded part of the brain to new use. They learned to walk as a child, now need to learn again.

It is now known that dementias shrink the hippocampus, and yet regular walking, the physical activity of moving forwards into the landscape where uncertainty lies, has been shown to delay the onset of some dementias by ten years. Walking increases the size of the hippocampus by adding 700 neurons a day, increasing in size by 15% over a year. Daniel Goleman and Richard Davidson in *Altered Traits* show how quickly this neuroplasticity can occur, the alarm signals of the amygdala can become damped by 30 hours of mindfulness training, cutting its activity by a half. The amygdala still receives incoming signals, but no longer over-responds or amplifies. This training also reduces inflammation, creating a new trait, as it influences all of our days and not just when we practice.

The brain learns by intense, repeated experience, and so neuroplasticity occurs, in short, when we are paying attention. A key question is this: how long does it take? Wendy Wood of the University of Southern California has long worked on habits and concluded in *Good Habits Bad Habits* that it takes about 66 hours of repeating a behaviour to make it automatic, with a range of 20–90 events or days. A rule of thumb emerges from a wide range of research on habit formation: it takes about 50–60 hours of concerted effort to set up or drop a habit. This could play out as half an hour a day for three months, thus over about 100 days; it could be 25 two-hour driving lessons. But the evidence is consistent on one thing: you have to do the new thing long enough to form new neural pathways and fix them permanently.

If you want to change a habituated routine, it needs to be brought back to the cortex, and amended with fresh learning. This needs fixed concentration. But changing ingrained habits is tough: you have to force the brain and body to behave differently. Such motivation might come from wanting to achieve something, say passing a driving test, or you could be drawn into something through an external signal, say to take up gambling. Dropping habits is doubly hard when associated with something perceived as pleasant, such as eating or drinking.

Yet, during the time required to automate, there will be many places where we might slide off the path and end up in a ditch. "Do you have the patience to wait, till your mind settles and the water is clear?" asked Lao Tzu in the *Tao Te Ching*.

The evidence also suggests that many deliberate interventions designed to help us break habits do not last long enough for us to form new neural structures. They do not lead to automation. Diets are rarely designed to last long enough; a one dry-month break from alcohol is too short to stick.

And anyway, you are being constantly reminded that this habit-break will come to an end soon, and then you can return to the old habits of eating or drinking.

Habit breaking is so often doomed, unless you feel strong enough to go at something new for 50–60 hours, 100 days, three months. One practice challenge is thus to identify the habit-releasers for improved health: new behaviours need forceful and fierce action. Another centres on the architecture of social and cultural environments that in turn shape choices about behaviours. All our lives we are responding to external signals and internal habits, and these social, ecological and economic environments shape behaviours and habits, for the good and bad.

One way or other, you will have to go on a journey into the dark woods, maybe even to another world.

The keys are these: belief that the new habit will work, resolve to continue long enough, and strength to silence the noisy inner voices saying *stop*. It is the attentiveness and immersion at the heart of the good life that can quieten the chatter, and provide the happiness and contentment to replace the old habits. This is where flow activities come in.

The 19th-century Buddhist nun, Otagiki Rengetsu wrote in *The River*:

> The world is, dust and dirt,
> Flows away, all is purified,
> By the waves, by the Kameo River.

Flow Activities

Out of emptiness and space comes life and the world. And perhaps, too, the best of a good life. Mihalyi Csiksentmihalyi said that flow related to our best moments in life, when we are so immersed that nothing else intrudes or matters. Stephen Kotler in *The Art of the Impossible* defined flow as, "An optional state where we feel our best and perform our best."

When the self and ego disappear, the sense of freedom increases. You could be riding a fast motorbike or weaving or performing surgery, you could be practising yoga or shepherding, deeply into fishing or skiing. Such attentiveness and immersion have also been called peak performance, non-ordinary states of consciousness, mystical states, altered states of high flow pursuits, and "little e" or "big E" enlightenment.

Flow can also look like a trance state. And it gives you a sense of freedom.

Contemplative and immersive states are common across most cultures: Pentecostal singing and dance, Sufi spinning, Buddhist meditation, Hindu oneness blessings, Christian contemplative prayer and walking. When the mind is quiet, the self is stilled. And so immersion and flow can also occur when you are preparing, cooking and eating food, walking in nature or gardening, being with other people we know and like, engaging with crafts, art, reading and dance. It is also true that there is nothing passive about mindfulness, meditation and prayer. These all are active interventions for the mind. Time stops, well-being is high. These are also the things mentioned earlier as being core activities in a good life.

This flow behaviour then does something interesting. The founder of the Flow Research Collective, Steven Kotler, describes how surfing brought him out of three years of Lyme disease illness and depression. Time slowed to a crawl, and his sense of self vanished as he merged with the ocean. This was the oddest part, he said, "I felt great. For the first time in years. The pain was gone, my head was clear, my mind sharp."

One late afternoon, I once walked from the point where the Blue and White Niles merge, inland from the raised riverbank toward the sprawling settlement of Omdurman. In those days, the wide plain was speckled with grave mounds and children ran with coloured kites. A Sufi crowd had gathered, mainly men in white robes edged with green and red, carrying the green tarique banner. At the tomb, there were drummers and wind instruments, and the men began to chant and spin, on one foot, spinning faster as the chant rose, *La ilaha illallah*, there is no God but Allah. The atmosphere became hypnotic as the rhythmic dancers circled, and the wall of crowd clapped and stamped, and the dust rose and the falling sun became bright red in the desert air.

In this way, the devotional dhikr creates a state of flow. It is celebratory and personal. It connects with Earth and God, it brings well-being to all participants. By culture, it is fixed as a habit.

Andrew Newberg and Mark Waldman of Thomas Jefferson University surveyed 1000 people to explore their varied experiences of enlightenment. They observed that this thing called enlightenment seemed a universal phenomenon, with exemplars found in cultures across the world. They defined it as, "An indescribable experience that alters the brain and our awareness of ourselves and the world in a way we find deeply meaningful."

They resolved it into two types: "little e" for moments when you are uplifted and blissful, and "big E" when your entire worldview is permanently changed.

Importantly, they found that enlightenment causes long-term structural changes in the brain, as we discover a different us. Enlightenment causes decreased activity in the parietal lobe, and thus a blurring of the lines between the self and the world. Frontal lobe activity also falls away after an increase arising from concentration and focus. Newberg and Waldman discovered long-term changes in the thalamus, causing colours to be more vibrant and us to feel more empathy with others. As the attention network goes up, so the inner chatter quietens. Pleasure follows. And in big E experiences, our sense of connection to the world expands.

One person wrote, "I watched all my problems vanish, not because any of them were solved, but because the questions themselves disappeared." Newberg and Waldman also reminded, even though the brain can change in beneficial ways, the external world has not yet changed. You may have the elixir, but the changes will require translation into the old world.

It does not matter which concept is preferred, nor really which architecture of principles suits best. If you want to talk about the source and ground of being, or of accessing god, it only becomes a problem if you find yourself saying, my way is best, it is the only way, and thus all others have no merit. What matters for the Earth system is that we are able to choose activities and behaviours that are low to zero in carbon emissions, that these activities make us happy, and then that we do this enough to feel we are living a fulfilling life.

Amy Isham, Tim Jackson and colleagues at the University of Surrey have recently shown how flow activities are linked to big questions on climate and the material economy. First, they found that people with material values have a lower tendency to select and experience flow activities. A consumption mindset involving striving for money and wealth seems to prevent flow. They also found that flow activities are mostly low in carbon emissions and other forms of negative environmental impact. These flow activities included art and craft, writing and reading, games and sport, walking and gentler forms of exercise, intimacy, religious and spiritual practice, being creative, helping and giving, and close relationships with others.

The importance of the lived experiences of the body has been highlighted in the revealing and ultimately uplifting book, *The Body Keeps the Score*, by Bessel van der Kolk. Trauma is held in the body, he shows, and is not responsive to words alone. It is produced by both physical and verbal threats, and yet the medical establishment tends to see people mostly as individuals rather than as social beings. If you have an economy that disrupts social relations, there will be consequences for individuals, and some of these may play out in people's own homes. Van der Kolk provides an accounting that is in itself traumatic. For every soldier in a war zone, ten children are endangered in their homes. These are the numbers: 1 in 5

Americans were abused as a child, 1 in 4 beaten by a parent, 1 in 3 couples experience or engage in physical violence, 1 in 4 people grow up with alcoholic relatives, 1 in 8 witness mothers being beaten or hit. Such trauma produces physiological changes, makes individuals hyper-vigilant, and thus some resultant behaviours are not the result of moral failings or lack of willpower.

All trauma, van der Kolk says, is preverbal. Some feelings are impossible to articulate, words fail people. We might say, we are losing our mind. His therapies thus focus on physical activities: these could be meditation or dance and drumming, the use of eye-movement desensitisation and reprocessing (EMDR), body-calming techniques, acupuncture. The most important thing is being with other people, and feeling safe and nourished with support. Stress hormones continually released by the firing amygdala, "wreak havoc with health." People do feel they are in a prison. In schools, there ought to be physical exercise, choral song, dance and theatre, drumming and mindfulness. Doing things with others creates togetherness, flow and attentiveness. For a while, the body and mind are calmed and in recovery.

As we shall see, these choices and consequences relating to attentiveness, immersion, focus and flow have wider social and political dimensions.

Nearby awaits a sunlit absence in what Martin Laird calls, "An unshakeable flow," referred to by the Trappist monk Thomas Merton as, "Our deep self," and by St Teresa of Ávila as, "The interior castle."

The light comes from within, wrote Taitetsu Unno, the attention bandwidth has widened, the world is brighter, sediments have settled away, and light is now the source of transformation.

> Thoughts will come,
> If you find yourself obsessed, sit back,
> Be still, let go of clever persuasive thoughts.
> *[Anon. monk, The Cloud of Unknowing, late 1300s]*

The Inner Chatter That Makes Us Unhappy

Mind and land are tied together. Set between the two is your body and its behaviours. Mind shapes land, and land shapes us in return. Once upon a time, this external world comprised nature and aspects of local culture. Now those external shapers come in different forms.

It is only recently that we took that ship, glittering and golden, and sailed into a storm. And so we find, we are in a bit of a pickle. The Earth system was damaged so quickly that the foundations to our own house have become shaky. External harms have had internal consequences. And so, some solutions do lie inside as well as in the structures of society and economies.

Late stage human evolution was marked by a wonderful expansion in the size of our cortex. This is where language and ingenuity reside, the capacity to plan and design, the very thinking that we all take for granted. At the same time, no one gave us an internal off-switch.

Thoughts are there when we wake at night, worrying over work or family. To use a modern metaphor, they are often constructed as videos of the past or of possible futures, episodes that run and re-run. These are a common source of anxiety and discontent. We wish they would stop. Thoughts are a wondrous source of creativity; they can also be a cradle of torment.

Mark Twain is said to have written in his later life (though something similar has been attributed to many writers), "I am old man and have known a many troubles, yet most of them never happened."

This mental chatter has often been given animal metaphor: thoughts are babbling monkeys, wild horses and wild hawks of the mind; they are a pack of hyenas, a swarm of bees, the hobgoblin of interference. And also a cocktail party, a flood of commentary, subtle whispers, mental poisons, judgemental films. You find yourself in a cage, travelling on a runaway train, running in circles. You feel pummeled by your own thoughts. In religious and spiritual literature, this raging of thoughts has been called the work of devils, taking on disruptive forms capable of deliberate action. This kind of wandering of the mind has been shown to be an unhappy mind.

This inner chatter of thought has a thousand arms.

St. Augustine said he felt "Deafened by the din of my mind," and that he wished to "Quell the tempest." The Vietnamese monk, Thich Nhat Hanh, called it, "Radio non-stop playing," and said you can't hear the call of life because of this noise. The reactive mind is like being caught in a phone box with a bee, wrote the Augustinian, Martin Laird.

The trouble is, this noise often also has direction as an inner judge, as Joan Borysenko put it, always simplifying into guilty or not-guilty, bad or good, them or us. "The thinking mind is a marvel, but it has limitations," also observed Martin Laird in *Into the Silent Land*.

Early in the 20th century, Welsh poet Ronald Stuart Thomas wrote, "But the silence in the mind, is where we live best."

Modern neuroscience has called the location of this chatter the Default Mode Network (DMN).

It is situated in and around the midline of the prefrontal cortex, in the verbal and language centres. We humans do not have, though, an internal mechanism to think this default mode off. There is no switch.

You have to choose something for the body to do. It is possible to see the effects in an MRI scanner when a patient practises meditation: the DMN drops out, goes temporarily silent. This has also been called *transient hypofrontality*, a term used first by Arne Dietrich of the American University of Beirut. Transient hypofrontality produces a sense of peace, reduces tension and anxiety, and creates a friendlier world. It reduces mind-wandering, it increases selflessness.

This is also related to what was called the Relaxation Response by Herbert Benson in the 1970s, when heart rate and blood pressure fall, and individuals are distracted away from their worries and concerns. This Relaxation Response causes the brain to help build grey matter, particularly in the hippocampus and thus improve memory. This induction of the response causes the expression of nearly 4000 genes, changing our internal responsiveness to stressors. This causes physiological changes by reducing oxygen consumption, lowering heart rate and blood pressure, reducing the risk of inflammation, improving sleep patterns, and increasing the release of serotonin and dopamine.

All of these feel good.

The Idea of a Blue Brain and a Red Brain

Let's return to a little more on brain architecture, as this is another feature common to all humans regardless of culture, income and age.

This model centres on the idea of us having a red brain and a blue brain.

The evidence comes from many sources: evidence from neuroscience and brain plasticity, from spiritual and wisdom traditions, from mindfulness and talking therapies, and from the lifeways of indigenous groups. The idea of a functional split for blue and red is not new. Daniel Kahneman called these type I and type II brains, one for thinking fast, the other for thinking slow. Kahneman's system I is red, system II is blue. David Goleman called them bottom-up and top-down modes: bottom-up is fast, involuntary and the executor or habits; top-down is slower, voluntary and effortful.

The red brain is located in the brain stem, an ancient evolutionary structure. It contains our survival functions: it is fast-responding, involuntary, automatic, impulsive, driven by emotions, and is the executor for habits and routines. This red brain is spiked into action by the amygdala, the sentinel for threats, which then drives the sympathetic nervous system (SNS). The red brain and SNS are all about fight and flight.

The blue brain is located in the cortex, having expanded in size rapidly during the later stages of our evolution. It is slower, voluntary, able to learn and plan, and contains centres for social abilities of empathy and language. This blue brain drives the parasympathetic nervous system (PNS) and is all about rest, repair and digest.

All humans have pretty much the same starting genetic legacy (red-blue structures, transmitters, hormones and mechanisms), but as a result of experiences occurring during life the patterns and amounts will change. Too much red brain activity is bad for health, causing inflammation and stress. The blue brain promotes conditions for repair, yet is also the centre for mental chatter.

Red is fast, hot, active; blue is slow, cool and calm.

You could think of the red brain as in the basement of your home, where the boiler is housed, producing energy, heat and action. The blue brain is in the top storey with an outlook to a blue sky and white clouds, a long view of the world. You need both, but you don't want to spend too much time in the basement.

Let's look at these structures in more detail.

The red brain is quite marvellous. It acts without thought, cascading into the sympathetic nervous systems and the hypothalamus-pituitary-adrenal axis (HPAA). You think you see a lion in the grass, and Bang! You run. You see something fall towards the baby and you leap. A car pulls out and you stamp on the brakes. In evolutionary history, to miss one threat may have meant disaster; to miss a false signal was not critical. Better to jump or run than not. The gatekeeper to the SNS–HPAA is the amygdala. This is your sentry, up on a watchtower with a loud alarm. It responds to incoming alerts. There is no moderation to the amygdala. It is a binary responder, either off or fully on.

The problem is that too much stimulation causes the amygdala to become over-reactive. Too much cortisol and too many repetitions cause it to take on hair-trigger status. Small incoming signals start to seem large. The air-traffic controller begins to over-respond. Now the amygdala is becoming physically damaged by the stress hormones, causing it to shrink in size. In the old economies of hunter-gatherers and small farmers, sites for millions of years of evolution, the mode of blue brain tended to dominate. The predator and poisonous snake were actually rare; threats from other human bands were very uncommon. By contrast, in modern econo-mies, the red alert mode is regularly activated.

Going red feels bad because it is bad. It weakens the immune response, wears out the cardiovascular system, the hippocampus goes off line, and neurons atrophy and die.

A chronically stressed amygdala is a smoke alarm that keeps on ringing.

At the same time, the hippocampus, the location for the formation of memory, becomes worn down. Memories become harder to form and retain, as the cortisol suppresses new neuron formation. The red brain is saying: we don't need any of this memory stuff when we're stressed and running away. We're just worried about survival. Feedback mechanisms are thus important. An environment producing too much stress leads to an over-reactive amygdala, which in turn results in high anxiety and a growing shading of past memories with fear and anxiety.

It is now known that too much SNS has negative impacts on the gastrointestinal system (ulcers, inflammatory bowel syndrome), on the immune system (colds and flu, slower wound healing), on the cardiovascular system (hardened arteries), and on the endocrine system (producing type 2 diabetes).

Stress manifests internally as the accumulation of effects over long time peri-ods, causing conditions such as post-traumatic stress disorder, depression, and cardiovascular disease. Over time, continuing red brain activity exacts a high cost, accelerating non-communicable diseases (NCDs), and causing atrophy of brain structures.

The NCDs of affluence are partly more common today because the red brain is over-activated. And the outside world of material late-stage capitalism has been changing the inner world. It is easier to habituate over-eating or drinking, to rely on buying things or on pharmaceutical interventions. Over time, such habits can evolve into pre-emptive strikes: promoting consumption before the anticipated discomfort.

Loneliness is an example. It is stressful being lonely, and we know together-ness with people reduces stress. Individuals with smaller social networks have been found to have both amygdala and hippocampus smaller in size. Lonely adults display elevated cortisol and epinephrine levels and have higher blood pressure. The more active SNS results in poorer sleep and lower immune function. It accelerates physiological decline with age. The lonely have become red brain dominant, and their lifespan is reduced. Loneliness has the same adverse health impact as smoking 15 cigarettes a day.

In these ways, the external social and economic conditions in which we live have biological effects.

The poet Rainer Maria Rilke wrote in his *Sonnets for Orpheus* (Book II, verse 29):

> Silent friend of those far away, sense
> How your breath expands space.
> Say to that earth of silence: I flow.
> Say to the rushing waters: I am.

Attentiveness, Flow and the Green Mind

Quieting the mental chatter switches on the blue brain and the workings of the parasympathetic nervous system. You feel good.

It might help to set out a binary categorization, dividing thoughts into these two types.

A-type thoughts are creative: from our inner store of memory and habits you are able to fashion something new. We humans like new things and activities.

B-type thoughts are responsive: internal or external signals are quickly formed into a story to which you find yourself responding. You often do not know how to stop these.

The idea is not to stop thinking, just to limit the B-type thoughts. We want them to recede into silence, some of the time. In Buddhist traditions, it is common to talk of first and second arrows. We cannot control the arrival of first arrows: these might be external events or words. It is the way of the world. Threats will come, and often do cause pain. But we can govern our responses to these first arrows. It is the firing of the second arrows that leads to added anxiety and stress. Indeed, it can become a habit to respond in a particular way to first arrows, whether this was a car pulling out in front or an unpleasant language threat on social media.

But it is hard. When something wrong and unfair, careless and painful, occurs, we feel we must respond. Long ago, Confucius wrote, "Before you embark on a journey of revenge, dig two graves." The second arrows can result in more pain than the first. There is wisdom in letting go.

A second problem is that we mistakenly take these B-type thoughts to be ourselves, and so become attached to them. You often cannot avoid first arrows, say when in the grip of pain, grief or low self-esteem. Even worse, the second arrows can start arriving before the first. You anticipate something bad, and maybe the first arrow never comes along anyway. Mark Williams and Danny Penman of Oxford put it this way, "We re-live past events and re-feel their pain, and we pre-live future disasters, and so pre-feel their impact." Joan Borysenko called it *awfulising*, the tendency to see the worst in a situation, and then append a flood of extra *what ifs* and *if onlys*.

All we can do is try to slow or stop the automatic flow of B-type thoughts through attentiveness and flow. When Martin Laird says *be the mountain*, or Bruce Lee says *be water*, they are recommending don't let the weather of thoughts and feelings overwhelm you. It is only meteorology, let it pass by.

The mountain is not concerned by the weather. In Tibetan, a popular phrase says thoughts are only writing on water: we should not take them personally as they soon will disappear. Today these B-type thoughts have become a major source of stress, causing anxiety and unhappiness. They have provoked our fight–flight responses, and are effective precisely because they are unpleasant. Your brain is saying, this is unpleasant, so try to avoid it in the future.

And so back to the switch.

You can select behaviours that calm the chatter of B-type thoughts, bring silence through transient hypofrontality, switch on the rest-repair of the blue brain, and thus bring about well-being and improved health. And the thing is, we already search out these behaviours on a daily basis, perhaps without realising. They are the fabric of our lives. They just need naming for what they do, whether we dance or run, knit or sing, walk or ski, listen to the birds or chat with friends, watch the clouds or pray.

Figure 7.1 contains a summary of the actions and states that can deliver episodes of well-being and happiness across our lives at any age. At the centre is the source, the luminous mind surrounded here by the circular form of the brushed enso to represent freedom of the mind.

These are the four key phrases: we can *be attentive, become immersed, be focused, become in flow*.

There is no single word that works best above all others, though many writers and analysts have preferred to use just one for simplicity. Mihaly Csikszentmihalyi wrote about *Flow*, Steven and Rachel Kaplan developed an *Attention Restoration Theory*, many have used water metaphors, many others have talked of the need to be aware and awake. Taitesu Unno wrote that attentiveness and focus both lead to deep hearing, which is all about wanting to produce a better world for all its people. There is no right answer; all are good.

If we look back, being highly attentive brought evolutionary advantage to us as hunter-gatherer-cultivators. Watchful awareness was central to the hunt, was vital for caring for plants and animals across the seasons, it was critical for memory creation so that we are able to return to sources of water and food. Hunter-gatherer-cultivators spend large amounts of time waiting, observing keenly, and preparing and eating food. But today, many modern lives are lived on simmer, in a state of red stress and anxiety, where in high-consumption living we find ourselves measuring our worth against others.

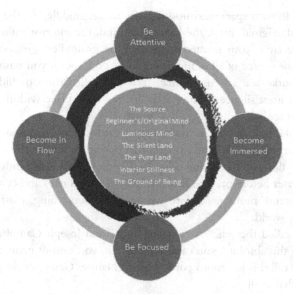

The Source
Beginner's/Original Mind
Luminous Mind
The Silent Land
The Pure Land
Interior Stillness
The Ground of Being

Be
Attentive

Become In
Flow

Become
Immersed

Be Focused

FIGURE 7.1 The Enso Circle for the Green Mind

It is these four activities and states that encourage well-being:

Be attentive: also aware, awake, attuned. This is you opening up bandwidth to all inputs, which could be externally or internally-sourced. It could be listening to birdsong or steadily repeating a mantra or prayer.

Become immersed: you are inside an activity or behaviour, fully absorbed. All the current and past worlds of memories and stories have fallen way. You are submerged, and have the sense of self and ego has disappeared.

Be focused: you are more narrowly fixed on an activity, and are probably applying skills learned in the past. Your focus quietens the brain's default mode network (DMN) through transient hypofrontality, the inner voices are silent and the silence expands.

Become in flow: you are inside the river, and it could be a quiet murmur or surging rapids. You could be at the level of peak performance, we have become skilled at something and now are flowing in automation. You might call this being in the zone. You are inside the sky where clouds glide gently. You sense unity and connectedness, to people and to nature.

This immersion and flow can feel like we are on a high vantage point, being on a hill looking down upon the distant valley, on a cliff looking at the silent sea. The *Bhagavad Gita* calls this *watchful insight*, implying also clarity of mind, gentleness, self-control and silence. As the attentiveness comes online, so the default mode network falls silent. The two work in opposition. Attention quietens the inner chatter.

We can enter this circle of the green mind from any compass point: north to south, east to west. The mind at the centre is surrounded by an *enso* symbol, an imperfect circle with space and nothingness in the middle. At the centre is the beginner's and original mind, the luminous mind, the interior stillness, the silent land and pure land, your ground of being, the groundless ground. It is sacred ground. It is the source of all of life. It could be god, too, if you wish.

Ben O'Rourke was an Augustinian friar and called this our hidden treasure, where there is great silence of the heart: "In the deep place within, there is calm at the sacred space."

In Zen Buddhism, this *enso* is the vast space enclosed by our mind. In Christianity, it is not surprising that baptisms occur in or with water, in the early centuries always in the flow of rivers and seas. In Islamic traditions, hands and feet are cleansed in water before prayers. Equally, many cultural rites and ceremonies simply involve intense moments before we go inside something, passing through a door to a new world.

Carl Jung called the self, "A circle of light," and Joseph Campbell said, "You cannot define this absolute, you cannot picture it, you cannot name it."

You could call this luminous core of mind Nature, Grace, or simply Spirit. Or nothing like that at all.

The Korean Zen master, Kyong Ho (1849–1912), wrote:

> Cultivate the mind without allowing it to be tainted by worldly things;
> Then plenty of good things will happen.

The Pro-Social and Pro-Nature Green Mind

Here is a tale about attentiveness, told to Ruth Sawyer by a baker from Italy, and appearing in her 1942 book, *The Way of the Storyteller*.

It is called "The Magic Box."

In one of the fertile valleys of the Apennines, north of Emilia, long ago there lived a wealthy farmer. His land was small, but the vineyards were perfectly pruned, the olives watched over and never suffered a frost, his grain golden, cattle sleek and sheep heavy with wool. Yes, everything prospered with him. He had two sons, one was an expert with the sword, the other easy-going, a bit of a rascal. The older marched off to the wars, and Old Gino said to the younger son, I am leaving you as fine an inheritance as any in the north. Look after it for your older brother. After he died, young Tonio settled down to easy living, went to the fair with friends, stayed at the inn, danced and fiddled. Often he slept until the sun threw shadows on the foothills.

Soon the farm suffered: bandits stole the cattle, wolves the sheep, the bad little oil-fly came in swarms over the olives, the grapes turned thin and sour. The autumn rain mildewed the harvest, and rats ate their share. Ten years passed, and one day Tonio saw soldiers at the inn, drinking and bragging. Suddenly he thought of his brother, and terror grew like a fire. He ran to a neighbour, who laughed. At the inn, they said let's dance first. The priest did not laugh and told him to light candles and pray for wisdom. On his way to the shrine of St. Anthony, he met a herd boy who tapped his forehead and said, "Master, you are so frightened, it has made you quite mad, like me." Then he whispered, go to the old gypsy woman in the grotto, over the pass. She makes magic, high and low.

In the end, Tonio climbed the mountain, and found the woman, as old and withered as a fig. She put a small casket in his hand, it was bound by brass bands. In the top was a tiny hole. She said,

> Every morning, while the dew is heavy, shake on grain of soil from the box in every corner of your lands – barn, pasture, vineyard. See that none is forgotten. Never miss a day, nor look inside. If you do, the magic will be gone.

That night, Tonio did not fiddle or dance. He was up early and took the magic box around the farm. There were cattle unfed, and corn fields uncut, scythes rusting in the yard, and men asleep in their huts. Holy mother! the men said, defend us. The farmer is looking about, just as his old father did.

Every day Tonio went, shaking a grain of soil here, a grain there. And every morning, he was seeing something that needed care. Soon the neighbours were wagging their thumbs at him, as they did towards his father, and the farm began to prosper.

And what happened: the older brother never came home, and Tonio married and his daughters looked after the farm with care and attention too. As he lay dying, he asked them to bring the box. They broke the bands, as his hands were weak by then. He looked inside, eager to find the magic. In the bottom were a few grains of soil, the kind that anyone can gather from the forest by the trails that climb into the Apennines.

Under the lid was written, "Look you, the farmer's eye is needed over all." And so.

> Everyone's journey,
> Through this world is the same, so I won't complain,
> I place my trust, in the dew.
>
> *[Ilo Sōgi, Zen poet, Japan, 1421–1502]*

In the famed Anglo-Saxon poem, The Seafarer said, "I have spent my life on the ice-chilled sea, on the exile's roads."

Out there, is the world of culture, structure, nature and economy. It still needs navigating.

We each have available, at any point in life, behaviours that can be called on. You can use the switch for troubling thoughts and habits; you can deploy attention, immersion, focus and flow; you can up-regulate the blue brain and parasympathetic nervous systems; and you can use guidance for the time and motivation required for the formation of new habits. But there is still a whole planet out there, in a bit of a state.

This opens up another door. We can take deliberate action with a purpose. Interventions can be in one of two domains: inside us, to change brain and body, and outside us, to change economic and natural environments.

But trouble has been brewing. The attentional commons is a term coined by Matthew Crawford in *The World Beyond Your Head*. These are increasingly enclosed by commercial interests. Advertisers and corporates have long understood the need to manufacture not just the goods, but also your desires and cravings. They have become good at making us want stuff.

In bounded circumstances, you understand the game: advertising on TV and at cinemas, on billboards and social media feeds. They aim to take your attention. More and more, though, new places are found for these signals, on sports team jerseys, through announcements on flights, on taxi, train and bus. If you can afford it, you fly to the sun hoping for a minimum of fuss and bother, and then plan to do nothing much at all. And when you get to the beach, you might find parts claimed by a hotel or resort. The sophistication of this invasion of attention is always greatest on private property, another reason why the physical commons remain important.

Flow has also been acquired by state and commercial interests. In *Stealing Fire*, Steven Kotler and Jamie Wheal point to the many facets of an altered states economy, where we are permitted or encouraged to pay more for immersive or silent experiences that involve some kind of altered mind state: the quiet business lounge in noisy airport, the iMax cinema, the acquisition of legal and non-legal drugs. Bee Wilson in *The Way We Eat Now* has observed that you cannot put a barcode on a carrot or tomato, nor on a tree or mountain path. Not yet, anyway.

It would be easy to let fatigue come upon us all. If you think instead of the wilds, you could sleep at night with the windows open, and listen to the timeless music of the marsh.

The Western Apache people have a particular phrase when talking about the importance of stories in places. They talk of mental smoothness, a mind that can be calm and focused. The smooth mind is a tightly woven basket, they say, yielding but strong, resistant to the jarring effects of external events. Such a steady human mind relinquishes all thoughts of personal superiority, it forgets itself and so conducts affairs in harmony and peace. Drink from natural places, the Apache girls and boys are told, then you can work on your mind.

"Wisdom sits in places," said elder Dudley Patterson to Keith Basso, "You will walk a long way and live a long time. And your mind will be smoother and smoother."

If you are entirely attentive to birdsong, to painting or knitting, to walking or sitting, rock-climbing or open-water swimming, to riding a bike or roller-coaster, then being in the moment switches off the chatter. This transient hypofrontality produces a state of immersion and flow that brings us happiness. This flow also implies an element of impermanence. Nothing stays as it is. Things come and go, we are born and later die. We should also try to let go of being attached to the idea of permanence.

High attentiveness brings the opportunity for something more. The close observation of nature and people as they are, at that moment, increases equanimity and compassion. Attentiveness is not selfish, it ends up increasing the connections between people and people, and people and nature. If you accept things as they are, at a particular moment, this also suggests that happiness can be increased independent of conditions or circumstances. You do not have to rely on something else to intervene.

The approach is modest. Send someone to fetch the flowers.

You could say that mind and land are connected. When we damage land, we damage ourselves. When our minds are harmed, we are taken away from the land. When each is repaired or restored, then the other benefits. Some of this is being implemented in secular mindfulness programmes. There has been, note Joel Levey and Michella Levey, "An inner revolution in mindfulness, meditation, mind-fitness and contemplative science." Mindfulness is becoming popular in schools to universities, from business to parliament.

In the 2021 Mindfulness Institute report on schools, one nine-year-old pupil is quoted as saying, "Mindfulness lets our brains cool down so that they can bubble up with ideas." It will be hard, though, to reach everyone. Many societal habits have themselves become cages. Shinzen Young has also pointed out that continued practice elevates the base level of attention and focus. We get better at attentiveness, finding it easier to start and also deriving more quality for the experience. We release an inner healer. Along this path lies both the good life and low carbon living.

Joe Hyams said to actor and martial artist Bruce Lee, "You want me to empty my mind of past knowledge and old habits."

"Precisely," Bruce Lee replied, "and now we are ready to begin."

Let's Summarise

This is how the brain and that elusive thing called the mind works.

Wrote the Japanese calligrapher, Tenepu Nakamura in 1948, "Even though the universe is essentially empty, it produces the most wonderful objects."

We all start at the same place from birth, with a beginner's or original mind free of memory and habit. We all end in the same way, able to take away nothing material from this life.

Habits are the brain's way of automating body behaviours and thought patterns. Learning habits takes time, but delivers great efficiencies. You come to do many things without thought, such as walking, talking, riding a bike or driving a car, and this allows the brain to divert resources for other functions. Habits are hard to form and are even harder to break.

The frontal and parietal lobes of the brain's cortex are the centre for all powerful and creative thoughts, yet evolution did not give us an off-switch. Throughout history, every recorded religious and spiritual tradition has addressed the troubling nature of the mind's inner chatter. Today, this is a still concern, playing a leading role in anxiety, stress and mental ill-health.

Flow is the off-switch. If you actively choose behaviours inviting attentiveness and focus, this leads to states of immersion and flow. These bring a sense of well-being by stopping the inner chatter, as well as activating the blue parasympathetic nervous system. The silence and calm improves well-being.

The brain displays plasticity. Its structures grow and shrink according to what we do and experience in life. We physically respond to our environments and are changed. It has become a dominant paradigm in the modern era to think that the brain cannot change and that it just decays with age (we are what we are, and there's no point trying to change).

The brain has two modes, red for fight-and-flight and blue for rest-and-repair, and these map onto our two nervous systems that have different functions, the red sympathetic and blue parasympathetic. What we might call the green mind that is pro-nature and pro-social uses flow activities to up-regulate the blue brain, and keep the red for only emergencies.

When you pay attention, you can let the old stuff go, and so the world might just brighten up. Then you need to engage with others to make the wider change.

Pay attention to this nothing.

[Meister Eckhart, Germany, 1260–1348]

~*~

8

ENOUGHNESS

Creative Slowth Is Better Than Infinite Growth

If you are capable of tranquility,
The world of dust, becomes the world of truth;
No matter what you do,
Forsake ideas of excess and inexhaustibility.

[Hung Ying-ming, *Living a Satisfying Life*, 1593–1665]

Abstract: Well. What now? Four crises, the keys to habit change and quietened inner chatter, the use of flow activities to reduce carbon. This chapter centres on the concept of enoughness, the sufficiency that creates slowth. The corporate model, for many, is always for more: more growth, more sales, more desire, more growth again. Yet, we always have lived within enoughness, the space between not-enough and too-much. Overshoot means going too far, wrote Donella Meadows, one of the authors of *The Limits to Growth*. The bad life has sharp teeth. "How big should it be?" asked Hermann Daly, a big question referring to the world economy. This economy anyway a wholly-owned subsidiary of the environment. In stable-state economies, interesting things happen, as they did and do in every indigenous culture worldwide. The Edo period in Japan was 260 years of slowth, and gave rise to an age of pilgrimage, poetry, painting and printing, to new temples and festivals of belonging, to great libraries. But worldwide, growthism brought rising inequality and environmental harm. Tim Jackson commented, "There is no growth on a dead planet." There are many ways to engage in personal and collective forms of growth: these are the infinity games. It is through contemplative crafts and art, the flow of activity and skill, that we find a good life in which work is about living well and long. These slow the climate crisis and also create divergence and diversity.

DOI: 10.4324/9781003346944-8

Why Enough Seems So Hard

It is big and it burns, this simple question for the planet. Why is it so hard to slow up and pause at enough?

This Earth has clear boundaries, and beyond the brim is darkness. The authors of the Club of Rome report in 1972 issued warnings. All their modelling for future decades ended up with a broken planet if consumption went on growing. That was 50 years ago, when carbon dioxide levels in the atmosphere were a safe 327 ppm, when too there were 700,000 more square kilometres of Amazon rainforest than today.

And so, how might we find a true kit for living?

In 2020, the oil company ConocoPhillips' Arctic sought approval for the Willow Project. It was to be located on public land in the far north of Alaska and intended to drill for oil at five sites, building 800 km of new roads, an airfield, a gravel mine, and produce a new flow of oil for 30 years. The permafrost in the region, however, was not playing ball. It had so melted and become unstable that the company said they intended to install giant chillers to refreeze the permafrost. Without a stable ground for the drilling rigs, it seemed they were in danger of not being able to extract the oil from rock below. There was no self-sense of irony or absurdity. The giant freezing units were there as a novel technology to aid faster fossil fuel extraction for both oil and gas.

At the same time, the largest soft drinks firm in the world issued their results and hopes. This one company has 5000 beverage brands and sells nearly two billion servings daily in 200 countries. Each day, the world population consumes 60 billion servings of all drinks, mostly plain old water from river, tap or well. In the Annual Report for 2019, Coca-Cola indicated, "One hundred and twenty-five years and we are just getting started."

Further along, their prime aim was still, they said, "To become more competitive and so accelerate growth." No doubt this explains, in part, why the average person in some countries consumes more than 200 litres of sugar-sweetened soft drinks each year. Perhaps one day, more of us will weep and mourn.

This conspicuous consumption, as Thorstein Veblen first called it, was no accident of the 20th century. It was designed that way and later magnified. The consumption economy created images of a good life attained through certain goods and services, a car, a dishwasher, a handbag, a holiday. Here was comfort and luxury, here was status, here was utopian imagery to suggest this could be accessible to all. Above all, in a world of hardship, suffering and death would soon admit defeat.

But something else was happening. This dreamland was also combined with the design of dissatisfaction, so that things could be promoted to fill the gap we felt. In 1929, the Head of Research at General Motors, Charles Kettering, told fellow manufacturers that they needed to keep the consumer discontent so as to sell more goods:

> If everyone were satisfied, no one would buy the new thing because no one would want it. The ore wouldn't be mined; timber wouldn't be cut. [The consumer] must accept this reasonable dissatisfaction with what they have and buy the new thing, or accept hard times. You have your choice.

If Kettering's corporate pals could provide kind and persuasive reminders to the public that they were to be dissatisfied with what they had, then part two was to show how to escape this suffering. Material consumption was quickly linked to happiness, and thus for all people: more sales, prosperity, joy. The power was thus to offer not just things, but empowerment and liberation. Consumption became an escape route from an imagined past of restrictive cultural and social norms. It was the new good life.

Yet, for them, fear lurked in the shadows. What if a person consumed just enough, and then stopped? What would happen to the new economy if you went instead for a walk, read a book, volunteered, meditated or prayed, played with your children, went dancing, painted a picture?

This would not do. So commerce was crafty.

They would tell about the top-of-the-range watch, barbecue or car not to encourage you to buy them, but to make you believe that something less expensive, but still more than you intended to spend, was appropriately frugal. This was the beauty of consumption: you could never get what you were told you wanted. Consumers were thrown into a land of perennially unfulfilled desire. In this last generation, dishwasher and air conditioner ownership rose from 10% to 70% of American households, tumble-driers from 20% to 90%. Barry Schwartz of Swarthmore College in Pennsylvania asked, "Does this mean we have happier people? Not at all."

And, meanwhile, an acrid haze began to build upon the plain.

Convenience and comfort were blameless desires, most of us would agree, but something important did not happen. Dissatisfaction did not disappear.

Companies became skilled at exploiting the readiness with which you measure yourself against others. You are happier with a $40k income if it has risen from last year's $35k than if it fell from $45k. You prefer a lower salary if it is higher than other people you know or work with rather than a higher salary that is below the average.

And then there is adaptation. You get used to things, then take them for granted.

Adaptation is useful in a world of misery: it allows us to cope. But as Barry Schwartz also said, "If you live in a world of plenty then adaptation defeats your attempts to enjoy good fortune." One famed comparison showed that two groups of people were equally happy a year after winning the lottery or becoming disabled in a car accident. They adapted. Robert Frank of Cornell University pointed out that you do get a rush of satisfaction when you get a new TV, fridge or phone, but soon it wears off.

Tim Kasser of Knox College in Illinois was one of the first to speak of the high price of materialism, the damage done by "winner-takes-all economies." He concluded that the pervasive negative correlates of material values are these: they make you unhappy, you end up with lower mental and physical health, you tend to believe others are malevolent, and so avoid being dependent on them, and you end up becoming non-generous, possessive and envious. At the close of play, he writes, "The successful pursuit of materialist ideas typically turns out to be empty and unsatisfying."

When faced with a huge array of options, there is also the creation of new worry around whether a choice might have been the wrong one. These regrets are corrosive and yet encouraged. And so there is a sliding into comparisons often without realising.

Victoria Husted Medvec and colleagues at Cornell University showed in the mid-1990s that Olympic bronze medallists were happier than silver medallists. Everyone sets out wanting to win, of course, but silver medallists ended up regretting that they came so close but failed, yet bronze medallists finished happier as they could see they came very close not to get anything at all. In this way, hope and expectations shape contentment with the outcomes.

Medvec called this counterfactual thinking, centred on *if only* and *if I had not* lines of thought. The upward counterfactual thoughts about missing gold centred on regret, guilt, blame and anger; the downward counterfactuals were about relief and gratitude. And so it comes to pass: each of us comes to be easily manipulated. You can choose what you spend, and appear to have free will, but you cannot choose what someone else does. This can then be made to influence your own behaviour.

If, of course, you stay in this game.

We saw earlier that cross-country and within-country data clearly shows that rising material consumption does produce more happiness at very low incomes. This is typically at less than \$10–15,000 of annual GDP per person, in the C1 and C2 countries where people are short of food, clean water, sanitation and domestic energy, and need access to health, education and transport services. But as GDP grows, happiness increasingly tails off. The fault line is partly with the way GDP is measured: pollution appears to contribute to progress, but volunteering does not.

When I talked with the Right Livelihood award winner, Chilean economist Manfred Max-Neef, in the late 1980s, he already knew we had a problem. He said over coffee, "You can only add to GDP, never subtract." At the time, he was keenly promoting the idea of *a basic income for all*, a national income publicly supplied and designed to close off the worst of inequality. In pointing to paths out of the dark woods of the Anthropocene, Jason Hickel recently said in *Less is More*, "If you grow your own food, clean the house, care for relatives, then GDP says nothing."

In 2018, 238 scientists called on the EU to abandon GDP as a measure of progress, and then in 2019, William Ripple of Oregon State University led a group of 15,000 scientists from 150 countries to call in the journal *BioScience* for the world's governments to shift from measuring GDP towards sustaining ecosystems and improving well-being.

> What a glorious luxury it is, to taste life,
> To be full, for even a year;
> Never has there been the wise person, who was rich.
>
> *[Yoshido Kenkō,* Essays on Idleness, *1283–1352]*

Finding Enoughness

Look around. Everywhere there are signs of enoughness.

Our whole lives are lived by balance and sufficiency. Our body temperature is tightly regulated. Too hot or too cold, by just three or four degrees, and death could follow. Within geological eras, the Earth is also in balance. There is enough oxygen in the air for life: too much and trees would combust, too little and, well, you know the answer.

There is a sweet spot in everything: diet, physical activity, clothing. Many sports require throwing, hitting or kicking a ball. But too much effort and power, and the ball misses the target; too little and it misses again. In the middle is enoughness.

This quest for balance has a long history. The Tao was seen as the life path on which things and activities were characterised by yin and yang. Neither was better than the other, both are needed for a life of poise. Tim Jackson in *Beyond Growth* has written of Aristotle's concept of *areté*, the search for virtue and how to be good at being good. Each virtue was seen by Aristotle as being flanked by two vices, on one side was deficiency, on the other was excess. An arête later came to describe a sharp mountain ridge where the land fell sharply away on either side.

Virtue was once about balance, not excess. To Amish people today, to be worldly is to be *a little on the fast side*, to stray away from norms that define meaning and well-being. They say you should take a long and slow stride, as there are two deep ditches along every road in life.

Doris Fuchs and colleagues at the University of Münster have framed this place between too little and too much as a consumption corridor. "All people want to have a good life," they write, and so the best consumption corridors are enabled by citizen engagement and deliberative democracy. The idea also makes clear that the pursuit of a too-much life hinders others from achieving an at-least-enough good life. This is, in short, living within limits.

And here we meet a fundamental differentiation: the trickle-downers say by thinking only of the individual everyone can be wealthy, and it is just a matter of time before the poor will benefit. The good-lifers say living within limits can mean thriving and flourishing for all. The latter is not zero-sum, where someone always has to lose in order for someone else to win.

A sustainable consumption corridor thus sits between a minimum consumption floor and maximum consumption standards. Another way to think of this space centres on enoughness and sufficiency. Thomas Princen of the University of Michigan wrote about the logic of Sufficiency: this is when we do not take out of the environment more than it supports, and we do not take into our economies more than we need. It has ancient roots: "People are perfectly capable of living within ecological constraints, and have a long history of doing so." Communities have self-limited fishing effort to protect fisheries, have rotationally harvested forests, have shared irrigation water, and have gardened for pleasure more than maximising profit and yield.

The idea here is that enoughness is not so much a thing that can be measured. It is more a guide and device that helps us think and act together and then put in place appropriate policies to shape and link actions.

In the background is a similar word: efficiency. It has long been an argument of those in favour of ever-increasing consumption that environmental efficiency will one day catch up and overtake total consumption. Producers will limit cost to become more efficient, and the planet will benefit. The energy required to produce one tonne of material output, or one item for sale, will come down over time, and so we can continue to consume without limits. And all will be well.

But it never worked out that way.

Private operators are rewarded when they displace environmental and social costs into the public sphere. In other words, they do not become more efficient, just better at passing on costs to those who cannot avoid them, or to the planet itself. Enoughness seeks to create the space for lifestyles and ways of living that does not fudge efficiency, and does not pretend that the route to happiness lies in too-muchness.

Thomas Princen wrote, back in 2005, "Now people are tired of being demeaned as mere consumers and employees. Now people crave real meaning in the material and social lives." Well, we could add, today, now citizens are fed up with promises of endless plenty, when just a few get rich and most get nowhere or just poorer. Efficiency sounds good, an equivalent to progress: make use of every minute of the day, don't do nothing, don't waste time by not working. These have not proven to be routes to good lives for all.

Sufficiency can be voluntary or obligatory, somewhere between having and not-having, and thus being content. But social and advertising pressures are huge. These have to be resisted, or you are drawn back into being made to want more than you need. Voluntary simplicity does have a role to play, but it cannot solve the larger problem.

Michael Maniates of Yale-NUS College in Singapore says a type of magical thinking is also at work. As we consumers become more concerned about limits, as we take actions, so he says we become side-lined into political irrelevance. The corporations, and governments, just become better at green-washing. A water company in the UK leads on a campaign called "save every drop," encouraging its domestic consumers to be careful and efficient. At the same time, it permits water leaks to continue, makes deliberate sewage emissions into rivers and seas, and so saves costs that can be paid to shareholders in the form of higher dividends. A twist, of course, is that this puts the blame back on consumers if water becomes in short supply.

You may conclude, if I stop consuming, this just allows others off the hook.

Or perhaps instead, if I reduce consumption, this allows the poorest to obtain more.

It is also magical thinking to believe that advances in eco-efficiency can deliver sufficient advances in reducing unit material use that everyone can consume without limits. Yet, today, the idea of growth still holds a vice-like grip on thinking about what comes next. You will find talk of green growth and clean growth, suggesting that we could become more environmentally efficient whilst still growing the old economy. You will find some saying we will just reduce carbon intensity as we grow. And that this will add up.

You will also find these are empty hopes. Decoupling has not occurred in ways that can offset the overall carbon and material footprint of consumption, despite the many claims for corporate responsibility. Where there have been efficiencies created in resource use, then more material growth simply swamps advances. There are two further problems: rebound and leakage. Rebound occurs when cost savings are recycled for different forms of consumption; leakage where natural capital is saved by one hand and spent by the other. Added to total consumption, these just swamp efficiency gains. Dominik Wiedenhofer and 16 colleagues at the University of Natural Resources and Life Sciences in Vienna recently reviewed 11,000 papers on GDP and the decoupling of resource use. Very rarely did any research address the fundamental incompatibility of seeking permanent economic growth on a finite planet.

The authors concluded that green growth is logically impossible when seen through the lens of GDP.

For at the heart of growthism has been rising inequality as well as environmental harm.

Markets and trade have been around for thousands of years, since at least the advent of agriculture and the emergence of the first city-states. But late-stage capitalism differed, with the imperative of constant growth and studied lack of understanding of planetary boundaries. But societal and individual habits will be hard to break, having had twenty decades to establish. Jason Hickel in *Less is More* makes this point about colonial appropriation, the age when natural, physical and human resources were acquired at very low cost: India and other colonies resourced Britain's industrial development, as the latter took resources from the former. The same goes for expansionist empires across human history. Between 1970 and 2002, Tim Jackson added, African economies borrowed US$540 billion from affluent countries and international banks. They have paid back $550 billion, but still find themselves owing a bill of $300 billion to the lenders.

Extractivism is a way to describe the effects of externalities. You acquire resources cheaply and make goods and services at low cost. Or you shift costs elsewhere for others to bear. But there is only the one Earth system, and we cannot externalize costs outside the planet boundaries, only move them about.

Enoughness has to define how we think and act. It also needs to bring people together into collective action and mass participation. Doris Fuchs wisely observes that sustainable consumption on its own is not enough. We also need new ways of living and new worldviews. Both of these may have very old roots too.

"There is beauty," says Clare Greenwood, "in the right amount of anything." She wrote of the Japanese concept of *oryoki* in *Just Enough*, about how to think about the right amount. She offered an example: if we carry on filling a room with furniture, then it will be cluttered and uncomfortable; if there is none, then we have to sit on the floor. Joel and Michella Levey in *Mindfulness and Mind Fitness* said something similar for music: if you over-tighten the strings, the instrument screeches; if you leave them loose, then you get just as much of nothing. This balance point is called the Goldilocks price by insurers. The bears' porridge was neither too hot nor too cold, the beds neither too hard nor soft. Daniel Kahneman and colleagues in *Noise* point out a high premium might seem advantageous to an insurance firm, but might result in lost business to competitors. Too low a premium may be equally costly. Somewhere in the middle is the perfect place.

Overshoot means to go too far, as Donella Meadows and co-authors wrote. It was to slide on the icy road when you put on the brakes, it was to have a ferocious headache the day after a party, it was to produce pollutants that air and water cannot assimilate, it was to fish so much that boats have to stay in harbour. There are so many examples, you wonder, how did none of this enter mainstream economic thought? And yet, this was long ago said by Mahatma Gandhi: "The world is big enough to satisfy everyone's needs, but will always be too small to satisfy everyone's greed."

And so: Imelda Marcos was found to own 3000 pairs of shoes when she and her dictator husband were ousted in 1986. A generation later, still a billion people worldwide each have at most one pair of shoes.

In the 1960s, the Chief Economist to the British Coal Board had been invited on an aid and advice trip to Burma. E F Schumacher saw something unexplained, and asked himself, "Why does everyone look so happy?" He stayed on for a retreat and visited temples, and came to make a plan for his 1973 book *Small is Beautiful*. Schumacher asked, "What is enough, who can tell us? There are poor societies which have too little, but where is the rich society that says, Halt! We have enough? There is none."

Even then, 50 years ago, he was able to observe, "It is clear that the rich are in the process of stripping the world of its once-for-all endowment." He called for a new philosophy for materialism, a shift from goods to people and from large to small, and for a new era of moral choice. Technology, he wrote, should have a human face. Then all fell silent. Now, still we wonder, what are the best ways for everyone on the planet to live well and happily, and yet not destroy our very own home?

The bad life has had sharp teeth. Neither too much nor too little works. A sixtieth of all the people who have lived for the past 50,000 years are alive today and have too little. Another one-hundredth today have far too much, and seem incapable of stopping. The Earth system might say, if it could, things have gone too far already. And meanwhile, swathes of economists and politicians worldwide have failed as yet to explain how carrying on growing forever is possible on a finite planet. Something has to give or stop.

Is there a higher road that works for all? Can we make a good life work, and ensure everyone has enough?

The investor and philanthropist John Bogle reported an encounter in 2007 at the party of a billionaire on Shelter Island, at the eastern fringes of Long Island. Storyteller Kurt Vonnegut guesses to his friend Joseph Heller that the hedge-fund host had made more money that day than Heller had made from ten million sales of Catch-22, his novel on crazed war-logic.

Heller replied, "Yes, but I have something he will never have . . . enough."

Even at midnight, it's only about the moment coming to all of us. At the instant of approaching death, if we have the chance, we will no doubt want to ask something like this: was it a good life? What were our most memorable events, and did we do well for others? None of the things we had accumulated will be much use at this point.

Taitetsu Unno wrote that pure land of Shin Buddhism *turns bits of rubble into gold*, and asked:

> After my death, what would I claim,
> As the most memorable events in my life?
> Could I come up with something to say,
> Yes, my life was a good one, a memorable life?

Why Slowth Is Good

What might a stable economy look like that does not breach planetary and health boundaries, one that favours the pursuit of a good life for every person? When at the World Bank, Herman Daly asked another big question referring to the world economy, and no one had an answer: "How big should it be?"

This economy, he said to colleagues, was, "A wholly-owned subsidiary of the environment."

The President of the World Bank, replied publicly: "That's not the right way to look at things." Just bigger than now, Daly was told, and then we might assume, more again.

Yet, in stable-state economies, interesting things happen.

Tim Kasser said this, by way of a recommendation: "Go for a walk. Read a book. Do volunteer work. Meditate. Play with your children. Talk with your spouse. Go dancing, work in a garden. Listen. Paint a picture. Go fishing." This was all, in his eyes, an agenda for a good life.

And development and environment specialists from the University of Oslo, Karen Lykke Syse and Martin Lee Muller have wisely observed, "When we cross-pollinate sustainability with the good life, the questions of the good life becomes anchored not only in our own mortality, but in the mortality of the Earth itself." Environmental philosopher Kate Soper calls this "alternative hedonism," a way to open up, "A post-consumerist approach to human flourishing."

The land of always-growth is not going to trickle down or be good for everyone. It has already led to conflict. Kate Soper also said, "The longer we continue with the growth economy, the more intense the competition for dwindling resources will become, and the more uncivil the methods to which richer societies are like to have recourse in defending their advantage."

An abundant flourishing occurred in Japan's Edo period, 260 years of economic slowth after a century of civil war. Between 1603 and 1868, the country looked inward for social and personal growth.

It was a time of peace and watching cherry blossom, relatively benevolent rule, and strong communal life. Shinto and Buddhist shrines and temples expanded to some 160,000 in number, and pilgrimages became highly popular, attracting half a million people a year to cross-country walking journeys. On the road was merchant and peddler, pilgrim and female singer, country doctor and subscription monks, and people just travelling for pleasure. Travel guides became hugely popular. Haiku poetry was invented, festivals dedicated, and new behaviours of engagement and togetherness normalised. The practice of *gantaka* became wildly popular, the giving-up of favoured food and drink until 100 or 1000 visits had been made to a nominated shrine. People became interested in stories: there were thousands of *satsuwa* tales with moral lessons, and *rakugo* drawn from social life. By the early 1800s, there were 900 active publishers in Edo-Tokyo, and 500 in each of Osaka and Kyoto, and 1500 bookshops. Great libraries were established across the country. Recycling was the norm: there were 4000 old clothes dealers in Edo. Nothing was wasted in this period of sustainable degrowth.

After the Meiji reforms of 1868, the next ten years brought 190 agrarian revolts, making the contrast with the previous period of stability so much more sharp.

Wintry gales might blow, but this slow ethic persists today. The 1400-kilometre Shikoku pilgrimage route was walked by Kōbō Daishi in the 9th century and covers 88 named temples and sacred sites on the perimeter of the isle. The route today attracts 150,000 visitors annually.

Ben Dooley and Hisako Ueno wrote in The Japan Times in 2020 of the traditional *shinise* shops that sell craft and home-produced food in Japan, often on these pilgrimage routes. There are estimated to be 33,000 *shinise* in Japan that are more than a century old, 3000 of which have been operating for at least two centuries since their establishment during the Edo period. Their values are called *kakua*, family precepts that look after employees, support the community, and produce crafts that inspire pride.

Naomi Hasegawa sells toasted *mochi* rice cakes from a cedar-timbered shop in Kyoto. The shop was opened in the year 1000 and operates not by trying to maximise a profit. It values persistence, and the production of pleasing *mochi*. Tanaka Iga began making Buddhist goods in 885 CE in Kyoto, the current family head, Masaichi Tanaka, being the 70th generation president. Some have added modern products. A metals firm that began making kettles in 1560 now also produces high-tech machine parts; a 300-year-old kimono manufacturer now also makes textiles for home furnishings and electronic goods.

From these *shinise* emerge a type of story about business service, about continuity and persistence, about acting in ways that do not damage the planet. When your timescale is a thousand years, then priorities inevitably differ.

When Helena Norberg-Hodge first travelled to Ladakh in the 1970s, she also found a 1000-year-old enclave of stability. It was a sere land of wind, high desert, remorseless sun and lengthy cold winters. The culture of this Little Tibet was founded on sharing, respect and gratitude, on always helping others. The severest insult was to call out someone who *angered easily*. In the mountain rain-shadow, villages were brilliant green oases lit with bright prayer flags snapping in the breeze. Life depended on channeling water long distances from glaciers high above, just as it is in the Pamirs and Hindu Kush to the west. Over the years, Norberg-Hodge came to see it was the Ladakhi attitude to a stable life that was most powerful. They sang in the fields, laughed together, the old were active until the day they died. There was beauty in their turquoise jewelry and attention to detail in the elegant carvings of house balconies. The wildness of the snow leopard was a mystical presence, as Peter Matthiessen found further east in the Himalaya, dangerous, fearless and silent.

Yet, the people said, *chi choen*, what was the point of worrying? They said there was more than one path, there was strength in difference.

The Ladakhi monk Gyelong Paldan said to Helena Norberg-Hodge:

Though we lack progress here,
We have happy peace of mind.

Sustainable Degrowth

It should be no surprise that the powerful and rich move costs to the poorer who have little or no voice. Authors and analysts have been trying to send a signal about growth: Donella Meadows called the Club of Rome report *The Limits to Growth*; a quarter century later Herman Daly wrote *Beyond Growth*; and another quarter-century again and Tim Jackson has written *Post Growth*, subtitled *life after capitalism*. For now, something new is being articulated.

The economic system of material and late-stage capitalism has been fingered: it created the mess. It cannot get us out by buying some splendid green paint. Tim Jackson said, "There is no growth on a dead planet," and so we're all going to need a new post-growth story.

We have already seen that GDP is a rotten measure of national success. There are two options: create new indices that capture the goods and bads in an ethical way, or simply ignore GDP by pointing towards the things that count for the maximum number of people.

One way or the other, the current economic system will have to change. It is an open question whether it happens before climate collapse or after. Meanwhile, we have Living Planet reports, Ecological Footprints, Genuine Progress indicators, and calculations of the number of planets needed to resource the economy. We have indices of Sustainable Economic Welfare, Environmental Performance, Better Life, and World Happiness. Each has helped us understand what has gone wrong, but being blunt, we have to conclude that they have not changed much yet. We are going to need new ways to think about value. We might need a new kind of currency altogether.

Nonetheless, creative options are emerging, there are calls for degrowth, agrowth and circular economies, for less is more, for living within the doughnut. Paul Ekins in the United Nations Environment Programme's sixth *Geo-6* report of 2019, called for *HP2*, a focus on a healthy planet and on healthy people, where priorities centred on "Nature's contribution to people" and "Preventing costs and creating value."

A circular bio-economy would be nature-based, bio-based and regenerative. It would invest in nature and people, and seek to build the regenerative assets of natural, social and human capital. It would invest in the global commons so that everyone has the potential to benefit. The priorities for circular bio-economies could include well-being and happiness, investing in nature, ensuring equitable distribution of prosperity, transformation of industry to eliminate externalities, and the re-imagination of living places with nearby nature.

Sustainable degrowth is a way of thinking that shines a light on the important things in life. It could be a way to dematerialise the economy. It emphasises the personal and social priorities in life whilst maintaining a steady-state economy, just as occurred in Edo Japan, in Iceland since the sagas, and in ten thousand indigenous and small farm cultures of the world. Tim Jackson calls this the Cinderella economy, a focus on the friendly and contemplative. Inês Cosme and Daniel O'Neill of the University of Leeds recently assessed more than 100 papers on degrowth and concluded actions should be both top-down from government and business, and bottom-up from communities and local action.

Four themes were common: the creation of a convivial society, redistributed income, reduced environmental impacts, and the emergence of new eras of stability.

The steady state of degrowth also raises new thinking about the commons. The industrial revolution brought enclosures of land and public institutions, so increasing inequality. Some countries have used their affluence to invest in new healthcare and educational commons, making them available to all citizens, for example, Costa Rica, Japan, Portugal, Norway, Denmark. Others previously created great commons, such as the UK's National Health Service in the 1940s, an institution that now employs 1.3 million people. Now new opportunities are arising when it comes to cutting carbon emissions.

We find public systems of goods and services tend to have lower carbon emissions than private.

If a country invests in renewable energy by solar photovoltaics, then they create a new commons whilst benefitting the Earth. Grameen Energy is training a female solar engineer for every one of the 70,000 villages in Bangladesh; electricity in Paraguay and Costa Rica is already almost 100% renewable; Kenya is building a solar rural network for all villages. Such investments release countries from some of the burdens of buying fossil fuel. Kenya, Senegal and India were spending a decade ago 40–50% of all their export earnings on importing oil. Now, they have the resources to invest yet more in renewables to produce collective benefits for home and planet.

Another course will be carbon capture and carbon trading. These are a necessary part of the race to zero but could create a new problem. Michael Mann and colleagues at Oxford have indicated there are real grounds for optimism is seeking unicorn projects that each could deliver annual 1 Gt net cuts in carbon emissions. These might include conversion of cheap renewables for hydrogen or ammonia fuel for shipping, trucks and aviation, next-generation high-efficiency solar cells, low-cost electrolytes, hydrogen fuel cells and engines, water treatment plants, and new forms of fermentation. The geothermal plant at Hellisheiði in Iceland is fixing carbon dioxide into rock. Sustainable and regenerative agricultural systems could fix several Gt of carbon into soils and above-ground perennial crop systems and trees.

On the other hand, such carbon capture on its own would let polluters off the hook. Even worse, they could pay others to capture carbon while they increase polluting. What is needed is both absolute reductions in carbon emissions, and carbon capture to get carbon out of the atmosphere as fast as possible.

We come back to the mix: individuals and what you can do easily, and governments and what policies can do. Growth has been a story of *more*, and never *enough*. The circular bio-economy and degrowth are about enoughness. And this will need support from new classes of social norms and taboos, similar to the creation of sacred mountains and forests, of mass pilgrimages to shrines, of new devotion and changed behaviours. Andy Couturier wrote that his Japanese friends had an abundance of less: "They seemed to have a lot of life."

Perhaps we modern humans are scared of having too much time. A sloweddown life offers a rich and palpable contact with the natural and social worlds around us. Stability is coming, in the form of a *Slowdown* wrote Danny Dorling. An era is ending, bringing the slowing down of almost everything. And stability could bring wisdom, the ending of the greedy era.

In *The Limits to Growth*, the steady-state economy was called the equilibrium society. This meant no overshoot, living within boundaries, and yet was neither stagnant nor stale.

And then there might be space for doing not very much. Jenny Odell has asked *How to Do Nothing*, and suggests this is an act of resistance against material consumption: "We might just find that everything we wanted is already here," and not in the "blasted landscape of neoliberal determinism." Being attentive creates new content, she writes. It is creative. But be aware too, in the attention economy, there are still many other interests seeking to capture attention to encourage another purchase.

> It's my snow,
> I think,
> And the weight on my hat lightens.
>
> *[Takarai Kikaku, Edo poet, 1661–1707]*

Personal Growth and Infinity Games

Each of our human lives, as far back as you dare remember, is about personal growth. Not the modern economic type, where growth means using up the planet, but change within a stable structure of body and community.

Personal growth is desirable across our lifeways. You are going to need to build and repair things, innovate to address new problems, develop new private and social businesses, develop new care systems for the elderly, share ideas on diets and dishes that work, write music and poems and dance and sing. You will find that these things add up to the good life, an art of living where moral and social progress is valued over selfishness. Many are already low in their carbon footprints. Fritz Schumacher asked in *Small is Beautiful*: "Everywhere people ask, what can I actually do! The answer is simple. We can, each of us, work to put our inner house in order."

Each life starts with an empty mind, and there are routines and habits to learn. Crawling and walking, language and social interaction. We have to find the food, carry the water home, light the fire and start the evening tale. This kind of growth is about learning, the enticement and excitement of the new, the satisfaction of achievement of something meaningful. It is about the currency of stories that grow with the telling.

At the age of 93, the cellist Pablo Casals was asked, "Why do you still practice?"

He replied, "Because I think I am making progress."

The American modern dancer Martha Graham danced into her 90s and was working on new choreographies in New York at her death aged 96. She had conformed decades earlier to social norms by formally retiring, but then suffered severe depression. She later said, "When I stopped dancing, I lost my will to live." After 45 years of tai chi, Henry Wang said, "I still look forward to my regular daily practice, which becomes more enjoyable and meaningful day after day."

The Okinawan developer of the martial art *karate*, Gishin Funakoshi, observed in his later life, "After seventy years of *kara-te*, I might finally be beginning to understand." Hokusai was in his 70s when he began in the 1830s his wildly popular series of woodblock prints of the views of Mount Fuji, each intended to be a story in itself. He wrote:

> None of my works before my seventieth year is really worth counting. At the age of 73, I have come to understand the true form of animals, insects and fish, and the nature of plants and trees . . . At the age of 90, I will have got closer to the essence of art. At the age of 110, each dot and line will be alive.

Bernard Lainé had been a sculptor all his life in Paris, and for one winter week I sat for him with a friend in his attic atelier.

It was on the Île Saint-Louis, the island in the Seine said to be eternally tranquil. He was wiry with sunken cheeks and sparkling eyes. He had the large hands of a crafter, a cigarette permanently on his lips. He seemed old in the silence, and yet nearly 20 years later Peter Turnley's essay of Paris contained a photograph of Bernard sitting in an old bath on the roof, reading and still with a wise smile. He looked the same age. My friend had brought me to photograph her mime troupe in studios at Bastille. Each day we climbed the seven flights up the dark stairwell. The rain fell on the Paris streets, far below, and soft light filled the studio. Through the open windows came music, a piano played, and there were bicycle bells ringing and doves cooing.

Bernard served green tea, and cast our heads in plaster. We breathed for hours through a straw. He talked of his life making things. One day we visited a Man Ray exhibition and came back with a book by Boubat. "Oh yes," he said, "My friend Édouard!" There was pea soup and bread for lunch on the Île, and one evening at an Algerian café *Kind of Blue* was playing, and still the rain fell on the slick streets. Bernard had fifteen heads on a table, moai-type casts, and said a lifetime lay ahead for any sculptor. Every day was different. At Dieppe, a group of the mime dancers rushed up and swayed on the dock and waved as the ferry cast away.

And the currency for that one week in 3000 lived, so far, across a life? Not much contribution to GDP, some sculptures, photographs of dancers, memory. And a story, which is currency itself.

In the survey of the good life discussed earlier, the category that appeared in every return was personal growth. More than a fifth of all returns came in this class, on average more than two per person in their ten choices. Regardless of our current deep embeddedness in the material economy, respondents put high value of opportunities for growth in both work and non-work domains, and across the whole of the life course.

To them, the good life was creative and satisfying work, it was learning and an active mind. It was crafting and engagement in painting, poetry, music, knitting and needlework, writing code. It was reading and listening to oral histories. It was gardening and growing things, managing and engaging with the ever-changing natural world. It was carpentry and repairing things, helping others to fix stuff. There was singing with friends, dancing, festivals, gigs, gaming. There was helping others, coaching sports teams, charitable work, giving talks. There was travel and learning about other places. Many of these activities were not what are sometimes called hobbies. They were about life itself, not simply filling the time and space of non-work.

Many of these are also what Stephen Kotler calls infinity games in *The Art of the Impossible*. There is no winning and losing when it comes to dancing or gardening, knitting or volunteering. There is no counterfactual thinking, no *if onlys*. There is only the path that lies ahead. And this will never end.

When we paint or walk, there is no failure or shame. There is no race that must be won. The author Bill McKibben has talked of his pleasure in cross-country skiing, where there is always someone ahead of you, to pull you on. There is also someone behind about to overtake; there is blue sky and snow on branches and the swish of skis on the snow crust. There is always change, we have anyway grown a little older each time we try. Bill finishes in the middle of the pack, quite anonymous, and says, "Literally no one cares how well I did, not even my wife. But for me, these are always great dramas."

Arne Naess was the founder of the Deep Ecology movement and told of these not-winning-games in a conversation with Alan Drengson. In Aikido, observed Drengson, the philosophy was to harmonise with the opponent, seeing the attacker as a partner. In Gandhian tennis, replied Naess, the sport was also about co-performance. Indeed, if the ball was returned in a way that was impossible to play, then the point was in favour of the would-be receiver. As a tai chi practitioner, I am not trying to win. I first saw the martial art being practiced in the early morning by quiet crowds in parks of China's cities in 1981. I practice to be attentive. I seem to be getting better, but there will never be a time when all will be learned.

Taitetsu Unno said of calligraphy, "There is no good or bad in calligraphy, only that it manifests the person writing it."

Just do your best, there is no need to be perfect or have to beat others. Remember, in the fictional town of Lake Wobegon, Garrison Keillor wrote that every parent believed their child to be above average. Unlimited riches in the neoliberal world are equally structurally impossible.

There is only performance, the slowth of being inside the thing we have chosen to pursue. The potter throws the clay, and every piece is different. The teacup not a perfect circle, the knitted jumper has smallest faults. In this way, there is always a hinterland to explore. Crafts and arts are never over, and so too give meaning, an additional reason for facing the future with anticipation.

Taisen Deshimaru said in *The Zen Way to the Martial Arts*:

> I am practicing because I am practicing,
> Create your own life, here and now,
> There is no victory, and no defeat.

An example of infinity games comes from anthropologist Wade Davis. In *The Light at the Edge of the World*, he describes harmony within cultures. In the annual ritual of the Akwé-Xavante of the Amazon, men and boys form teams of similar age to run the boundaries of their lands and rivers. The aim was to define the physical and social world in which they lived. Each team carried a large and heavy log, and the leading sides stopped often to allow others to catch up. Their expression of harmony came at the end. Everyone arrived together. Crossing the line first and alone was seen as failure.

What would happen, asked David Maybury-Lewis, should the race not occur, or if it never ended in a line? The culture would end, observed Davis.

These craft activities and infinity games involve attention, immersion, focus and flow, and these bring well-being. When you are young, with a beginner's mind, your potential is infinite. Who knows what you will come to learn?

Over time, though, as behaviours and thoughts become automated through habits, some options for growth open up, and others close. Some slip away without us noticing. Barbara Ehrenreich wrote of the collective joy of dancing in *Dancing in the Streets*, its archaic roots in ecstasy. Every culture, every person when young, dances, sings and chants, some to the state of exhaustion, and certainly to an altered state of mind. There is synchronicity and merger of whole groups, at music festival and gig, at carnival, at gospel ceremony and at the maypole, in the welcome of a Māori haka.

It is easy, as we age, to let horizons shrink, to let entropy overcome us. We lose energy, become smaller minded. Norman Fischer says in *The World Could be Otherwise*, "But our lives are larger than this. Great plans await each step of life."

And so, asked Mark Williams and Danny Penman of Oxford in their book *Mindfulness*, "When did you stop dancing?"

There are two phases to personal growth. The first involves high concentration and focus to build new neural networks and patterns. You have to devote resources to learn the new routines, the verbs and declensions, the colours of the palette, the song sequences of each bird, the multiple parallel sequences needed to drive a car, the dance moves. This in itself brings relief from the mental chatter, but it is hard work. After a certain point, automated flow kicks in. You can do the thing, use the skills, yet are immersed and watchful. Exploration continues, and the sense of satisfaction and contentment grows.

But then often comes a third phase: life intervenes, and you stop.

The 13th-century Persian poet and mystic, Rumi, founder of the Sufi traditions of spinning in harmony with nature to create true health, the term dervish meaning doorway, wrote:

> A secret turning in us,
> Makes the universe turn,
> Dance, when you're broken open,
> Dance, when you're perfectly free.

But as each life proceeds, you get busier, things become more complex and pressured. Activities seem to narrow, and you start to abandon the important stuff without realizing.

Inch by inch the territory is ceded, the book club, singing, exercise, dancing, all given up.

You say, I'll catch up next year. You think; I'll have that holiday or weekend away, and recalibrate.

The young love to dance, the immersion eliminates the inner noise, and the flow brings well-being. Nothing else in the world matters. You are in your body, you are with other people. You are smiling and laughing. In some cultures, Williams and Penman wrote, doctors don't ask when did you start feeling depressed.

Instead, they ask about why and when you stopped dancing. During the first pandemic lockdown year, Marea with Fred Again sang, "We've lost dancing."

So why did you stop dancing when you grew older, if it had been so much fun?

Contemplative Crafts and Habits Slow the Climate Crisis

The activities of personal growth bring contentment and happiness. Many are also characterised by the state of flow. They require attentiveness to the matter at hand, they result in immersion that down-regulates and quietens the default mode network. There is no need for flight or flight.

Contemplative habits also contain a secret sauce of well-being, they bring silence to the mind, a certain stillness, a slowing down in the face of an onrushing world. Most of the time, they happen to be low in material consumption too. So, it would be a surprise if modern commerce and politics was wholly supportive of such craft, ceremony and ritual. You are not spending enough. If you join a chorus to sing, walk for health, ring the church bells, attend a knitting class, run in a parkrun event, you would be happier, slower and find silence.

You also slow the climate crisis.

Many have written about the tea ceremony in Japan. It takes years to learn the rituals, the grinding of tea leaves and the pouring of hot water. Come in and sit quietly. Look around. There are ikebana flowers in the vase. The wood of walls and ceiling is jointed with precision; the paper door slides with a rustle. There is a view of maple and pine in the enclosed garden. The cups are porcelain clay, fashioned and fired, yet each differs. There are always subtle variations, in the weather, where we are in life. It is an open system with no fixed endpoint.

All that is required, for now, is to drink the tea. You cannot hurry. There is no shortcut to learning the skill and belief. We cannot trick our brain into learning something significant in minutes. Interestingly, it can become a journey of a lifetime.

Richard Sennett in *The Craftsman* spoke of the Isaac Stern rule. "The better your technique," said the violinist, "The longer you can rehearse without becoming bored." Getting good at something increases the motivation to continue being good, and this brings immersion and a sense of flow. There is space for imagination and reflection, and the steady evolution of skills.

Craft time is thus slow time. Craft, says Sennett, is "The desire is for a job well done for its own sake."

The making of a Japanese sword is a remarkable process. It takes several months, and by the end, each sword is said to contain soul and personality. The Komiya family has been making swords since 1786, and each day begins with everyone praying at the shrine in the yard. The wooden bellows are pulled in and out, the flames crackle, and the senior craftsmen beat the glowing iron, step back, dip the metal in water, hissing steam filling the air. The metal is heated again, folded many times, the impurities worked outwards. Their arms are speckled with burns, but the swordsmiths cannot stop. Their muscles ache and as they beat the sword it thins. A smaller hammer is then used, and a mixture of soil and charcoal powder is painted on the tempered blade. Its soul is folded in. The swordsmith inscribes the blade with their name, and now the sharpener works for weeks, giving balance to the blade.

At the end, the sword has a shine that seems to come from within. In this state of craft flow, a sense of self can disappear.

Time passes strangely, if at all. As attention increases, so the Default Mode Network is silenced by transient hypofrontality. It feels good because internal rewards are being made with the release of dopamine, serotonin and endorphins, oxytocin too if there are elements of social interaction. It is the parietal lobe that defines a boundary line between you and the world. It helps you navigate in space, defines you and the chair, you and the walls and air and light. Under extreme focus and flow, this lobe goes silent.

As these borders disappear, the brain concludes, as Stephen Kotler writes, "We are one with everything."

Alexandra Lamont and Nelinne Ranaweera of Keele University and the Royal Northern School of Music studied a 1000 amateur knitters and musicians. They were not experts, just committed to their creative crafts like you or me. It was found that engagement, social relationships, meaning and accomplishment all increased well-being and happiness. Knitting induced flow, and singing in groups reduced stress and increased happiness in all participants.

Again, many such crafts are low in carbon emissions. You substitute knowledge and physical labour for external energy, taking raw materials to make something. There is wool to knit, wood to carve, food to cook, paper and pigment for painting. There is no end to these roads, and thus no point where we would say we have won or the game is done. All the cognitive battles are personal: how to get better. We have been motivated, by self or by others, to get in the game. Learning then keeps us there.

Here's a story about a certain type of growth.

Walk this path to the edge of the village or town. Here are communal allotments, at the edgelands between houses and the railway line. Here are the small plots of enriched soil. Allotments bring people together; these jumbled plots are remnants of the Enclosure Acts, guinea gardens that waxed and waned, and became popular again. Up a dim lane overhung with ivy, by a field with rolled-up bales, a mix of design and wild chaos. Squash and maize, potato and sweet pea, raspberry cane over black plastic, riotous weed and waspy wildflower. One patch is all poppy and mallow, swaying red and purple in the warm breeze. Each shed differs, there is incinerator and stacked tool, raised bed awaiting the spade and fork. Shiny discs spin from bamboo poles to ward off birds; there is tall artichoke and wispy fennel, onion in razor-rows, a snapping flag. An axe handle is propped against a hut. It is seasoned, aged by wind and rain.

We have researched allotmenteers, and found gardening increases well-being and reduces material consumption.

Allotments have always been political spaces, statements of personal intent regarding the food system and local land identity. What the good life needs is more of this. More digging, more fitness, more healthy food. Allotments stimulate metaphors about the way we live, wrote Colin Ward, they are about looking after the earth and friendships, about self-reliance, planning for the future. Something more may happen. The silence opens up other ways of experiencing the world. The fabric of colour, sounds, deep values, all may change. It opens too new forms of behaviours, allowing you to lie back and listen to the wind in the poplars.

These lives, where nothing is really missing at all. There is wisdom in the fields, and you have not traded away the best years of your life.

Personal Growth: No Place for Perfection

In a slowed-down life, there is richness. Andy Couturier in *The Abundance of Less* wrote of the lessons in simple living in rural Japan, where people seemed to have a lot of time. People lived in rich and palpable contact with the natural world and slipped into timeless space when working at crafts of choice. There was intimacy with sounds, patience and endurance, a life often half in meditation. A craftsperson does good work for its own sake. Each was engaged fully, with hands and body with and brain and mind connected. Some were also engaged in the public good, crafting for people. Their activities produce few negative social and environmental side-effects.

Plato said, "All craftspeople are poets."

Crafting is also a process of self-transformation, as Peter Korn has written in *Why We Make Things*. It is not about just making things, it is making self. The creative effort and flow brings well-being and spiritual fulfillment. The conversation with materials puts thought into matter, and this dance of making means times distorts, actions and awareness merge and self disappears. When the object enters the world, it gathers associations of its own. Stories stick to it. In *Cræft*, Alexander Langlands observed that crafters of things flow into deep-time signatures. He wrote of the strimmer and the scythe. The function of both is to cut grass, yet only one allows the user to hear the birds and feel the wind. In ancient times, it was no wonder that the smiths were seen as magi. Through magic, it seemed, they knew how to melt rock and make metal. First iron alone, then blends for bronze, more mixtures to temper iron to steel.

The potter Bernard Leach and Sōetsu Yanagi, founder of the *mingai-kai* Japanese Craft Society, wrote in *The Unknown Craftsman* of pottery and painting, observing that these were fountainheads from which the creator could draw, and you always find fresh water springing forth. You never use it up. They wrote of the potter, Shōji Hamada, who said you hold crafts in your hand, but hang paintings high up on a wall. Hamada deliberately built his kiln so large that it was impossible to control all the internal conditions. He was, he said, at the mercy of an invisible power that emerges from grace.

Hamada said, "Objects are born, not made."

Leach wrote, "Good work proceeds from the whole person, heart, head and hands, in proper balance."

Yanagi and Hamada travelled to South Korea to explore the revered tradition of *hakeme*, where limewash is brushed on pottery to produce the rhythms of life, the wind and streams of cloud. There was beauty in irregularity, an art in imperfection where fine cracks were accepted along with glazes of uneven thickness. They visited a village where craftsmen made lathed wood objects and were astonished to see blocks of fresh pine, still sap-green, which when turned made a wet spray filled with the scent of resin. "Why do you use such green materials," they asked. "Cracks will come soon."

"What does it matter," said the craftsman, "I'll mend them." For him, the cracks introduced imperfections and uncertainties, letting each item become individual and of-itself. It also produced unique and beautiful art, observed Leach and Yanagi.

Such craft was something you never used up.

In late 2020, Kathryn Wortley wrote in The Japan Times of the ethos and culture of traditional crafts. Practitioners took time and showed patience across the years, building their living experience into each item. They were low in carbon.

Suzanne Ross said of *urushi* lacquerware, "It is not something you can master in a lifetime. It masters you."

And Valentine Brose said of the art of bonsai, "I learned to be humble, and you are never finished learning."

Such slowth means great diversity.

There is the time, when the hurry to consume is removed, to invest in doing existing things better as well as inventing new ways. Writer and artist, Hannah Kirschner, writes of life in the mountain town of Yamanaki in *Water, Wood and Wild Things*. A 14th-generation saké artist and tiny bar owner serves hundreds of varieties of sake, each flavour with its own type of cup, wooden, lacquer and glass; blond and floral. You slow down, and subtle diversity emerges. There is the tea ceremony, the vase in the *tokonama* alcove celebrating the current few days of each of the 72 seasons of the year. Attentiveness and focus increases when someone says, I've been practicing this for 20 years. Hannah Kirschner says, "This is one of the rare places I am a fully present, disarmed from distraction."

Slowth also means more time to be attentive in other spheres. Victoria Sweet calls for *Slow Medicine*, in which health professionals take time, think about the social context of individuals, and consider the value of non-medical interventions. This is medicine with kindness, where the focus is on growing, caring, tending, fussing and fiddling. Rather like being a gardener. The general practitioner and doctor, Gavin Francis, also writes of the value of time.

A patient says, "I've got three problems"; the GP asks, "and what's your fourth?"

All recoveries are unique, there are many landscapes of illness, so all we can do, he writes, is interpret experience, and work on what might be possible for the patient and health professional in co-producing a better future. Peter Beresford and Susan Carr have talked of the need for participatory health policy, a paradigm shift towards experiential knowledge and approaches for engagement. This turns experts away from telling people or patients what to do, and more towards being guides and gardeners.

In these open systems of learning, there is no endpoint, no arrival, no summit. There is continuous dialogue, between body and hand with materials for crafts, between body and world for tai chi and other martial arts.

There is also no place for perfection. You may produce something very good, perhaps almost flawless, but tomorrow, well, all could be different. And this, too, is why the years of life do not matter. There are always opportunities for learning and exploration, if we believe it is worth the effort. Or indeed, we are permitted or encouraged so to do by the social and economic environment around us. In Japan, the success and happiness of the oldest-old is partly because of the cultural emphasis on doing something cognitive every day. A hobby or crossword, origami or managing the garden.

In many a care home of the affluent west, where the old often feel abandoned, the opportunities for such growth seem more limited, perhaps even suppressed.

The importance of lifelong learning is embedded in the remarkable folk-bildung movement in the Nordic countries. These were launched in Denmark in the 1860s. The aim was to transform the lives of individuals and communities. People were to author their own choices, and so the educational approach emphasised well-being and knowledge, and focused on collective action and citizenship. Above all, the mode of learning was not so much about achieving personal targets as developing life itself. In Finland, this school-for-all at public expense is tied to the greater social commons of three-year maternity leave, food and health care in schools, and free travel for the young.

Personal growth is thus the creation of a new currency, a redefinition of wealth in the midst of affluence. It brings days of sweet contentment, the return to a golden age.

An extreme form of craft could be doing nothing, watching the world, falling asleep on a long summer's afternoon, or in front of the winter fire. There is great craft in making these attentive choices. Creating habits of immersion stops the dust of affluence.

Musashi was the most famed Samurai of all. He was said to have fought 60 sword bouts by the age of 30, defeating some without even drawing his sword. He spent the rest of his life in the early Edo period as garden designer, calligrapher, poet, ceramics maker and painter. He wrote in *The Way of Walking Alone*, "Do not ever think in acquisitive terms, do not be intent on possessing valuables in old age." Body and mind must be free to flow: "Let the mind swing peacefully, not allowing it to stop, even for a moment."

The Kannous Sutra says such a life becomes:

An unending sea of blessings.

~★~

9

PUBLIC ENGAGEMENT AND NEW POWER

The Race to Net Zero

Gambatte kudasai: Please, do your best.

[Traditional farewell to pilgrims, Japan]

Abstract: Now for the new power of public engagement. We cannot solve these crises alone. Yes, please do your best, goes the farewell to pilgrims in Japan. But the dilemma is this. We know the problems, we have identified some solutions. Now how to get more than 200 countries serious about the race to net zero? How, too, to create the social movements that move themselves. This chapter sets out the principles and mechanisms of engagement, the changes that come from new rather than old power. Public engagement means giving up of some authority and control, so this is going to be hard for some. A new typology of public engagement is discussed: how to go from passive and consultative modes to co-creation and transformation? Yet, there is good news. Many novel institutional models have been created, rather quietly and forming new structures of social capital. Devolved health and care institutions, climate action groups, farmer field schools, participatory civic budgeting, crowd funding, social prescribing, web-sharing platforms, microfinance for women's groups: all are models of new power than can help build regenerative cultures. Five guidelines are set out for slowth and regeneration.

DOI: 10.4324/9781003346944-9

New Power

So. There is this dilemma.

For you, me and everyone else. The problems are known: too much consumption producing rising carbon and other greenhouse gases in the atmosphere, fast-growing inequalities and unhappiness, economies destroying nature, and the spread of non-communicable diseases.

The solutions are also known. This is unusual. For many big problems, years of work are needed to discover and evolve technologies and practices, methods and social movements. We know what reduces emissions, what increases carbon capture. There is more talk about these as workable solutions.

But still the two hundred and more governments are going slow, much corporate finance is tied to fossil fuels, poor foods are being made and sold, forests are being felled for the soya to feed the animals in inhumane conditions. And there is another hole: a lack of global institutions with enough public support to marshall and guide 7.8 billion people to amend what they want from life, and what to do about it.

So, where to start? If you think this is only down to individuals, why should you act first and suffer? If you think this is for governments to act first, then someone needs to persuade them through social movements and agitation, or become them.

The individual transformation, the heroic journey, the low-carbon good lives, all need togetherness and social action. And now at warp speed. There is a window, but it is closing.

Where are these new social movements, this new local, national and international leadership, where the bravery rarely seen? We may find, if we search, some novel institutional models built around new power beginning to point the way.

Old power can be defined by central control and ruthless competition, by creating winners and losers, and by experts who hoard and protect. New power is defined by participatory combinations of many knowledges and worldviews, a sense of collaborative agency, calls to action to improve the world, and inclusion of experts by both learnt and lived experience who share, facilitate and learn more themselves by doing so.

A core theme to the examples of success centres on building a movement. Movements have narrative: they tell stories with values and morals. Henry Timms and Jeremy Heimans in their ground-breaking book #newpower have observed that "A movement is not a movement unless it moves without you." They indicate there are these ways to build a crowd on a common platform: don't try to connect to everyone; build a brand (idea) with a call-to-action; use the voices of members to ensure salience; lower the barriers to participation so that people can just join; think about the types of public engagement, and aim eventually for transformational; and harness stories and flashpoints by acting on behalf of whole platform of the engaged.

Perhaps then we may find the innovations to co-create regenerative cultures worldwide. We will need to start with place, seek diverse perspectives, design for circularity, reduce consumption, prioritise health, happiness and longevity, and utilise forms of new power to spread social justice.

It has been well established over the past 30–40 years that there is a considerable and positive participation premium. Public and private organisations who reach out to engage with their customers, consumers and members, with their staff and students, with wider public groups, all find that people generally want to engage and have something interesting to contribute in the form of ideas, time and money. They enjoy being creative and contributing to a greater good. They wish to help to make improvements to people's lives and the wider state of nature.

Equally, they do not appreciate forms of "we-washing" (analogous to green-washing using sustainability claims) or manipulation of engagement for the sole aim of promoting the interests of professionals or experts. But effective public engagement involves the giving up of some authority and control and the accepting of the inherent value of multiple knowledges and viewpoints within particular systems and cultures. Public engagement thus involves, at its best, thinking about new forms of power and relationships, and building a creative knowledge and understanding distributed across social groups.

If you want to be taken seriously by a crowd or community, taking them seriously is a good place to start, say Timms and Heimans. These kinds of outside-in approaches ensure that all are focusing on what their members, users or customers care about. The best public engagement operations do not mind not being in control, they try to ensure their consumers or members are. They build on the advocacy of the public to become a lighthouse around which others steer. Such movements move themselves.

This is the thing. Public engagement is not that easy.

It means giving up some existing status and power, often acquired after years of training and learning knowledge and norms. It means recognising and then abandoning a paradigm that defines expertise in a particular field as inhering only in certain individuals or groups. It means adopting a mode that values the knowledge and understandings of all people in a system. Public engagement, done well, requires effort and challenges everyone to think and act differently.

Is This Too Hard?

This is especially hard when years of training have emphasised a mode of knowledge and technology transfer from those who know to those who do not. This has been called a "deficit-model": valuable knowledge is transferred to fill a gap for recipients who are assumed not to know something, and thus this should be welcomed by them. Over time, knowledge sits with power and becomes self-referential, often missing how the world and the public have already changed.

Yet, paradigms do come to shift, often slowly, sometimes suddenly. Thomas Kuhn in *The Structure of Scientific Revolutions* defined a paradigm as a coherent set of bounded beliefs and theories about how a segment or sub-system of the world works. Most investigation and exploration from the sciences, social sciences and humanities sets out to fill the gaps within an existing and accepted paradigm. Then, out of the blue for many actors, something shifts. Someone thinks something new, develops a method, theory or technology, and boom. The paradigm flips. Some feel threatened, their careers and habits too tied to a particular set of explanations and evidence to change.

Some opponents become blockers, seeking to prevent change or discredit the new. Thomas Kuhn noted that a common feature of a paradigm shift is, "Fierce controversy, international name-calling and dissolution of old friendships." The physicist Max Planck had earlier stated this: "A new scientific truth does not triumph by convincing its opponents and making them see the light, but rather because its opponents eventually die and a new generation grows up that is familiar with it."

This Planck's Principle is often true, but change can be faster. Old power institutions or individuals can be overturned very quickly, perhaps more so today with fast and broad communication platforms and networks. And this is what can happen.

In 2015, Jeremy Heimans created *GetUp! Australia*, an online platform to make it easy for individual citizens to email their legislative bodies. Tens of thousands of individuals responded, writing for the first time to their elected representatives. They raised issues they cared about. One government politician appeared on national TV to condemn the campaign, saying, "There are hundreds of emails arriving in Senators' offices. They are beside themselves . . . This is highly irresponsible, this is spam." GetUp! remains issue based and not supportive of any particular political party. It has one million members, more than all national political parties put together. "Our work is driven by values," they write, "not party politics".

DonorsChoose is a crowdfunding web platform connecting the public to public schools (in the USA, a public school is publicly-funded, in the UK it is a private operation). DonorsChoose was started by a history teacher, Charles Best, in a Bronx school. It is a crowdfunding platform with explicit values and aims to reach those students and schools missed by existing institutions of support. It has allowed teachers to raise funds for what they felt their students needed. They have since funded 1.7 million classroom projects with donations from 5.2 million citizens, giving US$1.2 billion of support to 86,000 schools and benefitting 18 million students. This has been meeting needs missed by existing structures.

But responses were not entirely supportive. One professor of political science at Colombia University was dismissive: "We have vested school boards, mayor and superintendent offices with the authority to make decisions about schooling," because, he said, "Only we understand." Put another way, he was saying he did not trust the interference of teachers or the public in these matters. Some school boards have since banned crowdfunding, as it draws attention to the failing of their own structures and responsibilities.

And this is what can go sideways, inside institutions, when you start engaging the public. Old power tries to reassert itself. Here is an example from NASA.

The National Aeronautics and Space Administration Johnson Space Center in Houston launched its Open Innovation programme in 2010 in response to feedback from the US Congress that they were not being sufficiently innovative (a code for "being noticed"). NASA picked fourteen strategic and developmental challenges, and laid them out on an open innovation platform: 3000 people in 80 countries responded. NASA found that the crowd solved problems in 3–6 months instead of NASA's norm of 3–5 years. They found these analyses were impressively accurate, and breakthroughs led to renewed internal enthusiasm. But another internal faction quickly emerged, who viewed citizen science as a waste of time and a threat.

Some turned saboteur, others ignored crowd ideas when presented. Professional privileges and knowledge had been hard-won, and many did not want to give them up. Their instinct was to hoard information, not expose it to the scrutiny of an unqualified crowd, as they saw it. But Open Innovation continues to grow: the supportive group founded a space apps hackathon, bringing together 25,000 people from 69 countries. Similar open innovation platforms have been developed in some large private companies, and these too have resulted in internal divisions.

These kinds of barriers will need to be broken if worldwide progress is to be made on solving the crises of climate, inequality, nature, and food ill-health.

Wrote Hildegard of Bingen in 1152:

> I a fragile vessel, speak these things,
> Not from me but from the serene light.

Typology of Public Engagement

Public Engagement can lead to the formation of social capital and the development of regenerative cultures, where natural, social and human assets are built. Christian Wahl observes that "A regenerative future requires the capacity to listen and learn from diverse perspectives."

This is what Patricia Wilson of the University of Texas calls "ensemble awareness": the quality of presence, relational awareness, and effectiveness. She uses the concept of the social field, in which transformative outcomes are sought in both the outer and inner journeys. Deploying public engagement effectively does mean changes in both attitudes and mind. It implies generative patterns of practice, where creativity leads to new ways of seeing the world and acting in it. The story is reframed through public engagement, and people were able to say, "We did it ourselves."

Participation often can mean finding something out and proceeding as originally planned. The problem is not that people are hard to reach; it is that professionals tend not to question their own approaches. When they do, public engagement can mean developing processes of social learning that change the way that people think and act in the world.

When little effort is made to build local interests and capacity, then people have no stake in maintaining structures or practices once the flow of incentives stops or policies change. If people do not cross a cognitive frontier, then there is unlikely to be sustained change that might be counted as an improvement.

Public engagement is a multi-dimensional concept that can incorporate communication, co-creation, dialogue and the creation of social capital. Its deployment and methods and approaches have differed widely, representing variations in values and principles.

The first typology to express these variations was created as a *ladder of participation* by Sherry Arnstein to indicate how improvements to civic planning could occur. Since then, a number of spectrums, ladders and typologies have been developed and refined. All have sought to indicate a spectrum from passive public engagement (people are told what has been decided) to consultative modes (people are asked set questions), to collective and interactive (people work together in joint analysis), to transformative (where the worldviews and behaviours of all actors change), and finally to self-mobilising and connected (citizens take action independently of external institutions). A typology of public engagement is depicted in Table 9.1.

Passive-Informative is the easy default; Self-Mobilisation is hard. The former is old power, the latter new. An early characterisation of public engagement was the fashion for what was called public understanding of science, in which there is an assumed gap between science and society. The Royal Society called for a better understanding of science in business, government and schools, and concluded that "scientists must learn to communicate." This was the partial deficit method, assuming that groups of the public lacked something (knowledge, understanding, technology) and they just needed to be told about it by experts for their lives to improve. Good communication can do this, but it is not guaranteed to cause positive change.

One way to look at this typology is not that these types are mutually incompatible, nor that one is inherently best (or worst). One type may lead to another. It may be that engagement starts with communication, leads to the development of an audience, which then leads to more transformative engagement. This escalator approach can allow for the building of confidence in actors. It can allow time for individuals to relax into a participatory mode. Public engagement can then become a journey.

Peter Beresford is a leading researcher and commentator on public participation in health and social care and has noted, "This participation is not independent of concerns for representation, democracy, social rights, community development, provisions for unrepresented or ignored groups, and then state reactions as conflicts and competing agendas develop."

There are also many things experts and professionals will know that local people do not. Public engagement can thus be a way to co-create and share knowledge and change behaviours that lead to improvements. Knowledge is dynamic, developing and changing over time. In local circumstances, there are things we might know well: the changing of seasons, the timings of plants and the behaviours of

TABLE 9.1 Typology of public engagement

Types of public engagement	Characteristics of each type of public engagement
Passive-Informative	Public engagement is unidirectional, with information being pushed outwards. Information belongs only to professionals and experts, and thus external systems are unlikely to change. This type can be positive where the information is useful and interesting, but it can also be manipulative.
Consultative	Public engagement centres on asking question to gather information about the world. Organisations and professionals are in listening mode, and professionals are under no obligation to take on board people's views. Feedback is not guaranteed, and this PE may include bought forms of participation (for food, income or other material incentives).
Co-Created	Organisations and professionals work together with people in co-created platforms for joint analysis, development of action plans and formation or strengthening of local groups or institutions. Learning methodologies used to seek multiple perspectives, and groups determine how available resources are used. This increases total knowledge, data and perspectives, though often neither are changed by the experience. Public knowledge can feel like a threat to the status of professionals and experts.
Transformative	Public engagement changes worldviews and values, causing permanent shifts in choices and behaviours of individuals and groups. Organisations and professionals change, and bonding and bridging social capital founded on trust, reciprocity and institutions is built.
Self-Mobilising and Connected	Citizens and the public take action independently of organisations to change systems. This might emerge as deliberate resistance to existing structures; it might lead to the development of contacts with external institutions for resources and technical advice they need, but retain control over how resources are used. This can mobilise large numbers of people and can lead to systemic changes in power and outcomes.

animals, the quality of local leadership, the potholes in roads and drainage patterns from the hills in storms.

Paul Mason has written in his *Post Capitalism* that "Capitalism is a complex adaptive system that reached its limits of capacity to adapt." It no longer can cope with the costly externalities and damage; it no longer seems to have the institutions to address the great gaps in inequality that it has opened up. Yet, at the same time, collaborative production continues to rise and spread, forms of production not responding to markets. Wikipedia and its 27,000 volunteer contributors and editors is an example of a global, open-access knowledge commons.

No one in a system can know everything: only engagement leads to sharing and mutual understanding.

The collector and teller of stories, Ruth Sawyer, wrote in the 1940s, "This is living art, this is creative art . . . Never give it direct, and never hoard it, or all its virtue will be gone."

Types of Learning

It has been argued that public engagement has become harder today precisely because of breakdowns in social capital, civic disengagement, the loss of volunteering and hyper-local institutions, and the fracturing of social networks. Poverty, hunger and the need to hold multiple jobs in the gig economy all set the scene for many people, both in affluent and poorer countries. No one, it could be said, is against meaningful praxis and transformation, but the time for engagement may simply be unavailable when people are seeking to retain one or more jobs.

Public engagement is about individual and collective learning. We learn something from someone who knows, we come to learn together with others. And the process can be transformative. New options in life appear, new behaviours and actions may follow. In the early 1970s, E F Schumacher asked in *Small is Beautiful*: "If education is to save us, it would have to be education of a different kind; an education that takes into the depth of things." Chris Argyris and Donald Schön observed that addressing sustainability and social justice needed new systems of learning, especially on "how to learn how to learn."

We can distinguish three levels of learning, in which there is a transition from learning information to meta-learning, and then to epistemic learning.

> First order learning: we learn about things within a particular boundary without challenging assumptions or values (work within a paradigm);
> Second order learning: we as learners critically examine beliefs, values and assumptions, leading to a shift in the way we see ourselves and things in the world;
> Third order learning: epistemic learning, in which a shift occurs in the operative ways of knowing and thinking that frame our perceptions of the world, involving thinking about the foundations of thought itself.

One of the best examples of the institutionalisation of these approaches to learning was at Hawkesbury, now part of the University of Western Sydney. This 20-year experiment on methodological pluralism sought to transcend the limitations of positivism and reductionism. Said Richard Bawden: "Together we could learn how to see the world differently and in the process discover just how difficult a transformation this is." The Centre for Systems Integration and Sustainability at Michigan State University then emerged under the leadership of Frank Fear and

Richard Bawden: their aim was also to shift to effective "ways of inquiring about the world."

Can this change the world, enough at least to address the contemporary crises? Well, we know forms of public engagement can be transformative. Let us turn to three examples: bibliotherapy, storytelling to young people, and a university course on happiness.

Bibliotherapy in prisons has become a successful educational programme. It was established by Robert Waxler of the University of Massachusetts, Dartmouth. The *Changing Lives Through Literature* programme sought to engage prisoners with many convictions and a high likelihood of reoffending on release. The aim was to use stories and reading. These were men, to begin with, who had little voice in their communities, and no obvious means to imagine or create new lifeways. With the support of the judiciary and probation service, the programme engaged them in reading and group discussions. The rate of reoffending fell by a half, and the programme subsequently spread to eight other states in the USA, with programmes for men and women, adults and juveniles.

Individuals discovered renewed interest in family ties and education, enhanced self-esteem, and created different possibilities for their futures. Characters in stories became both inspiration and teacher. Robert Waxler observed, "All stories, at their best, have an ethical standing." Such activity is extremely dangerous, he added, "It can change lives, both inside and out."

One prisoner wrote, "There were books I could not put down. They kept me interested in the positive aspects of it all." It is of course simplistic to use reoffending as the only measure of success. It had changed lives, but perhaps not the society to which the prisoners would have to return. Yet, several thousand prisoners have completed the 8–12 week programme. Imprisonment and probation are costly; public engagement is much cheaper.

A variant on this bibliotherapy is the University of Leeds' *Writing Back* programme that is tackling loneliness through intergenerational age groups. Students are paired with older residents in Yorkshire, having been identified by public libraries. The project was established by Georgina Binnie, who had exchanged letters with her grandmother during her time at university. This method is used to link 200 students a year to the elderly and lonely. Letters are exchanged every two weeks, and participants are encouraged to write about hobbies, travel, poetry and music, also to share photographs. Evaluations show that 95% of both students and older participants said the project had improved their well-being. Many long-term friendships have been formed.

Malcolm Green is a storyteller and environmental educator and created a story platform with primary school children in a deprived part of Newcastle to awaken curiosity about nature. The problem situation was this: the nesting kittiwake colony on the nearby Tyne Bridge had led to local calls for their enforced removal. Some people were delighted by their presence; louder voices insisted they were a nuisance. As part of the voices of the River's Edge project, Green spent a term working with children who had hardly ever went outdoors and certainly did not visit the countryside. The project centred on observational storytelling.

The children were taken to watch the birds, and then were asked, "What kinds of questions would you like to ask the birds?" They were asked to wonder what was home like for the birds; where did they fly to; what would it be like for young birds in the middle of the ocean; how do teenage birds spend time together?

The children made maps, wrote their own stories and poetry, imagined they were birds, and then performed a story at a public festival in Gateshead. They had inhabited the emotional world of the birds. It came to pass that the Council dropped plans to eradicate the kittiwake colony. On an outdoor trip later that year, the children noticed gulls on a pond. A new child in the class said they must be kittiwakes. Another replied, "Don't be silly, they won't have come back from the sea yet." Storytelling was a way to understand and act.

Bruce Hood of the University of Bristol's School of Psychological Science launched an undergraduate course on the Science of Happiness. Its aim was to educate and involve students in what is proven to make us happier. The course was based on a similar exercise at Yale University and was open to students of all disciplines. Students obtain credit towards their degrees but do not complete exams or coursework. Instead, they engage in well-being assessments, join happiness hubs and write a group report. They also have to carry out practical tasks: performing an act of kindness, chatting to a stranger, taking time to savour an experience, exercising, sleeping well, and writing a thank you letter.

Evaluations of the course subsequently found that a significant proportion of the thousand students taking the course had ended up themselves feeling happier.

One student observed,

> It's made me feel more conscious of my happiness. I've thought a lot about success and happiness. Lots of people think they will be happy if they are successful. We can turn it around: if we are happy we are more likely to be successful.

On a nearby branch, a hototogisu cuckoo once was singing:

> To whom shall I show it,
> To whose ears shall I bring it –
> This dawn in the mountain village,
> This music of the hototogisu as it greets another day?
> [*Lady Sarashina Nikki,* As I Crossed a Bridge of Dreams, *1008–c1059*]

Citizen Science Improves Health and Environments

Recent years have seen a flowering of methods, approaches and projects involving citizens as scientists. These projects and programmes are helping to engage large numbers of people in gathering unprecedented quantities of good quality data. The best forms of citizen science represent socio-technical co-evolution and innovation, bringing changes in values, norms and behaviours, resulting in increased awareness, learning and trust, and then adoption of new solutions by the public. Citizen science has the power to change behaviours.

The European Citizen Science Association has published Ten Principles for Citizen Science to summarise best practice. It recognises variations to continuing engagement over time. It still defines citizen science as where "citizens are actively involved in research, in partnerships or collaboration with scientists or professionals," and where "there is a genuine outcome, such as new scientific knowledge, conservation action or policy change." The US Citizen Science Association goes a little further, and states, "We value all Citizen Science projects, from scientist-driven to community-driven, from contributory to co-created."

Both the ECSA and CSA state public engagement in citizen science should actively include citizens, projects should have generic science outcomes, and both professional scientists and citizen scientists should benefit from co-learning, personal engagement, social interactions and satisfaction with the potential to influence policy.

Citizen science has been deployed in many ways. Community volunteers have been collecting data on poverty, nutrition and health in the Philippines, monitoring water in Peru, implementing climate-smart agriculture in Brazil and using low-cost rain gauges in Sub-Saharan Africa. In the UK, it has become common for large numbers of the public to report sightings and first seasonal arrivals of rare or migratory insects, birds and mammals, such as stag beetles, bats and swifts. The voluntary British Trust for Ornithology in the UK uses five initiatives to increase public engagement on data gathering for birds and their habitats: Bird Atlas, Bird Track, Breeding Bird Survey, Garden Bird Watch, and Garden Nesting Survey. The public engaged in these projects have produced more than 30 million bird records.

Here are some other exemplars.

The Belgian CurieuzeNeuzen project (curious nose, meaning in local dialect "nosing around") began in 2016 and has engaged 5000 residents of Antwerp directly in air pollution monitoring, with a further 35,000 deliberating on the results, causes and potential solutions; the project has gone on to engage the public in eutrophication monitoring and the impacts of urban heat and drought on gardens.

The *eBird* citizen science initiative from the Cornell Lab of Ornithology and National Audubon Society was launched in 2002, and by May 2021 had grown to one billion bird observations, with 400,000 public participants now logging 100 million bird observations annually. This has resulted in the production of 150 peer-reviewed papers. Additional public engagement projects from the Cornell Lab include Urban Birds, Nestwatch, Feederwatch, and a Great Backyard Bird Count.

Drain detectives in Victoria, Australia, who monitor water flows and pollution to beaches, the 50 citizen scientists over two years using photographs, observations, water sampling and smartphone data platforms.

The seven-year Mildew Mania project has been run by Curtin University in Western Australia and works with 220 schools and 16,000 students to establish trap crops of barley and wheat in school grounds. Primary-age children grow the crops between June and October and monitor them for powdery mildew. If the disease appears, samples are sent to Curtin researchers to map the spread of mildew strains across the state, helping timely solutions to be developed for farmers.

An innovative agricultural programme in northern Ecuador led by Steve Sherwood and Myriam Paredes and involving international organisations and universities was focused on knowledge-based and socially-oriented interventions for pest management in potato cultivation. Farmers were known to use more than 50 fungicide and insecticide compounds, 85% of which were classified as highly toxic by the World Health Organisation. Farmers did not use personal protective equipment for cost and availability reasons, and by observation, it was clear that farmer exposure through the skin was high. National data put recorded pesticide poisonings at some of the highest rates in the world. A non-toxic fluorescent powder that glowed under UV light was added to back-pack sprayers, and the team returned at night with UV lights and video cameras to identify exposure pathways.

They found pesticides on young children, on clothing, throughout each house, in beds and kitchens. Says Sherwood, "More than any other activity, the participatory tracers inspired farmers to take action themselves." Farmer field schools were then used to create low to zero-pesticide approaches for farmers through participatory action learning.

Science and Technology Backyard Platforms (STBs) were established in China to increase the sharing of knowledge and skills between scientists and farmers. STBs bring agricultural scientists to live in villages and use field demonstrations, farm schools and yield contests to engage farmers in externally-developed and locally-developed innovations. Over six years, STBs resulted in yield increases of crops and reductions in input costs. Now 71 STBs operate in 21 provinces, covering a wide range of crops. Some 50,000 small farmers have been engaged and benefitted. Evaluations show the value of in-person communication, the emergent socio-cultural bonding, and the increased trust developed amongst farmer groups and scientists.

The MyShake project was developed by Richard Allen and colleagues at the Berkeley Seismology Lab. It is a citizen science project to provide earthquake early warning in earthquake-prone regions. In three years, it has built to 330,000 users engaging via an app. Earthquakes are being detected and located, and the magnitude is established in 5–7 seconds after origin time, and alerts are delivered to smartphones in another 1–5 seconds. MyShake creates personal alerts for protection (drop, cover and hold on), and rapid institutional action, such as automated slowing of trains, opening and closing of pipeline values, and readying of emergency response personnel and equipment.

Mad studies began in Canada and have spread to the UK and elsewhere. The aims centre on resistance and understanding and seek to create alternatives to simple medical models of mental ill-health. The value and emphasis is on first-person knowledge, building alliances and creating platforms for radical change. Peter Beresford and Jasna Russo state Mad Studies offer, "Insights for advancing understanding of experiences of madness and distress from the perspectives of those who have had those experiences, and also explores ways of supporting people oppressed by conventional systems."

It is now clear that this richness of citizen science can create significant economic value. Elli Theobald and colleagues from the University of Washington, Seattle, quantified the value of 1.3–2.3 million citizen scientist volunteers on 388 biodiversity projects at US$0.7–2.5 billion annual contribution to the economy.

> How can I possibly sleep, this moonlit evening?
> Come my friends,
> Let's sing and dance, all night long.
>
> *[Ryokan, 1758–1831]*

Emergent Institutional Models for New Power

To what extent have these systems of innovative learning and research involving public engagement been institutionalised? And have they led to new expressions of power? There are two possible processes: forms of public engagement could lead to emergent institutional models. And new institutions might be developed in order to increase public engagement.

Participatory Budgeting is a fabulous social innovation in public and civic engagement at the city or regional level. It began in Venezuela and Porto Allegre in Brazil and has since spread to many countries, including India, Indonesia, Kenya, Mexico, Peru, Philippines, Poland, Portugal, South Korea and the USA. It is centred on a simple idea: civic authorities engage their public to decide how to allocate and spend public money within their cities, regions or neighbourhoods. Participatory Budgeting in Brazil, for example, tends to be a year-long process, in which citizens exercise voice and vote. They negotiate amongst themselves and with government. More than 120 cities and communities in Brazil alone have adopted participatory budgeting, some now with 20 years of experience.

The simple idea of giving local people the power to decide on priorities for public expenditure in their own places has led to multiple outcomes. At its best, participatory budgeting produces a school of democracy, where citizens learn to deliberate, study government functioning and engage in democratic practices. It is, however, least likely to be successful in single-party states. Participatory Budgeting programmes are strongly associated with increases in health care spending, decreases in infant mortality and growth in numbers and membership of civil society organisations.

Participatory Budgeting seems to produce benefits in six areas: stronger civil society by increasing the number of groups and their links with government; improved transparency by increasing citizen knowledge; greater accountability and shared interaction; improved social outcomes especially for underserved communities.

"Participatory Budgeting has generative effects within civil society," conclude Michel Touchton and Brian Wampler of Boise State University. There should, they said, be more of it.

The Place-Based Climate Action Network in the UK was established by the University of Leeds. Its aim was to provide an institutional architecture situated between national government and individual households. The model of independent commissions supported by commissioners drawn from the public sector, business, charities and universities has been adopted by cities, counties and sub-regions, including Aberdeen, Belfast, Croydon, Edinburgh, Essex, Kirklees, Leeds, Lincoln, Surrey, Yorkshire and Humber. This close engagement with the concerns and interests of the public in their regions has ensured the opportunities for addressing the climate crisis are embedded in a wider set of social, economic and cultural structures.

Climate action is not just about developing and adopting low-carbon technologies, it is also about inequality, social justice, behaviour change, volunteering and sharing economies, and visions for new ways of living. Many Commissions began with a target of Net Zero for 2050, echoing national and UN targets, but after the Glasgow COP26, ambition has increased. The Yorkshire and Humberside Commission has set a target of 2038 in their late-2021 report; the Essex Commission has published more than 100 recommendations for local government, businesses, schools and other institutions across the county.

PatientsLikeMe is an online community of 930,000 people with 2900 conditions and diseases. The mission is to improve the lives of patients through knowledge derived from real-world experiences and outcomes. Begun in 1998 by one patient, it has grown to become the world's largest community-health management platform. Each individual is seen as a citizen scientist. Data is collected and quantified and provides context on lifestyle choices, social and demographic conditions, and the effects of treatments on individuals. This gathering and sharing empowers patients with their own agency. It is thus said to be transformative.

Jos de Blok established *Buurtzorg* community health care in the town of Enschede with four nurses in 2006. He had a simple idea for the social enterprise: eliminate bureaucracy and back-offices and managers, and give teams of nurses the authority and responsibility for providing care to groups of patients in particular neighbourhoods. Teams are connected through a web portal that enables nurses to share information and knowledge, and extend and receive support. Today they have grown to 500 self-governing teams, 10,000 nurses and 4500 home-help workers, caring for 100,000 people. They have only one stated objective: delivery of the best and most appropriate care.

The patient-centred model of community nursing has been found to be good for patients, carers, general practitioner doctors and other health professionals. The service is more responsive to specific needs of patients, with community nurses able to make operational and clinical decisions. Buurtzorg has high patient and employee ratings and is now the most satisfied workforce of any Dutch company with more than 1000 employees. On average, Buurtzorg nurses end up using only 40% of the care hours they are allocated per client/patient, realising considerable local and national savings for health care. Healthcare systems across the world are trying to adopt primary care service models that will reduce health costs and keep as many people out of hospital as possible. There are teams now in the Netherlands, Sweden, Japan, the USA (Minnesota) and France.

As we saw earlier, there has been remarkable innovation in rural social capital across the world, particularly in emerging economics. A successful public engagement innovation has seen the spread of informal microfinance systems embedded in local groups. The largest numbers of groups have been formed in Bangladesh (1.8 M groups), India (4.16 M groups) and Pakistan (0.12 M groups). The leading innovator was Grameen Bank in Bangladesh, later joined by the third-sector organisations Bangladesh Rural Advancement Committee and Proshika. All groups work primarily with women, and members of groups save each week in order to create some of the capital for re-lending.

A major change in thinking and practice occurred when banking professionals began to realise that it was possible to provide micro-finance to poor groups and still ensure high repayment rates. The systems work on trust, and payback rates typically reach 98%. Grameen has nine million members in groups spread over 81,000 villages: 97% of its members are women. BRAC has five million members and takes a deliberately integrated approach to poverty pockets, especially in wetlands, on riverine islands and for indigenous populations. Through a single platform they provide agricultural and skills support, education, legal services, health care and loans. More than 130 of its women members have been elected into government structures.

It is important to note, nonetheless, that public engagement has often emerged as deliberate resistance to existing institutions, norms and political structures.

In the health and social care sector, the PatientsLikeMe and Shaping Our Lives platforms began as user-led projects. The latter is now a Community Interest Company led by disabled people. The key aim of Shaping Our Lives is to give stronger voice to individuals with multiple identities and disadvantages. They note that 70% of the UK health spend and most of adult social care is for people who are disabled. They seek to consult, engage and co-produce.

It is also true that public participation may not need external institutions to be making deliberate efforts to mobilise the public. Resistance may occur in different ways, such as after the first #MeToo social media posting by Alyssa Milano in 2017. The Stolperstein project ("stumbling stones") takes this resistance into the space of public memorials. Begun in 1992 by Gunter Demnig, concrete cubes with brass plates inscribed with the names and life dates of victims of Nazi persecution and extermination have been installed on streets. Some 75,000 have now been installed in 1200 cities and towns, making these Stolpersteine the world's largest decentralised memorial bringing daily engagement by walkers.

In *A Paradise Built in Hell*, Rebecca Solnit has called aversion to public engagement, "elite panic," and considers it to be grounded in recent social change. Those who have advanced their personal interest during the period of neoliberal economics cannot believe other people will not also behave selfishly. They do not understand why people should be involved or engaged, nor do they understand why being kind is a way to build mutually beneficial relationships.

This elite panic, this fear that civilisation is only skin-deep, fosters a sense that people cannot be trusted. Frans de Waal called this a veneer theory. Rutger Bregman recounts the famed Ata Island episode, when six Tongan boys in 1965 were stranded by a storm on an island far from home. It was a decade after William Golding's dystopic tale, *The Lord of the Flies*, had been published. Put the people in charge, and things will fall apart seemed to have been that message.

The Tongan boys, by contrast, divided tasks, set up their own government, grew food, built rainwater harvesting structures, made a gym, and designated a cooling-off zone on a cliff for those who needed to get away. They tended a fire permanently and came to be rescued a year later.

But if you stand up for human goodness, this can invite accusations of naivety and weakness. It is easy to be a cynic. Old power is defined by central control and ruthless competition, by creating winners and losers, and by experts who hoard and protect. New power is defined by participatory combinations of many knowledges and worldviews, a sense of collaborative agency, calls to action to improve the world, and experts who share and facilitate.

Rutger Bregman further said, "For most of our history, we did not collect things, but friendships."

Good Guidance

Participatory approaches leading to greater engagement could help change institutions, their sectors and even have consequences across the world. One approach would be to launch more citizen assemblies, youth assemblies and parliaments, extend city-wide participatory budgeting, establish more place-based collective institutions for change, ensure there are more compacts between institutions explicitly seeking inclusive, innovation and impact. "It is not enough to have facts on your side," noted Henry Timms and Jeremy Heimans. Transformative public engagement will need to be part of the picture too.

For people to invest in collective action and social relations, they must be convinced that the benefits derived from joint approaches will be greater than those from going it alone. External agencies, by contrast, must be convinced that the required investment of resources to help develop social and human capital, through participatory approaches or adult education, will produce sufficient benefits to exceed the costs.

Elinor Ostrom in *Governing the Commons* put it this way: "Participating in solving collective-action problems is a costly and time consuming process. Enhancing the capabilities of local, public entrepreneurs is an investment activity that needs to be carried out over a long-term period." For initiatives to persist, the benefits must exceed both individual costs and those imposed by free-riders. There will also need to be new forms of social capital.

These structures can be called platforms: stages for actors to perform together. This will include for wider public deliberation, such as the successful *An Tionól Saoránach* Citizens' Assembly of Ireland, established by government in 2016, with each topic-focused assembly comprising 99 members plus the chair. David Orr of Ohio's Oberlin College, put it this way:

> Now we have to learn entirely new things, not because we have failed in the narrow sense of the word, but because we succeeded too well . . . What must we learn? We must learn to embrace a higher and more inclusive level of ethics.

Norman Fischer in *The World Could be Otherwise* tells of care volunteers at their Zen Hospice in San Francisco: "We tell them they will receive more from their service than they give." The patients somehow act unintentionally as guides to the volunteers, who find themselves opening up and feeling whole. This small world of a hospice, on the outside looking like only a place for death, ends up being a gift.

Public engagement needs this type of language and tone of kindness and generosity. It is about togetherness, forming bonds and trust to solve problems, and building natural and social capital. Universities need public purpose and social action.

To many organisations, it looks too hard to do. It undoubtedly means giving up some power and much certainty. If public engagement is to involve people, then their perspectives, ideas and views matter. They will changes processes and priorities by being able to express wishes and wisdom, and thus help create new knowledges. Public engagement brings a positive premium. It changes minds, institutions and environments. It can lead to improvements.

Here, then, are five simple guidelines for transformative platforms to create low-carbon good lives:

> *Choose* methods and approaches carefully using the typology, and ask: what can be done to escalate engagement and hear more voices?
>
> *Think* about how public engagement can transform existing systems, structures and institutions towards greater sustainability and equality; the process will change the people involved, so tell them in advance what could now be achieved.
>
> *Be flexible* and open to changing research and education systems, and modes of institutional working: if you involve people, they will bring new ideas.
>
> *Ensure* the principles of co-production are part of all research and education activities. Power structures may constrain, but they may also change with increasing clarity and breadth of voices working together.
>
> *Tell stories* about impact and outcomes, and ask: what was surprising, and how did the world and its institutions change?

I can say at once,
That I want nothing to be spared.

[*Guðrun Osvifsdottir*, The Saga of the People
of Laxardal, *9th to 11th C, Iceland*]

~*~

10

TRANSFORMATION

Achieving the Low-Carbon Good Lives

How terrifying it will be,
When the wealth of all this world stands waste.

[*The Seafarer* (anon), c970 CE, Old English]

Those days are past,
All the pomp of the kingdom of earth . . .
Fallen is this host, good times have passed by,
Just as now, each person in the middle world.

[*The Wanderer* (anon), c970 CE, Old English]

Abstract: The book concludes on transformation. The end is not near, if new ways to organise are found, new stories told, new aspirations for the good life shared and promoted. Low carbon behaviours produce the good life, not just giving up the high material consumption that caused these crises. So come now, let's dance. Peace is not freedom from problems; they will always come. It is a wide space and quiet mind. So put a stone in your pocket, a reminder to be more present, and slow down and reconnect. And tell stories, for this cannot be done alone.

DOI: 10.4324/9781003346944-10

The End Does Not Have to Be Near

One way or another, everything will change.

It will be a journey out and back, into the dark woods, over the sea, down a rabbit hole, escaping the problems of past habits. We can be sure of this. We live, and later to each will come to a moment when the things accumulated in life will then be worthless. We all started in the same place, with a clear, bright beginner's mind. There were unlimited possibilities, both good and not so good.

In these recent years, many have added up to not-so-good for the Earth. You've heard this kind of quote before: someone out there says, "I don't care, I'll be dead by the time anything bad happens." It is conceivable now they won't be.

About now, the path ahead splits into three. We do have choices.

Along one route, the material economy goes on growing forever and nothing bad happens. The path winds upwards through meadows and pastures, and the sun always shines.

Along another, the Earth suffers system collapse. The ecosystem services we relied upon cease to function. This path comes to the edge of a cliff, and the only way is over. Everything social and economic changes.

Along a third way, we find the means to act quickly to change our ways of living. It is hard to give up ingrained habits and hopes, but low carbon and attentive good lives seem attractive, as do happy longevity, greater equality and stable populations and economies.

"The end is not near," wrote deep ecologist Arne Naess, but actually it is for some futures. Path one is physically impossible, path two quite undesirable. There are many types of elixir along the third way. But from this vantage point, it could look like there will be only thorns and growling beasts.

Now is the time to go deeper into the forest. Guidance would help, stories to open the way. If sufficiently attentive, we may find they are already there, held in the land itself. Eva Tulene Watt was a White Mountain Apache born in 1913, and she explained to anthropologist Keith Basso how their lives had hummed with life and vitality, how the land was dense with guidance on how to live well:

> Those stories are tracks, lots of tracks,
> Those stories are people's lives.

The second encyclical of Pope Francis, *Laudato Si*, was subtitled, *Care for our Own Home*. He observed that "The world is a joyful mystery, not a problem to be solved." He also set out a congruence of mind and land: "The ecological crisis is a summons to profound interior conversation."

It is a wondrous place, this home, but people across the world are starting from different places. Jack Kornfield wrote, "A beauty calls to us, a wholeness that we know exists."

Places are made sacred by the repeated actions of people over time. This offers hope. In northern Lithuania, 100 kilometres inland from the Baltic, the Hill of

Crosses was first celebrated as the site of an 1831 uprising. Over the years pilgrims brought cross, crucifix, effigy and rosary, and today more than 100,000 crosses are crowded on this single hill on the plain. There is a monument from the Pope, and people recall the times when Soviet authorities tried to destroy the site. Large numbers of pilgrims visit the hill each year, adding to the many crosses.

In 1987, people in Estonia launched the four-year-long Singing Revolution, a way to celebrate culture and identity as well as opposition to the Soviet rule that began in the 1940s. People began singing poplar songs in public places, and this in turn led to greater engagement and togetherness. Then in August 1989, two million people in Estonia, Latvia and Lithuania formed a human chain along the Baltic Way, attached by holding hands on a route of 690 kilometres. Independence movements in all three countries grew, leading to formal independence declarations in 1991.

The elders of the Haudenosaunee confederacy (once given the name Iroquois) consider the effect of each decision they make to the seventh generation, roughly 175 years hence. Looking back, there have been 200 Tadodaho spiritual leaders, a long linked line, holding hands again. The incumbent in the 1980s, Grandfather Leon Shenandoah, said to Joan Halifax in *The Fruitful Darkness*, "We think of those not yet born, whose faces will be coming from beneath the ground."

Starting With Regenerative Culture

Every culture's actions bring side-effects, some positive, some negative. At some point, when too big, too dominant, the negative side-effects can come to outweigh the positive, and systems begin to fail. When natural, social and human assets are lost and not rebuilt, trouble is coming.

As the prospect of dusk falling as these great crises of climate, inequality, nature and food ill-health continue to roar, we are going to need mass collective efforts of redesign and regeneration. Sustaining things as they are is not enough. The world needs improvement. We need it. This is what life usually does. It is self-creating and self-generating, a system characteristic called autopoeisis by the great Chilean biologists and systems thinkers, Humberto Maturana and Francisco Varela. They wrote, "The world will be different only if we live differently."

Their phrase, "bringing forth a world through living as relating," well describes the human condition.

A new form of world awaits. All it needs is active redesign and collective action.

Thinking in this way needs a tale to tell it slant. "When is a tree?" asks designer and writer Stuart Walker of the University of Lancaster. What is a nightingale for? When is an electronic product? Does it speak to the mine and spoil tip, and deadly working conditions, was it later discard and landfill, was it only convenience in your hand? *Thinking Like a Mountain* was one of Aldo Leopold's most famed essays. If you kill the wolf, he reasoned, and the green fire of her eyes goes, then the deer fast increase in numbers, the vegetation is stripped away, the soil erodes in the rain, and the mountain soon slips into river and sea. To think like a mountain is to think of the whole system, the watershed, the city part, the trees on a tower block, the tall prairie grass and Leopold's mountain.

David Wahl asks, "How do we collaborate in the creation of diverse regenerative cultures," and the key word here is "we." For this can only be done collectively, through the transformative power of self-help groups, collaborative networks, civil society organisations, neighbourhood groups, self-built homes and communities, community wind farms, wildlife groups, and so very much more. These are all what Paul Hawken called the planet's immune response.

Public engagement to create this social capital that is good at regenerating nature is *The Great Work* that Thomas Berry had in mind. There is transformative power at the heart of social innovations. And as we have seen in earlier chapters, this social innovation may be place-based, as was customary in human cultures. It may also be web-based today, with the potential to grow very fast.

Where might we look for inspiration for projects of design? As we have seen in earlier chapters, we can look back at other cultures, we can look sideways at contemporary innovators.

We can look to natural systems too, and try to mimic key principles of design for more regenerative cities and settlements, buildings and economies, for ways of living too. We have already talked of the circularity of regenerative agriculture. The same principles of building natural assets, seeking synergistic interactions, helping one component help another, all these come into play in biomimicry and biophilia.

Stuart Walker said to be attentive, and you will see abundance, beauty, a world aglow: "Under the warm and wide embrace of a summer sun, one and one almost never make two."

As Wesaw earlier, both biomimicry and biophilia are about one and one making three or even four. Rob Hopkins was onto something when he created The Transition Handbook some 15 years ago. A journey was needed, it has to be collective if it was to impact the whole Earth; it contains elements of personal action and persuasion. This thing cannot be done alone, and this is much harder than escaping to try to live well within false walls and borders. The Transition movement has grown in size and coverage and has maintained the central ethic of inclusive change to improve lives. Rob Hopkins and Naresh Giangrande called for action to create "nourishing and abundant futures," and less of the narrative on how dreadful the future could be.

So, what can biomimicry and biophilia offer?

Biomimicry is about bio-inspired design and products: glue from mussels, green chemistry, optically-clear tough glass from abalone, anti-bacterial compounds, and much more. Biophilia is a recognition that nature gives many services and benefits, it is home after all.

Biophilic Urbanism, as Phillip Tabb has recently written, is about living in conditions of immersive nature, about daily doses of nature, and addressing structural inequalities. In the early years, the term biophilia was deployed to imply some kind of genetic link to the environments of human origins. This today is no longer a part of the argument. All we need to know is biophilic principles centre on greater access to nature, especially in cities, to sustainable and healthy lifestyles, and to the building of responsibility and contributions to the natural world.

There are many instructive examples. Phillip Tabb describes the Makoku floating school in Lagos lagoon, made of wood and bamboo with solar panels and rainwater harvesting, the children coming by canoe. There are Singapore Park Connections, 300 km of linear garden and loop in the dense city, trail and nature corridor that are less garden in a city than city in a garden. In Milan are the Bosco Verticale, 900 trees on terrace and balcony of two residential towers. In New York, the High Line Park, 2 km on the former NY Central Railroad spur, 9 m above the street, a promenade and town square in the sky. The human-made island city of Mexcaltitín de Uribe on a lake in western Mexico, a walkable and social place with church and square at the centre, and fishers setting off for prawn and oyster from houses by the water. In Denmark and Germany, the regenerative villages of Helsinge Hareby and Kronsberg built around agrihoods and public green space.

Herbert Giradet has long been a leading thinker and writer on how to turn nature principles and these kinds of examples into whole regenerative cities. Regenerative here means continually restoring and improving ecosystems, maintaining renewable energy, creating restorative relationships, and adopting new lifestyle choices for circular economies. Do this and the Petropolis that evolved from Agropolis could become Ecopolis. "Humanity," he reminds, "is building an urban future, yet urbanism in its current form is threatening the very future of humanity."

We have tried this before in the UK, in the forms of the Quaker village of Bourneville for factory workers, the garden cities of the early 20th century, and the 28 new towns established between 1955 and 1975. Yet, as Giradet says, "Most did not live up to expectations." It was not the fault of the designers. They were just overtaken and swamped by the consumption economy and late-stage capitalism. Regenerative models, as we have said, also need active design components for public engagement and social justice; otherwise, inequality comes to destroy the model too.

A fine example of the deployment of regenerative design is New York City's water supply system. It does not have one. This city of nine million consumes six million litres of water daily. It has the largest unfiltered water supply in the USA, and it is all done by the land, forest and farms of the Catskills and Delaware watershed. There had been a choice. Spend US\$6–10 billion on a treatment plant, and by engineering guarantee quality. Or \$100 million a year on watershed management over 4000 square kilometres, and by social action, outreach, education, joint initiatives and land acquisition, also guarantee water quality. Go upstream, in this case literally, and create healthy ecosystems, and downstream there will be supply of services that is cheaper than engineering.

We could say something similar about the 46,000 glaciers in the Himalayan mountain range that are sources for the great rivers of the Indus, Ganges, Brahmaputra, Mekong and Yangtze. These deliver drinking water for free to

1.3 billion people. What happens when the glaciers disappear as the climate heats?

Go on, said Valerie Brown and John Harris, be one of those wise sorcerers we can celebrate. We somehow have to found "a fresh tradition to an era of collective thinking." This Great Work, this grand transition, cannot occur by individual actions alone. There will be collaboration, co-production and continuing performance. It will take time to get it right. This is not, it should be said, the territory of the mass mind, where many people come to think the same, believe in a single route to success. This is the way captured by populists and demagogues. We will need many voices, perspectives, contributions to a common understanding of where we need to go.

But take care. Valerie Brown and John Harris also note: when big new ideas come along, there will be distortion, denial and disciples.

Kick-Starting Collective Action

This isn't going to be easy. This great collaborative chorus needs a pattern language, the transformative science and action of citizens. It will be convivial and build on flow activities; it does not need to be imposed. But it will have to resist, too.

And here is the reality too. Political scientist Ian Budge, in his book *Kick-Starting Government Action*, reminds us too of this: one-third of over 200 world states are not democracies. They are military regimes and autocracies. They are not designed to listen to their people.

He notes, individuals can change behaviours and choices, and they can organise into groups and movements, but you still have to push governments into effective action. But there is no time to waste on this: "Individuals acting together as well as individually can make a real difference."

And political scientists can point to many successful movements and groundswells: the Temperance movement, social crusades, support for public commons of health systems and pension provision, organisation to end commercial whaling, and so much more. Such collective ceremonies change mind and brain, and the sense of agency is increased by togetherness. The world can change, if enough people come together. It needs a spark of leadership to get things started and then more to keep it going until there is some permanent system shift.

Ian Budge recommends this: inventive political action, innovation in non-expert groups of citizens, civil disobedience, working through religious traditions, and tying the climate crisis to inequality. He writes: "We have to rely on territorially-based governments to carry it through." But they need a serious shove from their citizens and many others across the world.

But public engagement clearly cannot solve all social, economic and environmental challenges requiring unprecedented action and change. All things are connected, has written Naomi Klein. Now we face a climate crisis, the growth in inequality, declines in global biodiversity, and the spread of non-communicable diseases of affluence (type 2 diabetes, obesity, coronary heart disease, many cancers, mental ill-health, and loneliness). Some 300 million people have become economic and social migrants, forced away from their homes. A total of 800 million still are hungry across the world. Great technological, medical and communications advances have been made, and the world population is heading towards stability in total numbers. Many believed the 2-year COVID pandemic and lockdowns were a rehearsal for the climate crisis. In the early 1970s, the *Limits to Growth* team led by Donella Meadows produced scenarios that suggested planetary limits would be breached within 30–40 years, with consequences for all human societies. Now everything must change.

How then can we find more innovations that support co-creating diverse regenerative cultures worldwide? We will need to start with place, seek diverse perspectives, design for circularity, reduce consumption, prioritise health, happiness and longevity, and utilise forms of new power to spread social justice.

And how might public engagement help? It can lead to the formation of social capital and the development of regenerative cultures, where natural, social and human assets are built. Christian Wahl notes: "A regenerative future requires the capacity to listen and learn from diverse perspectives."

The development of the national K-Diet in South Korea is an exemplar. Korean citizens had already engaged in a remarkable experiment on policy evaluation, following 30 years of authoritarian military rule. Some 700 citizens were formally involved in evaluating the effectiveness of 43 ministries in Policy Evaluation Committees. Then in 2016, Korea launched the Seoul Declaration for the Korean Diet (K-diet), following a nationwide consultation. It resulted in the nomination and listing of the top 100 representative dishes for the K-diet. Through public choice, these emphasised dishes low in red meat, and high in vegetable, legume and fish, high in seasoned tofu and fermented vegetables. Whole grains and vegetables are considered real food, and kimchi is always served to make dishes a proper meal. There has, of course, been some entry into the world food economy, with increases in consumption of alcohol and sugar, yet only 3% of Korean adults are obese, even though GDP increased 17-fold over two generations.

So, in the face of international pressures for convergence, Korea has collectively been able to maintain culturally-important diets, and thus have low food-related carbon footprints too. In this way, the K-diet has been the route to promote dietary goods and not just focus on stopping bads. The Seoul Declaration had affirmed the K-diet, the authenticity and value of traditional foods and dishes, and above all of public discussion about food and health.

Successes elsewhere have been few with food. One comes from Oklahoma City in the USA, where Mayor Mick Cornett in 2007 launched the "City on a Diet" campaign, with the aim of all citizens, including him, collectively trying to lose one million pounds (450,000 kg). It was financed by a time-limited 1% sales tax that raised $700 million and focused on healthy food, walking and cycling, 140 km of sidewalks and tails, 300 km of new bike trails and lanes, design of a new central park, senior wellness centres, a public white water facility, and promoting public transport. Five years later, the target was reached, with 47,000 people actively engaged in weight loss. The city was now in the top ten fittest cities nationally. It would be easy to be cynical. Did the weight stay off, did people revert to earlier behaviours, did food manufacturers change their practices? But equally, leaders have to start somewhere.

Other changes have arisen from outside national policy settings. The Slow Food movement was launched up by Carlo Petrini in Italy in the mid-1980s as an anti-dote to fast food. The aim was to change the whole structure of food systems through the choices of growers, cooks and eaters. The pursuit of slowness and silence was a recognition of the value of taking time, of not falling into the trap of wanting food without the ceremony and culture. Choose local foods, celebrate locally-distinct dishes, eat together, and already the system changes. Slow food gave rise to Slow City and Slow Living movements, each producing an architecture to give value to the taking of time and local deployment of skills and knowledge.

Universal Basic Income

But, in all this talk of engagement and regeneration, how to break out of the great traps of inequality and overwork? We can see the good life, but it lies too far off for many people.

As we saw earlier, income inequality makes everyone less happy. It is not a net zero sum game to reduce poverty. Everyone gains. It could be we already have the answer, but too many governments are frightened of their own people to introduce it, or even to give it voice.

This is the idea of universal basic income. And it comes with heavy irony, for early experiments came on the back of income and profits from fossil fuels. And from some surprising polities.

Oil began to flow from Alaska in the 1960s, and the Republican governor in the mid-1970s, Jay Hammond, established the Alaska Permanent Fund for oil profits. From 1982 onwards, the annual dividend was divided into equal shares for each Alaskan resident. This is seen as a wholly unconditional right, not a privilege. The payment had fallen to $1600 per person by 2019, and the Alaska senate approved an increase from the Permanent Fund to $4000, topped up from general public funds to $5500 per year. This is $450 payment per month to each person, to do with it what they wish.

In some cases, the income from oil goes into the public purse. The Wyoming Mineral Trust fund uses income from mining to eliminate income tax in the state, and the Texas Permanent School Fund distributes $800 million from oil each year to public schools. The most famed, perhaps, is The Norwegian Government Pension Fund Global, also called the Norwegian Sovereign Wealth Fund. This was established in 1976 and delivered twice the returns on every barrel of oil than did the UK. It is used explicitly to fund universal public services (free education to university level, generous maternity and paternity leave, free social care for all, pensions, free health care), and still had grown by 2022 to US$1.2 trillion in size. On a smaller scale, the Shetland Charitable Trust Fund with assets of £23 billion from North Sea oil distributes income to charities and local groups on the isles.

What happens when you give money to people, especially those in poverty?

They will waste it, drink it, spend frivolously, buy drugs, say the neoliberals who believe it was selfishness that brought them success.

But actually no, the evidence says the opposite. People are careful, they spend first on essentials, and thus do not need to go into debt for food, energy, transport to find work, seeds for farming. They then save, often more than half of the payments. They know about being careful with money and choose what is important to them. It is thus odd: these are schemes where governments at various levels give money and say to people: you choose; we will not interfere. What is not to like, it could be asked of those neoliberals who emphasise personal choice.

There have been city and neighbourhood experiments in Cook County, Illinois; in Atlanta, Georgia; and in Santa Clara, California: several thousand residents and young adults were given between $500 and $1000 per month for one or two years. There have been experiments in Kenya and Rwanda: 30,000 people in 120 villages were given $30 a month, some guaranteed for ten years. In London, homeless people have had lives improved with payments of only £50 per month. When a number of Native American tribes began to build casinos on their sovereign territories, most elected to return profits equally divided to all tribal members. Like oil, the source of income may be morally troubling, but the use of profits is not. They are transforming lives and hope.

And these have been the findings from deployments of universal basic income: falls in hospitalisation, decreases in domestic violence, falls in mental ill-health, and increases in health across the whole community. Only when people have enough can they begin to think about reducing carbon footprints, investing in nature, thinking about a better future and the good life itself.

Related experiments are beginning to occur on the four-day working week. The pandemic and lockdowns changed working practices and led to rethinking. Perhaps not everyone needs to be at the place called work all the time. The stress of commuting may be cut, the carbon of travel also reduced. Maybe too, productivity increases if workers are trusted more, given more time for not-work. The experiments for a changed working week have followed, particularly on the 100–80–100 model: 100% of pay for 80% of time, in exchange for a commitment to maintain 100% productivity. In mid-2022, a six-month pilot was launched in 70 companies with 3300 staff. One chief executive noted that the new frontier for competition for staff is quality of life.

To begin with, good is about getting away from the bad.

Cutting Carbon

The ethicist, Peter Singer wrote that if you do good, you feel good. He asked, "What is the most good you can do?"

Your task, if you are already on the way or now intrigued is this: help others, tell the story, show how it can be done. Don't stop at the moral high ground. Use it to help others. Be the trickster, the hare or coyote, the ravens of Loki flying across the world each day. Moths will fly to the night light when it shines.

Richard Carmichael has written of behaviour change for the UK's Climate Change Committee. There is the performative dance, and we as individuals seek to adopt new low-carbon behaviours, and then persuade others to join them. Governments need to act to create incentives, structures and signals to aid these shifts, but they need a thoughtful push. Words are easy, but achieving targets or simply reneging, is much easier. We as individuals have autonomy, yet are shaped both by personal history and by current cultural, economic and social circumstances.

Governments have the capacity to make decisions, but clearly have not (yet) taken these four crises seriously: climate, inequality, biodiversity and food/health.

What can national and local governments do that individuals cannot? How about cycle infrastructure, grants for e-bikes and electric vehicles, subsidised rail and bus use, reduced charging costs, supporting house insulation and installation of heat pumps, air miles taxes, encouraging working from home, supporting research for meat analogues, change rules for public sector catering, changing rules about the use of ultra-processed foods.

We might then respond with more physical activity, more sustainable energy consumption, energy-efficient housing, new diets and transport preferences. And much more too.

This is the dance.

The cultural historian and priest, Thomas Berry, said, "Only intimacy with natural surroundings can save us." He called for a new movement called The Great Work of a people. It should include everyone, no one is exempt. It would be "our sacred story." It would be a bit of a meander, a quiet transition, a brooding one too, in the silence of morning mist. Above all, he said, "Such a transition has no historical precedent."

One way or another, this transition will need mass collective action. In March 2022, people in Tokyo were urged at short notice to cut their electricity use, for a number of circumstances were threatening widespread power cuts. It was a surprisingly bitter day, renewable energy production was low, and nuclear power stations were offline due to an earthquake two weeks previously. Mass consumer action worked, people were cold, wrapped in blankets, and electricity supply continued. But individual consumers cannot build new power plants, or speed the shift to wind power. There is a dance here between individuals and governments.

The model set out earlier proposed that One Tonne of carbon per person worldwide would be a good target to secure a safe place for planet and humanity. It would also be fairer. In the C1 countries, 1.3 billion people need to consume more to have enough food, clean water, housing, education, health services and transport. The other 6.5 billion, in the C2, C3 and C4 countries, need to start cutting now, by a half in the next decade, by another half again, then a last half by

2050. At the same time, the more we can do to capture carbon from the air into trees and soils the better.

Our collective primary focus must be on contraction of current carbon emissions, and convergence towards safe space of something like one tonne per person. The Race to Zero campaign coordinated by the UNFCCC has been mobilising a coalition of "real economy" actors and institutions to encourage commitment to hard targets. By mid-2022, this comprised 1,049 cities, 67 regions, 5,235 businesses, 441 investors, and 1,039 universities, collectively covering some 25% of global CO_2 emissions and over 50% of GDP.

Elsewhere we have: between 2000 and 3000 transition initiatives in 50 countries, spreading to towns, villages and communities worldwide after the launch by Rob Hopkins in Totnes in 2006. We have the B Lab initiative begun in Pennsylvania in 2007, certifying corporations if they benefit social and ecological systems: there are 5000 certified B-companies in 82 countries with a workforce of some 400,000 people. And then there are all those thousands of ecovillages too, and all eight million rural social groups.

There are many intentional cultures today: voluntary simplicity movements, slow cities and slow food networks, ecovillages, religious groups, retreat centres, transition movements. For all these, the great shift is underway, where growth is not growth, and all are seeking to create social models of being. Some will fail, but that's what happens when you have a large number of experiments.

This, you can see, is the way we get to stable green economies: by a combination of personal choice and government, business and institutional action. None of these might be perfect, but they are steps on route to collective change in many locations. You and I today cannot set policies that will change a national power generation and supply system entirely to renewables. You cannot personally ban fossil oil, allow open cast mining or ban it, set incentives for electric vehicles. Our influence is two or three steps removed, though can grow through organisation and action.

Yet, we know there is a strong link between the domains of a good life to health and happiness.

Nature improves mental and physical well-being. Healthy food from sustainable sources improves both personal health and sends market signals to farms around the importance of food production that increases biodiversity and ecosystem services.

Greater social capital in the form of togetherness improves health and happiness: members of social groups are happier. Regular physical activity increases health and wards off many non-communicable diseases.

Personal growth is a key part of engaging with learning and new activities and skills. A coherent spiritual and ethical framework is a wrapper for the meaning of life, giving further strength to choices and behaviours.

We do have these personal choices about ways of living, even though these are shaped by larger economic and social structures, by the constraints of history, by clever people trying to get us to buy more of their stuff. And these choices, added up, can lead to changes in policies. If you were to cut meat from your diet, intensive livestock producers would get the message. Animals will benefit, as so will the rainforest of the Amazon. If you ensure you only source renewable energy for your home, suppliers will get the message too. It will become easier for them to invest in wind, solar and other renewable infrastructure.

Many national and local governments have agreed on emissions' targets. This is a good start. But there remain many ways to hide the truth. By offshoring harmful consumption for carbon, you can make domestic numbers look better. By importing renewable electricity rather than investing in your own, you make domestic energy patterns look better. By investing in carbon capture and storage, you may ignore the need to cut emissions. If consumers continue to buy meat from animals raised intensively and fed on soya derived from fields that were recently rainforests, then good work is soon undone.

Some governments, especially populist ones, actively promote the pretense that decisions are only for individuals. It is not for government to interfere. If you make bad choices, well, the fault is yours. Not many have taken a coordinated approach to encouraging all citizens to cut carbon footprints. In the spring of 2022, the All-China Environment Federation issued a guideline containing 40 "green and low-carbon behaviours." These ranged from plant-based foods, recycling used clothes, buying sustainable products, use of energy-saving technologies, water and energy conservation, house waste sorting. The China President had previously set targets in September 2020 to ensure national carbon emissions peaked by 2030, and that net zero would be achieved by 2060 (too slow, but progress on previous announced targets).

The UK Climate Change Committee reckons on a split of about 65% of responsibility for cutting carbon lies directly with government alone, and 35% with individuals. This may be a fair assessment of the system, but it underplays what can be done without a huge amount of pain. And on this third path, there are many options.

You can't do everything yourself, but you can do one new thing now. We each have just enough time to select one change, see how it goes, get used to the new habit and behaviour, and then select another. Join an alternative food movement, volunteer for a local wildlife organisation, change your diet, help others choose walk across the threshold. As noted earlier, many options centre on non-material consumption. Go on, consume a bit of birdsong or cloud-watching. And let that splash and silence expand. You never know, something quite different and intriguing might just happen as green minds expand and our connections with nature and the world are changed forever.

The Currency of Story

Could the concept of *enoughness* still form the core mythology for a healthy planet and a happy life? Where we have done this before, the currency of wealth was storytelling with a purpose.

If objects have a story, they become magic vessels for memory and meaning. We are also more likely to keep them. If landscapes contain stories, then we find that wisdom sits in places, as Keith Basso has written of the Western Apache people. Natural places themselves become moral guides, especially if the stories are shared.

Ruth Sawyer wrote this in the *The Way of the Storyteller*, "The best of traditional storytellers have been those who live close to the heart of things: to the earth, the sea, wind and weather. They have been given unbroken time in which to feel deeply."

So, let's turn this around.

If we live in a stable culture, then stories are the currency of growth. They expand with the telling. They do not take away from someone else or from the Earth when they are used.

In many cultures, you might wear a band of precious metal around your finger, and your partner wear another. In others, bands of gold could be fixed around your neck or upper arm. Try this thought experiment, said Mike Bell of the University of Wisconsin: I will offer you an identical ring to the one you are wearing, plus $100 in cash. Will you exchange, do you swap the rings? There is money available, after all.

Could you live with the new ring and its lack of memory? In *The Wild Places*, Rob Macfarlane distinguishes between the kind of map with grids and names and the importance of certainty for navigation and the prose maps of wildness and memory. The latter warp and change, emerge from our own experiences. They become ships in a storm, a whale on the move. They take on, "A natural shape," a new and valued currency for our lives.

Richard Sennett said of the craftsperson, they are engaged in a constant discussion with materials. So who knows what then might happen?

Giving significance to objects, things and places is an important part of the good life. It was our common human reflex, until consumerism walked in the door. We create memories by doing things. This gives us a sense of worth and value, reminding that we made good choices or did good things (and not so good too, of course). And thus material things and places are not anonymous or neutral to human experience. They come to guide our choices and behaviours. A thing with memories is also less likely to be thrown away and replaced with a new one.

A place with memories is less likely to be allowed to perish, replaced by concrete and tarmac. A veteran tree under which you played, or where your family sat down for a picnic all those years ago, this tree is likely to secure your protection from the axe.

Memory is the magic of protection. Now you need a trickster to get you in the game.

And perhaps the story will be very short. Many other cultures have structures similar to koans in their stories. You have to work to fill the gaps. The Koyukon of Alaska begins with, "Wait, I see something," the Western Apache of New Mexico begins and ends stories with, "It happened at." The Dinesh people finish a story with, "Take it, I give it to you."

Here is an example.

Wait I see something.
I sweep this way and that way and this way around me.

This Koyukon song is all about their regard for nature, their love of being out-doors, living in remote camps, travelling over the great expanses of land. The song reminds them about being attentive to fleeting moments, mountains against the sky, reflections on water, a bird's song in the quietness. If you don't speak carefully about nature, they will say especially to children, "Don't talk, your mouth is small." The song itself is about grass tassels moving back and forth in winter wind, making little curved trails on the snow.

A single place name for the Western Apache accomplishes the communicative work of entire sagas. The narrator opens up thinking, encourages listeners to train their minds. They also do not want to speak too much. Each is guidance to be, "right here, at this very place." Here are two story examples:

"It happened at:
Water Flows Down a Succession of Flat Rocks."
"It happened at:
Men Stand Alone Here and There."

At one named place, a whiteman's cow was killed, and the Apache policeman needed cartridges and food, so he said only hello and goodbye, and the arrested man was released. At another, an old woman cried out in an attack to allow her granddaughter to escape, and she was killed instead. At a third, a stepfather attacked his daughter and was slayed by her uncle, and the body thrown in a pit. Travel in your mind to this and that place, bring what you know, let it work on you. You will feel better if you do.

All tribes and groups have hundreds of stories covering every plant and animal species. Some govern use, some are prohibitions. They add up to a web of manage-ment guidelines and moral rules. Joan Halifax says, "Stories are threads that draw us back into the fibre of the Earth."

The Timbisha Shoshone lived stably over centuries in the hottest lowest desert in the world, protecting water sources, moving up the mountain slopes when the heat was searing, camping under bristlecone pines that sprouted 5000 years ago. The Tuvan nomads of the south Siberian steppe moved camps four times a year, staying within their own boundaries to raise their livestock, fish the lentic lakes and hunt with eagles. By timbral and polyphonic throat-singing, epics have been sung for more than a thousand years. The longest transcription of the oral *Manas* epic of the clans of steppe contains 500,000 lines of verse.

During stability, stories emerge that come to span great eras.

Iceland was an empty land until discovered by Irish priests and then Norse explorers in the late 800s to early 900s CE. New farms were laid out, and open-air Thing assemblies were established so that all the people could gather and make collective decisions. A tradition of sagas then emerged, as the people made sense of their land of ice and fire. Soon every place had a story, where something happened to named characters. No longer was the landscape lava and grass, nor cliff and moss. An early vellum document recorded the names of every farm, and so we know there were 4560 farms a thousand years ago. There are still 4500 today. Farms named in the *Lándnamabók* are still present in today's landscapes. In 1897, W G Collingwood and Jón Stefánsson undertook a *Pilgrimage of the Saga-Steads*, and drew and painted 150 detailed scenes, the very places where events in sagas occurred.

Two things are very clear: the scenes are recognisable from the written records from nine or ten centuries before; they are also close to identical today. Many aspects of Iceland's landscapes and land use have remained stable for a millennium.

The sagas were written a 1000 years ago, and yet we can still recognise their locations. This cannot be said for many other affluent countries, where farm numbers have fallen sharply, landscape features removed, the commons enclosed and field sizes increased. Economic growth brought many things, but it took away the old stories.

And if we have no story, we might find we are no longer strong enough to care.

We all start life in the same place, with a beginner's mind. We all will end able to take nothing. In between, there is a life to live, a path to walk on an unfolding way. The river flows downward to the sea, the land itself gives up secrets. There is scent on the trail. The empty cup is slowly filled, with habits and skills, with memories and knowing.

And there, on the Earth's curved back, we might find the most wondrous things.

The good life was called "a detestable phrase" by American political philosopher John Rawls. But I'd like to think it has merit.

Antoine de Saint-Exupery flew the night desert, the seas, the pampas and the highest passes. He wrote in *Wind, Sand and Stars*:

> No sum could buy the night flight,
> With its hundred thousand stars, its serenity,
> That fresh vision of the world after a difficult phase of flight,
> Those trees, those flowers, those smiles,
> With the life restored to us at dawn, that chorus of small things,
> Which are our reward, money cannot buy them.

~*~

CODA

Let's Dance, Together

In Oxford Circus, I was passed a little handwritten note.
It said, "I can't get arrested because I am only ten,
But thank you for doing this for me."

[Jay Griffiths, *Why Rebel*, 2021]

Change is a snare, permanence is a snare.

[Joseph Campbell, *The Hero With a Thousand Faces*, 1949]

Abstract: And so. Let's dance. Together. There has never been such a great collective project of deliberate change. It could fail. But then the current economic system is already failing both nature and people. The many ways of living a low-carbon good life could deliver more happiness, greater health, longer lives, a whole and recovered natural world. Change is a snare, permanence is a snare, wrote Joseph Campbell. A gate, a threshold, lies ahead. Beyond: immersion in something new.

DOI 10.4324/9781003346944-11

I once was in a canoe on an upland lake at the east end of the Himalaya, up in the Shan Hills. The Intha people live on top of the one hundred square kilometres of pristine water, on floating villages, in houses on stilts. The boatmen row while standing upright, one leg coiled around the oar, and all around there was deepwater rice, fishnet and eel trap, floating garden and farm. We stopped at an untethered monastery, and the monk smiled in the morning sun, a host of cats sitting in the sunlight on the polished floor. The flow of thin light glittered on the water, the sky was cobalt blue. All was hushed and still.

Silence rites and rituals are often found more easily on a journey, on far travels, for we have already deliberately elected to leave all the other stuff at home. Now space expands. Preferences could change, habits too.

Now could follow a great unlocking of value from such structural change.

If we people could reduce harm to nature, there will be more ecosystem services to go around. If we reduce harm to health, and more people can live well and long, then health service costs will fall. If we prevent pollution and biodiversity loss and avert the non-communicable diseases of affluence, then money will be available to be spent on other things. And people will be happier and live longer. Nature will have been given the chance to recover.

So come now, let us dance.

It is true there will be losers. If you are in a company making billions from sales of pharmaceutical treatment for mental ill-health, then the last thing you want is for mental ill-health to be largely prevented. You want a pipeline of people to treat. If you are a producer of ultra-processed foods, and consumers shift to sustainable and healthy foods, then you will need a new business model. If you are a fossil fuel driller and distributor, then get out. There is only one future for you.

They have had a helluva run, built on harm.

If every kilo of fossil fuel were now to remain in the ground, then the cleverer fossil companies will have seen this coming and will have changed direction, investing in solar and wind power. The brightest, too, will be helping to create new economic systems based on sustainable and prosocial values, not trying to protect late-stage material capitalism. They will be thinking of wealth and affluence in different ways.

One day, Chief Henry of the Koyukon people said to author Richard Nelson, "I have had a good life. I have camped many times beneath spruce trees, roasting grouse over my campfire. So there is no reason to pray that I might live longer."

For two generations, high material consumption has been held as a sign of success. We were drawn to aspire and hope for a life that looked like that. But now we know, all this consumption has not made people happier. It increased inequality.

For a good life, focus on the goods, and then the bads will disappear as they quickly become irrelevant. The world will change for the better. There will be snowdrops in flower, and the grass will glisten. We will need both speed and patience. These crises will not pause, they are getting worse. At the same time, it takes time for each of us to drop a bad habit, and take up another. Mathieu Ricard wrote in his book *Happiness*, "Patience turns the mulberry leaf into satin."

Tokugawa Ieyasu, the first leader of the Edo unification of Japan, observed this about the cuckoo, about life itself:

> If the hototogisu,
> Will not sing,
> Wait.

So, does the new arithmetic add up? Could eight to nine billion people live within planetary boundaries and be happy and live long? The answer now is a resounding yes, if we focus on healthy food, nature, togetherness, physical activity and personal growth.

All of these are low in material consumption. If we choose them and consume sustainably using greener options, then planet-sized problems will indeed be reversed. We will find the climate crisis, biodiversity loss, plastic pollution, unhealthy air, dirty water, empty oceans, deal coral reefs, and retreating glaciers can be stopped.

Some natural assets will take a long time to recover fully. For others, it may already be too late. But we only have to say: these matter, a lot, to all of us, and now we'll make different choices, behave in new ways, consume differently. That's the way to start.

This may well need the quick wit and intuition of the trickster Hare to start up something new, perhaps the craft of Coyote, the thought and memory of Raven. Joseph Campbell said, "How are you going to change the situation unless you break some rules?"

Now fill in the creative space that follows the start of these great transformative stories. It happened at this very place in the world we live in. A good life will bring happiness and life satisfaction, and at the same time take each individual on a path of positive choices to reduce individual carbon footprints. A good life will not cost the earth.

And what of those old ways of living, the material consumption that caused all those ills? Well, a question: have you crossed the river and its rapids, and are you still trying to drag the raft while ashore? It is time to let those old habits go.

It is said, we always have 83 problems. And also an 84th: the wish all the other problems would go away. And then furthermore, we feel disappointed when they don't. What is that annoying sound? It is no more than sound waves reaching your ears. It is only irritating when we layer the sound with meaning. Said the old teacher to Dosho Port, "When the time comes, shake out your sleeves and just leave."

But what is left?

"A single wisp of cloud in the sky above the great plains."

Peace is not freedom from problems; it is a wide space and a quiet mind. So put a stone in your pocket, to remind you to be immersed and more present. And then to choose another low-carbon option. And tell someone else.

Wait, I see something. It is the restoration era. It is slow and happy.

Now, all of us, do our best.

The moon shone brightly,
Lighting every corner of the earth,
We sat there all night, looking at the sky;
At dawn, we went to bed.
[*Lady Sarashina Nikki,* As I Crossed a Bridge of Dreams, *1008–c1059]*

~★~

CHAPTER ENDNOTES

Preface: Transgression

1. **And what a place to start, at the Earth's great interlocking crises**: The interlocking crises are covered very well in recent books by Naomi Klein (*This Changes Everything; On Fire*), Kate Raworth (*Doughnut Economics*), Danny Dorling (*Slowdown*), Richard Layard (*Can We be Happier?*), Tim Jackson (*Post Growth*), Bill McKibben (*Falter*), Kurt Anderson (*Evil Geniuses*), Jason Hickel (*Less is More*), Paul Mason (*Post-Capitalism*).

2. **"What I know is that a good life is one hero journey after another," said Joseph Campbell**: The idea of the monomyth originates with Joseph Campbell in his 1949 classic, *The Hero with a Thousand Faces*; see also his *Myths of Light* and *Pathways to Bliss*. The folk tales of Alice in Wonderland (via Lewis Carrol), Cinderella and Goldilocks are well-known and widely told. Janet Yolen (in *Folktales*) has counted more than 500 variants on Cinderella told in folk tales across Europe. The hero's journey into nature is discussed and practiced by Bill Plotkin in *Soulcraft*, and across our life course by Stephen Gilligan and Robert Dilts in *The Hero's Journey*. Hero can be a troubling term, seeming to imply male and not female (hero, heroine); yet most uses today imply hero works at a meta-level to cover all people.

3. **The trickster helps when it comes to these necessary tilts and transgressions**: Trickster tales can be found in *Trickster Makes This World* by Lewis Hyde, and in many accounts of specific tribes, indigenous groups and cultures of the land and sea: see especially *A Story as Sharp as a Knife* (Robert Bringhurst), *Make Prayers to the Raven* (Richard Nelson), *Wisdom Sits in Places* (Keith Basso), *Medicine Man* (John Fire Lame Deer and Richard Erdoes). The classic koan on the dog is discussed by John Tarrant (*Bring Me the Rhinoceros*) and by many authors in James Ismael Ford and Melissa Myozen Blacker's *The Book of Mu*. The tales of Monkey King and Salish Coyote are from Joseph Campbell, and the Yoruba tales of Èṣù-Eshu is in both Campbell and Hyde. See Italo Calvino, Lewis Hyde, Janet Yolen, and Richard Erdoes and Alfonso Ortiz for further tales of tricksters worldwide. It has been noted, in the past, wolves get caught in traps but Coyote never does. They have been known to dig them up, turn them over and urinate on them. See also William Hynes and William Doty (1993): *Mythical Trickster Figures*. There is Brer Rabbit (actually a hare), Maui of the South Pacific, Yurugu the pale fox of the Saharan Dogon, Susa-no-o of Japan, Loki of the Norse, Raven of the boreal forests, Coyote of the deserts, the swan maiden from cultures across the world.

Chapter 1: Dust and Air: A Dangerous New Economic Worldview

1. **There once was a god called Enlil**: The first written story of Gilgamesh can be found in *The Epic of Gilgamesh* (trans. Andrew George) and *Myths from Mesopotamia* (trans. Stephanie Dilley).
2. **Andri Snær Magnason has recently asked, how do you say goodbye to a glacier**: The story of the disappearing Okjökull glacier in Iceland is from Andri Snær Magnuson's *Time and Water*, and is also discussed in Episode 1 of the podcast *Louder Than Words*.
3. **"What kind of times are they," wrote Berthold Brecht**: Bertold Brecht's stanza "in the dark times" was from the motto to the Svendborg Poems (*Svendborger Gedichte*), written in 1939.
4. **It seems things can be both bad and getting better**: An indicated earlier, the key texts for the climate, biodiversity, inequality and non-communicable crises are in a number of important aggregator texts (see note 1 for Preface). For scientific reviews of the climate crises, see the extensive reports and data at the IPCC website. The key findings from the IPCC 6th ARC Report were reported in early 2022: the six hottest years ever had occurred in the six years after the 2015 Paris Agreement. The data on school attendance by girls and boys is from the World Bank Open Data platform: https://data.worldbank.org/
5. **An advocate of this emerging neoliberalism was a 1920s Russian émigré to the USA**: You can find good discussions on Ayn Rand in *Evil Geniuses* (Kurt Anderson), *Falter* (Bill McKibben), and *Doughnut Economics* (Kate Raworth).
6. **We have three sets of ten years available**: Johan Rockström's variation on Moore's Law goes like this: three successive 50% cuts of emissions over three decades get the total down to 12% of the starting amount: personal communication. The key, though, is the first of the three decades, when a halving is needed. The target for 2030 is a half of current emissions.
7. **When he was alive, Hans Rosling often used this story**: Hans Rosling's *Factfulness* contains an account of the many reasons why we find ourselves being wrong about the world, and why things are better than many think.
8. **The OECD has calculated that between 2020 and 2060**: See Paul Mason (2015): *Post Capitalism*.
9. **Herman Daly was an influential economic insider at the World Bank**: You can find Herman Daly's excellent account of the early days of sustainability at the World Bank: see *Beyond Growth*.
10. **In one corner of south India, the non-government organisation SPEECH**: The story about my friend John Devavaram at SPEECH (the Society for People's Education and Economic Change) and later the Resource Centre for Participatory Development Studies came to an abrupt end when he suddenly died in 2018. He had been a brilliant innovator in participatory methods and approaches to help transform the lives of rural people, making especially significant improvements to the lives of women. We will always miss him.
11. **Shunmyō Masuno is head priest and garden designer**: at the 450-year-old Kenkō-ji temple on the north coast of the island of Honshu in Japan.
12. **Let's be blunt. Some economists lost a piece of moral core**; on the lost moral core of economics and late-stage capitalism: see Andrew Simms and Joe Smith (*Do Good Lives Have to Cost the Earth*), Tim Jackson (*Prosperity without Growth* and *Post-Growth*), Kate Raworth (*Doughnut Economics*) and Kurt Anderson (*Evil Geniuses*).
13. **Why would rationalists with only self-interest devote time on pilgrimages**: On pilgrimages, see Sibley (2013) for the Shikoko 88 temple pilgrimage, Joan Halifax for a lovely account in *The Fruitful Darkness* of the hard and harsh pilgrimage to Mount Kailas in Tibet when she carried the ashes of Tenzing Norgay, and Peter Stanford's (2021) book on *Pilgrimage*. The closer you come to the mountain, wrote Joan Halifax, the more it disappears: "During pilgrimage, there is no place to escape from yourself, even when

the wind takes off your skin, and the birds have drained away from the sky." The Trappist monk Thomas Merton writes about the nature of pilgrimage in *Mystics and Zen Masters* (1961), and William Scott Wilson walks the Kiso Road in Honshu, part of the 550-kilometre Nakasendo Way, in *Walking the Kiso Road* (2015).

14. **From the South Sea, did a guest appear**: The poems of Li Po and Tu Fu can be found in the translation by Arthur Cooper (Li and Tu, 1973).

Chapter 2: Ten Thousand Good Lives: Sustainable and Kind Ways of Living

1. **A hundred thousand sparrows**: The lines from Yang Wan-I (*Cold Sparrows*) are from David Hinton's *Mountain Home*.
2. **And then when we return, bringing back something essential**: The Jack Kornfield quote is from his book *After the Ecstasy, The Laundry* (2000).
3. **Once you get started, all is fine. You succeed or fail**: The Joseph Campbell quotes are from talks and chapters in *The Hero with a Thousand Faces*, *Myths of Light* and *Pathways to Bliss*.
4. **Take good care, of what is good**: The quote in this chapter is from the 13th to 14th century Meister Eckhart's *Meister Eckhart's Book of Secrets*.
5. **One-eyed Oðinn had two ravens, Huginn-Thought and Muninn-Memory**; For expositions on Norse mythology, see my *Sea Sagas of the North* (2022), and source material in the *Poetic Edda* and *Prose Edda*, Kevin Crossley-Holland's *Norse Myths*, Hilda Ellis Davidson's *Gods and Myths of Northern Europe* and *Scandinavian Mythology*.
6. **For the north-western Athapaskan and Haida cultures, Raven is a central figure**: Raven mythology is from *Make Prayers to the Raven* (Richard Nelson); see also Robert Bringhurst's *A Story as Sharp as a Knife* (2011). The Coyote story was told to musician Kevin North and me by Pauline Esteves of the Timbisha Shoshone tribe in what is now called Death Valley. The *Early Morning Song* is from Steven Crum's *The Road On Which We Came*. A very good annual summary of the state of lives of indigenous peoples and cultures is published by the International Working Group for Indigenous Affairs (IWGIA). See Dwayne Mamo (ed). *The Indigenous World* (2022), 36th edition, 850pp.
7. **One of the great Haida sagas is *Raven Travelling* , some 1400 lines long**: Gary Snyder wrote about raven and other bird tales of the Haida in *He Who Hunted Birds in His Father's Village* (1979). He wrote: "a curse on monocultural industrial civilisation."
8. **Back in Siberian Tuva, every place has an *ee* , a spiritual guardian**: An account of ways of living in Tuva (Tyva) in Siberia can be found in my *The Edge of Extinction* (2014). See also the brilliant book on throat singing by Ted Levin *Where Rivers and Mountains Sing*.
9. **Under the same broad sky, many small farmers say their goal in life is contentment**: Gene Lodgson's book is called *The Contrary Farmer*. He wrote many hundreds of reports and articles on being a contrary farmer before his death in 2016. Kentucky farmer, Wendell Berry, has written widely of the fundamental life and values of the small cultivator, the importance of community, and the consequences of the destruction of nature (see *The Unsettling of America*, 1997).
10. **I first travelled to Holmes County with David Orr of Oberlin College**: You can also find a full account of the ways of living of the Amish communities in Ohio in my *The Edge of Extinction*, in Amish farmer David Kline's *Great Possessions* and Donald Kraybill's *The Riddle of Amish Culture*.
11. **It is also a common trait: people give to each other, and to nature. And as gifts move, their value increases**: Key texts covering the nature of gifts and giving include *Braiding Sweetgrass* by Robin Wall Kimmerer, *The Light at the Edge of the World* by Wade Davis, and *Ancient Futures* by Helena Norberg-Hodge. The Tibet field observations are from Patrick French's *Tibet, Tibet*. The stories of Australia Aboriginal and

First Peoples perspectives and lifeways are from Deborah Bird Rose's *Dingo Makes Us Human* and *Nourishing Terrains*, and Tyson Yunkaporta's *Sand Talk*. On hunter-gatherer cultures, see Richard Lee and Richard Daly's comprehensive *Cambridge Encyclopaedia of Hunter-Gatherers* and Vicki Cummings and colleagues' *Oxford Handbook of the Archaeology and Anthropology of Hunter-Gatherers*.

12. **The Kula Ring is a cultural structure as wonderful and impressive**: The Kula Ring of the Trobriands was first researched and written about by the anthropologist Bronislaw Malinowski in the 1920s.

13. **Patrick French said this, despite the prevailing political and economic conditions**: see his *Tibet, Tibet* (2003).

14. In the 1950s, Lorna Marshall gave Kung San! Women: this story about the cowrie shells is in Lewis Hyde's **The Gift** (1979). It is worth noting, as Hyde points out, not all gifts regenerate the soul. Think of the Trojan Horse, or the poisoned apple given to Snow White.

15. **In the 1930s, Genzaburō Yoshino wrote *How Do You Live?*** This is the classic book about how to live well and kindly in Japan, read by almost all school children. All the people in the world of the story are bound by an unbreakable net, and when things go wrong, the boy Copper is reminded, "A tree does not recognise it you are miserable."

16. **Saichō was the founder of Tendai Buddhism in Japan**: The Saichō verse is from John Steven's *Mountain Monks* (2013).

17. **On happiness and GDP**, see these papers: (Pretty, 2013; Pretty et al., 2015, 2017; Pretty and Barton, 2020). See also Richard Layard's book *Can We Be Happier?* The annual *World Happiness Reports* (2013–2021), and Andrew Clark's *The Origins of Happiness*. It is also a common trait: people give to each other, and to nature. And as gifts move, their value increases. Happiness data is from the World Happiness Survey.

18. **Life expectancy data** is from the World Bank and Our World in Data. The data in Table 3.1 are aggregated from many open sources.

19. **Matsuo Bashō's haiku "Long conversations . . ."** is from *Bashō: The Complete Haiku* (2008).

20. **A total of 15 million people in the UK, some 23% of the population, now suffer from long-term health conditions**: our summary of NCD costs in the UK is contained in Pretty et al. (2015, 2017).

21. **In no country has obesity or type 2 diabetes fallen once it rose**: On obesity as a problem, it is worth just pausing and thinking about people's circumstances. Bessel van der Kolk in *The Body Keeps the Score* notes that problems can be used as solutions. Not all obesity is a failing of the food system or of individual choice. Obesity is a good thing, if it makes abused children feel safe by increasing the likelihood of being ignored. It is well-known, too, that many police officers and prisons/corrections officers deliberately put in weight when joining their employment as it reduces the likelihood of being physically pushed around.

22. **The good life is at the core:** See *The Book of Chuang Tzu* and *Tao Te Ching* by Lao Tzu (two translations listed in references). I conducted The Good Life survey in October to December 2020, with 190 respondents from 27 countries contributing: from Argentina, Australia, Austria, Bangladesh, Canada, China, Costa Rica, Denmark, Germany, Finland, France, India, Indonesia, Israel, Italy, Japan, Kenya, Netherlands, New Zealand, Pakistan, Singapore, South Korea, Sri Lanka, Sweden, Switzerland, UK and USA. The survey was open-ended, and respondents were asked to list ten or more components (activities or things) they considered central to what might be defined as the good life. These could include behaviours and choices, as well as hopes and aspirations. Respondents to the survey were 46% women, 54% men; by age groups, 1% were aged less than 18; 5% were 18–30; 64% were between 30 and 65; and 30% older than 65 years. The total of 1915 returns from 190 people were coded according to common content and values, and were found to fit into twelve domains. Nine of these twelve were related to personalised aspects of the good life, and three covered features of the surrounding local to national social and political framing conditions in which people lived.

23. **The good life is at the core**: Alastair Campbell is journalist, political strategist and sufferer from depression for most of his adult life. He wrote in *Living Better* (2020) of his transition towards a good life:

> Now I am obsessed with sleeping long, and well. A night in bed before 10pm is a night well spent. My diet is vastly improved, I exercise pretty much every day. Read books not newspapers is a new motto, which has definitely helped by mental health. Listen to music, not the news on the radio, learn to enjoy silence. All good, all good.

24. **Resulted in the foundation of the Right Livelihood Award**: The annual Right Livelihood Award was established in 1980 to "honour and support those offering practical and exemplary answers to the most urgent challenges facing us today." See Joel Magnuson's *The Approaching Great Transformation* (2013).

25. **The low impact communities contributing to understanding of the good life**: These were Erraid Community, Ewe House, Findhorn Centre and Ecovillage, Lancaster Cohousing-Forgebank, Lancaster Coop, Landmatters Cooperative, Lauriston Hall Housing Coop, London Catholic Worker Farm, Lancaster Cohousing, Monkton Wyld Court, Newton Dee Camphill Community, Old Chapel, Old Hall Community, Othona Community, Threshold Centre, and Windsor Hill Wood.

26. **Karen Litfin of the University of Washington**: See Litfin's *Ecovillages* (2014). Also Tendai Chitiwere (2018), Rob Hopkins, Dawson (2006), Miller (2018). The Global Ecovillage Network's shared purpose is "to link and support ecovillages, educate the world about them, and grow the regenerative movement": https://ecovillage.org/. It links 10,000 communities worldwide. One of the first ecovillages created was Sólheimar in Iceland in 1931. At its height, the kibbutz movement centred on communal living and social justice in Israel was home to 7% of the national population. Today some retain these principles, others have launched into the corporate world. Caroline Lucas, the sole Green MP to the early 2020s in the UK, has called for "human-scale settlement, harmlessly integrated into the natural world, a peaceful, socially-just, sustainable community."

27. **The way people live in low-impact communities and ecovillages is often called an attentive lifestyle**: For more on low-impact communities and ecovillages, see Sarah Bunker and colleagues' *Low Impact Living Communities*. See also Ken Worpole's *No Matter How Many Skies Have Fallen* (2021) and Anna Neima's *The Utopians* (2021).

28. **Dorothy Schwarz and Walter Schwarz, the then Guardian environment correspondent, travelled to low impact communities and ecovillages**: There are two fine books from Dorothy and Walter Schwarz: *Living Lightly* (1998) and *Breaking Through* (1987). Walter Schwarz died in 2018.

29. **Eight out of ten of the 75–100 year-olds**: On gero-transcedence, see Lars Tornstam (2011).

30. **"Cling if you want, to things"**: The Meister Eckhart quote is also from his *Meister Eckhart's Book of Secrets*.

Chapter 3: The Climate Crisis: The Safety of One Tonne Each

1. **There were a sizeable number of people who had been drawn to believe those corporates and kings when they said there was much doubt**: For an account of the ways of living in the Atchafalaya basin and the inland and coastal swamps and marshes of Louisiana, see my *The Edge of Extinction* (Chapter 11). In *Under a White Sky* (2021), Elizabeth Kolbert notes how the coast of Louisiana has lost 2000 square miles since the 1930s, and how 31 place names of parishes have been formally retired, because they are under water.

2. **"Tell all the truth," wrote American poet Emily Dickinson**: Emily Dickinson's phrase "tell it slant" has appeared as the title of a number of books. She did not give many poems titles. She was a recluse, spending many years without going outside her

house. And yet she was creative and daring in her work (see Dickinson, 2016, *Collected Poems*). She was born in 1830 and died in 1886.

3. **Carbon Emissions and Consequences:** The carbon and carbon dioxide data is drawn from open source datasets at the World Bank and OECD. There are many sources for the impacts and predictions for the climate crisis: see especially IPCC website for links to the overview reports approved by large numbers of scientists worldwide.

4. **Some 2500 years ago, Lao Tzu (also known as Laozi) wrote:** For Lao Tzu, see *Tao Te Ching* (trans D Lau) and Stephen Mitchell's *Tao Te Cing: An Illustrated Journey*.

5. **Figure 3.2: the per capita oil consumption and GDP data** for 181 countries is from The World Bank. Oil consumption is presented here as a proxy for the growth in all fossil fuel consumption (coal, oil, gas).

6. **One early morning, the helicopter rattled over the Shandur Pass in northern Chitral**: You can find a summary of the work of the Aga Khan Rural Support Programme at their website. The helicopter flight from Chitral to Gilgit was with Javed Ahmed in 1991. Emily Lorimer's book, *Language Hunting in the Karakoram* (1939) contains a fine account of life in the Hunza Valley in the 1920s. In his *The Transition Handbook* (2008), Rob Hopkins recounts a visit to the Hunza Valley in 1990, observing that "if on Earth there is a garden of bliss, it is this."

7. **"A day will no doubt come"**: The Po Chú-Í poem is from David Hinton's *Mountain Home*.

8. **Converging on One Tonne Per Person:** In this chapter, both Carbon alone (C) and Carbon Dioxide equivalent (CO_2eq) data is presented as they tell different stories. Throughout the book, CO_2eq is taken to be the standard for data reporting, as it includes the effects of the other major greenhouse gases, methane, nitrous oxide and CFCs. I have combined greenhouse gas data with social and economic data for all 220 countries, and these are used for the figures and tables in the chapter.

9. **Contraction and convergence:** This first proposed by Aubrey Meyer in the early 1990s. It was adopted as a valued way of framing the challenge: some increase carbon consumption, most reduce by large amounts. Whatever the model for equal shares in reduction, the overarching need is to transition to net zero emissions as soon as possible.

10. **At the same time, the iron cage of consumption**: Max Weber used the term iron cage to illustrate how Western societies (of capitalism) trap individuals into systems judged by efficiency and control by others.

11. **"How should I follow a rugged road?"** The quote from Korean Zen Master Taiwŏn is from *Korean Zen: Garden Chrysanthemums and Forest Mountain Snow* (trans. Haight, 2017).

12. **The Essex Climate Action Commission**: I am chair of the Commission, one of a number of place-based commissions in the UK: see https://pcancities.org.uk/.

13. **The 30 for 30 data:** adapted from these key references on carbon by behaviours: Institute for Global Environmental Strategies, Aalto University and D-mat ltd. 2018. 1.5-Degree Lifestyles: Targets & options for reducing lifestyle carbon footprints; www.iges.or.jp/en/pub/15-degrees-lifestyles-2019/en; Ivanova D et al. 2017. Mapping the carbon footprint of EU regions. *Environ Res Letters*, *12*(5), p. 054013; Ivanova D et al. 2018. Carbon mitigation in domains of high consumer lock-in. *Global Environ Change*, *52*, 117–130; Ivanova D & Wood R. 2020. The unequal distribution of household carbon footprints in Europe and its link to sustainability. *Global Sustainability*, *3*; Ivanova D et al. 2020. Quantifying the potential for climate change mitigation of consumption options. *Environ Res Letters*, *15*(9), p. 093001; Project Drawdown. 2020. *The Drawdown Review*. www.drawdown.org/drawdown-review. In this carbon schedule using the excellent research by Diana Ivanova and Daniel O'Neill at the University of Leeds. I have used median scores for carbon footprints by activity to offer guidance for individual choices. Each could vary according to your own circumstances. Diet savings, for example, will depend on the details of your starting and new diet, on your specific food preferences, on how and where your food is grown and raised. I am less concerned with high levels of exactitude, more with the schedule that offers choices for each of us, and

how can swiftly move to a final point of no more than One Tonne per person. Just this: do your best. Start now. And then we can dance.

Chapter 4: The Inequality Crisis: Togetherness Is Better Than Selfishness

1. **There was a clearing by the track, by the spruce and birch**: Individual chapters in my *The Edge of Extinction* are devoted to the Innu of Labrador, the nomads of Tuva (Tyva), and the small farm cultures of Ohio. Amish farmer David Kline wrote of the Amish way of living in *Great Possessions*, and Donald Kraybill's analysis of the Amish people is in *The Riddle of Amish Culture*.

2. **In the old days, the Timbisha people**: Pauline Esteves is an elder of the Western Shoshone/Timbisha people who live in what is now called Death Valley. A chapter in my book, *The Edge of Extinction*, is devoted to life in the western deserts. The Timbisha were granted a reservation in the year 2000, small compared with the wide home lands over which they originally roamed for ten thousand years.

3. **Brian Hare and Vanessa Woods published a book centred on the value of cooperation**: See Hare and Woods *Survival of the Friendliest* (2020) for a fine summary of their theories and analyses of the impacts of domestication on both animals and early humans.

4. **The Rise of Selfishness**: For more on Milton Friedman, Grover Norquist and Ayn Rand, see *Evil Geniuses* by Kurt Anderson, *Falter* by Bill McKibben, and *Doughnut Economics* by Kate Raworth.

5. **It looks like this era of selfishness has to come to an end**: See also Jamil Zaki's *The War for Kindness* (2018). Zaki said: "We should see selfishness as a sickness." And also: "Stories help us imagine other people's lives. They make others less distant." The Charities Aid Foundation runs an annual survey to produce the World Giving Index. The top ten countries in 2021 where citizens give the most to charity are: Indonesia, Kenya, Nigeria, Myanmar, Australia, Ghana, New Zealand, Kosovo, Uganda and Thailand. www.cafonline.org/about-us/publications/2021-publications/caf-world-giving-index-2021.

6. **Figure 4.1 Inequality of household income**: The country-level Gini coefficient and health expectancy data is from World Bank datasets.

7. **Twenty years ago, Thomas Insel and Larry Young of Emory University wrote**: Attachment theory was developed by Thomas Insel and Larry Young (2001).

8. **So how did we come to permit loneliness to spread so fast and far**: On loneliness, see Julianne Holt-Lunstad and colleagues (2010, 2015, 2017) and John Cacioppo and colleagues (2003, Hawkley and Cacioppo, 2007). George Vaillant's *Aging Well* shows how friendships and lifelong relationships play a key role in healthy ageing and increasing lifespan.

9. **Social capital has been in decline in the affluent and industrialised countries**: See the famed Robert Putnam study *Bowling Alone* (1995). Gautam Rao's research on Delhi schools is in Rao (2019).

10. **Roller skating decline in the USA:** In the film, *United Skates* (2019), one skater says, "Skating was our hope." Now it is harder to express civic pride and identity in physically-active settings that once had brought well-being and connection. See also Basu (2014).

11. **Volunteering is a further form of giving to others and to nature**: On the personal and health benefits of volunteering, see ILO (2011), Nicole Anderson and colleagues (2014) and Francesca Borgonovi (2008).

12. **"At dusk I came down from the mountain**: The Li Po poem is from *Poems*.

13. **A fresh era of collective thinking was launched by economist Elinor Ostrom**: See Ostrom's groundbreaking 1990 book, *Governing the Commons*.

14. **In the mid-20th century, Walter Goldschmidt studied the two Californian communities**: The famed Walter Goldschmidt studies on rural Californian towns and

agriculture structure are in Goldschmidt (1946), and follow-ups were conducted by Michael Perelman and Kevin P. Shea (1972) and Linda Lobao (1990).

15. **The enclosure of the commons began in our own country**: The story of the draining of the great wetlands called The Fens in the UK is a well-known part of taught history; the destruction of a whole hunter-gather people is not. Reports and observations are details in Ernle's *English Farming* (1912), and aspects of the whole one hundred years of conflict appear in my Sea Sagas of the North. E P Thomson's considered analysis of the pivotal role of the Waltham Black Act is in his *Of Whigs and Hunters* (1975). Calling the wilds and commons "wastelands" started at this point, and this terminology is still widely used today.

16. **Further changes to social structures were promoted by the conditional policies of structural adjustment adopted by international finance institutions**: On the impact of structural adjustment and Training and Visit (T&V) policies, see Benor et al. (1984) and discussion in paper on social capital by Pretty and colleagues (2020).

17. **Michael Cernea at the World Bank already had concerns in the late 1980s**: On early participatory approaches, see especially World Bank sociologist Michael Cernea's *Putting People First*, and Robert Chambers' *Farmer First* (1989). On adult pedagogy, you have to start with Paulo Freire's *Pedagogy of the Oppressed* (1970) and Rolf Lynton's *Tide of Learning* (1961). Our 29-author recent analysis of the growth of social capital worldwide is in Pretty and colleagues in *Global Sustainability* (2020).

18. **In Australia, the 30-year-old Landcare movement**: The personal and health values of the Landcare movement in Australia have been documented in KPMG/Landcare Australia (2021).

19. **"How can I possibly sleep, this moonlit evening?"** The Ryokan poem is from *Dewdrops on a Lotus Leaf* (1993).

20. **Oliver Scott Curry and colleagues at the University of Oxford Kindlab**: On the superb work of the Oxford Kindlab, see Oliver Scott Curry and colleagues' (2019) paper "Is it good to cooperate?"

21. **Kindness is our common state and response to threat. It is selfishness that is the outlier.** Rutger Bregman in Human Kind details the work of the Delaware Disaster Research Centre, which studied 700 examples of disasters worldwide since the early 1960s. In the overwhelming majority, people were actively prosocial. They did not go into shock, they stayed calm, and sprung into action to help others. Each year for the past decade, the Charities Aid Foundation has produced a World Giving Index (www.cafonline.org/about-us/publications/2021-publications/caf-world-giving-index-2021). It ranks countries by asking three questions: in the past month, have you i) helped a stranger, or someone you didn't know who needed help; ii) donated money to a charity; iii) volunteered your time to an organisation? More than three in ten adults around the world donated money to charity in 2020. The top ten countries in 2020 were Indonesia, Kenya, Nigeria, Myanmar, Australia, Ghana, New Zealand, Uganda, Kosovo and Thailand.

22. **Nelson Mandela spent 18 years on Robben Island**: Viktor Frankl wrote of his experiences in concentration camps in his 1959 book, *Man's Search for Meaning*. The reports on Nelson Mandela's approach to surviving prison are online, as is his invitation to the former guard Christo Brand to attend his presidential inauguration.

23. **"Just off to hoe my melons one day"**: The poem by Wang Wei is from his *Poems* (1973).

Chapter 5: The Nature Crisis: Regaining Earthsong and Attentiveness

1. **I once saw a rook's parliament on a cold clear winter morning**: The rooks' parliament was in a tree on the slopes of the Stour Valley along the Suffolk and Essex border in the east of the UK. Jack Kornfield wrote of the blackbird circle in the Himalaya in

After the Ecstasy, and aviator Guy Murchie's observations about birds and plans from his 1954 book, *Song of the Sky*. In my book, *The East Country*, I talk about how crows came one dawn to the flat roof outside the bedroom of my home the day before my father died, after a month-long illness. They had never come so close, nor had they ever faced towards the window and called so directly. There is a Japanese folk saying: "When the crows caw, the weather will change" (in W S Wilson, *Walking the Kiso Road*). Crows, rooks and other corvids are tricksters, opening doors.

2. **"This tiny bird," wrote singer and folklorist Sam Lee**: The stories and folk tales are in Sam Lee's *The Nightingale* (2020).

3. **"Tell me what it is you plan to do"**: Mary Oliver's piece on a wild and precious life is from her book *Upstream*.

4. **"Earth was home," Naess said**: You can read a fine range of works of Arne Naess in *The Ecology of Wisdom*.

5. **When I had lunch with the President**: I had lunch with President Jimmy Carter in Atlanta, Georgia in February 2007. This is the same Jimmy Carter who invited E F Schumacher to the White House in 1977. Solar panels were installed on the roof, and discussions centred on how to value deeper values over consumerism. Then came the cowboy on the frontier, and Ronald Reagan won the next election by standing for endless expansion. See Morris Berman's *Neurotic Beauty* (2015).

6. **This is what we keep finding: the interconnectedness of natural systems**: See Suzanne Simard's *The Mother Tree* (2021), and also Merlin Sheldrake's superb *Entangled Life* (2020). Fungal hyphae are the most extraordinary form of life. They can branch and fuse, pass through two doors and come together. They use volatile chemicals for signaling. They grow by getting longer, not by adding new cells, and can detect the light of stars. Fungal mycelia are about 2.4 billion years old, and some 90% of plants depend on mycorrhizal (MCR) fungi for nutrients, food transfer, signaling and support. MCR hyphae make up a third to a half of the living mass of soils. Modern agricultural fields using fertilisers and pesticides have low MCR; organic and regenerative fields have high levels.

    Foresters actively undermined Simard's work for years, as it challenged their norms. They called her "Miss Birch" for supporting an understanding of the role of the so-called weed, the birch. Said one man: "Well, Miss Birch, you think you are an expert." It was a battle for ideas, junior forester tended to be supported, but not much the old (mostly men). Simard had found in one block there were 250 trees connected to each other, the oldest individuals were hubs with the most connections. And what do foresters do in forests: they cut the largest trees and thus disrupt the working of whole systems. When the Douglas Fir was infected by Spruce Budworm, Simard found they sent half their carbon into roots and then to mycorrhizal fungi for safe storage, also passing a tenth to ponderosa pine neighbours, helping them to fight the infection.

7. **We looked into the shadows. A pair of red foxes strolled from the trees**: The story of the fishing foxes first appeared in my *Only The Earth Endures* (2007), a lecturing and field trip to Armenia organised by Vardan Haykyazan.

8. **All is Sacred Nature:** On the sacred treatment and designation of land, there are many fine works. On hunter-gatherers, see Richard Lee and Richard Daly (*Cambridge Encyclopaedia of Hunter-Gatherers*) and Vicki Cummings and colleagues (*The Oxford Handbook of the Archaeology and Anthropology of Hunter-Gatherers*); on settled city-states and civilisations, see Nicholas Ostler's *Empires of the World*; on current indigenous cultures, see the list of references under the heading below "Ten Thousand Good Lives."

9. **In each old place, a song deeper than silence still was coming radiant**: In Barry Lopez's final book, *Horizon*, he is the elder who for a while is then young boy of his youth, swimming tirelessly in a pool of glittering water. An now, he says: "the throttled earth, the scalped, the mined, the industrially-farmed, the polluted and endlessly manipulated." And so he says, "In the years ahead, I will listen to the elders." Author Albert Camus wrote, "The world is beautiful, and outside it there is solution," and so Lopez points to the Navajo ceremony called *Beautyway*, their faith in renewal of the world, the

beauty and harmony that pervades, and the highest level of collective thinking. This is where solutions exist.

10. **Places are made sacred by the repeated actions of many people over time**: In *The Art of Pilgrimage*, Philip Cousineau says pilgrimage should be a challenge to every-day life. It is "attentive travel," and a desire for new forms of belonging. The point of pilgrimage is to improve yourself by overcoming difficulties. What you might say at the start, he recommends, is something like: "I hope it does not go as planned." There are 6000 pilgrim routes in Europe.

11. **In Blackfoot culture, rock is a trickster**: See Wissler and Duval (1908). *Mythology of Blackfoot Indians*.

12. **Today we find some have dared row out**: On Japan, see Andy Couturier's *The Abundance of Less*, where he returns to families met years before and is able to track their personal and social changes. On small farmer cultures in the USA, see Gene Logsdon's classic *The Contrary Farmer*.

13. **Partha Dasgupta's comprehensive review**: Partha Dasgupta's report for the UK government, *The Economics of Biodiversity*, was published in 2021. The featuring of the commons was notable and important, coming more than sixty years after Garret Harding's notorious paper in the journal Science called "The Tragedy of the Commons." David Orr of Oberlin College has written wisely on the value of ecological literacy (1992).

14. **The leading thinker and writer on community-based conservation**: See Fikret Berkes: *Advanced Introduction to Community-Based Conservation* (2020).

15. **A study in England and Wales for the UK government department Defra in 2022**: This was led by Brett Day at the University of Exeter.

16. **And so emerged the *shanshui* mountain-water tradition**: An account of the emer-gence of the shanshui movement, and the later public reclaiming of the Huangshan Mountains of Anhui, can be found in Chapter 2 of my *The Edge of Extinction* (2014).

17. **Before my bed, there is bright moonlight**: Li Po's poem is from the 1973 collection *Poems*.

18. **The 13th-century monk and writer**: The book of the work of Eihei Dogen (Dogen Zenji) is translated by Tanahashi (*Moon in a Dewdrop*).

19. **We came up with the term *green exercise***: At the University of Essex, we invented the term *green exercise* in 2004 and have published many papers and books on the subject. There is a rich and wide literature on the evidence of the health benefits of natural places. Early work was led by Ernest Moore and Roger Ulrich on views from the windows of prisons and hospitals, and Rachel and Stephen Kaplan on attention res-toration. The term biophilia has proven to be useful to apply to the design of buildings and places, but genetic mechanisms are not really needed to explain the health benefits of nature arising from our behaviours and habits that induce attentiveness-immersion-focus-flow.

20. **Further research has filled gaps**: For life course studies, see Peter Elwood and col-leagues (2013); more than 1200 papers from the longitudinal Dunedin Multidisciplinary Health and Development Study begun in 1972 by Avshalom Caspi and Terri Moffitt (see Caspi et al., 1996, as an example); Deborah Danner and colleagues (Milwaukee nuns, 2001); George Vaillant's *Aging Well* (Harvard cohorts).

21. **Here is one more example from history. Asclepius was the Greek god**: On the several hundred Asclepian temples of Greece: see Esther Steinberg's *Healing Places* (2009). For Theocritus and Hesiod, see *Idylls* and *Works and Days*.

22. **John Ji and colleagues**: The oldest old studies in China, see John Ji and colleagues (2019). The UK Monitor of Engagement with the Natural Environment (MENE) is at www.gov.uk/government/collections/monitor-of-engagement-with-the-natural-environment-survey-purpose-and-results. For measures of the health benefits of nature, see Catharine Ward Thompson and colleagues (2019) and David Rojas-Rueda and colleagues (2019). On equigenic environments, see the work of Richard Mitchell and colleagues (Mitchell and Popham, 2008; Mitchell et al., 2015).

23. **It is, though, hospices and cancer treatment centres**: the health infrastructure of Maggie's Centres (for cancer treatment): see www.maggies.org/. Of city parks and Biophilic design, see Stephanie Panlasigui and colleagues (2021).
24. **The data on cars/autos by country** (Figure 5.1) are from Our World In Data at https://ourworldindata.org/.
25. **The term *non-exercise physical activity***: NEPA research was developed by Elin Ekblom-Bak and colleagues in Sweden.
26. **Copenhagen calls itself a green tiger**: The Institute of Happiness is led in Copenhagen by Meik Wiking. See his books on hygge and happiness (Wiking, 2016, 2017).
27. **In 2019, the UK government's *Twenty-Five Year Environment Plan***: The UK government's 25-Year Environment Plan can be found at www.gov.uk/government/publications/25-year-environment-plan. The Fields in Trust calculations are at www.fieldsintrust.org/. The UK's Chief Medical Officer's annual reports of 2013 and 2019 contain recommendation on preventing health costs. If we spend to prevent NCDs, then the greater (personal and social) costs of treatment would be saved.
28. **The UK health system has extended prevention-pays into a new intervention model called Social Prescribing**: For a summary of green social prescribing references, see Pretty and Barton (2020).
29. **Recently conducted an evaluation of four programmes**: The data on the study of green social prescribing and NBIs/MBIs are from Pretty and Barton (2020). The annual World Happiness Reports are at https://worldhappiness.report/.
30. **Leslie Marmon Silko describes her homeland of north New Mexico**: Leslie Marmon Silko observations are from her delightful, *The Turquoise Ledge*. Antoine de Saint-Exupery wrote about the training for dangerous night flights in his autobiography, *Wind, Sand and Stars*.
31. **Well, how well do you feel connected to your place, asked Walter and Dorothy Schwarz in their book**: The connections to nature questions were formulated more than 30 years ago by Dorothy and Walter Schwarz in *Breaking Through*. Tyson Yunkaporta's observations, that's why we're here, are in his excellent *Sand Talk* (2019). The Ryokan poem on wild peonies is in *Dewdrops on a Lotus Leaf* (1993). Another form of connectedness come from the forests of the North West of America. Bears carry 150 salmon per day into the forest to eat. The roots of trees take up decaying protein and nutrients, with the bear-salmon providing some 75% of a tree's nitrogen requirements. Over time, bears prefer certain trees, and pass this preference to youngsters. Across generations of salmon and bear, the fir forest prospers.

Chapter 6: The Food and Agriculture Crisis: It's Nourishment, Not Calories

1. **Agricultural yields have increased steadily for 60 years**: The data and trends on total and per capita world food production are derived from the UN FAOSTAT database at www.faostat.org.
2. **Here is a short story from years gone by**: The Roughnecks American Football team was established by oil families working out of Great Yarmouth, and played at the north Suffolk village of Corton. The four US airbases were at Alconbury, Bentwaters, Lakenheath and Mildenhall. The coach was Lawrence Elkins, a former professional running back with the Huston Oilers and the Pittsburgh Steelers. See also Ron Davis's book on Lawrence Elkins (2019).
3. **Wild edible species still form part of the diets of a billion people**: For more on the widespread values of wild foods in contemporary agricultural landscapes, see the 2010 review by Bharucha and Pretty; and IIED (2020) *Indigenous Peoples' Food Systems Hold the Key to Feeding Humanity*.
4. **They also noted that these additional costs were disproportionately born by marginalized and underserved communities**: You pay three times for food: see

my *Agri-Culture* (Pretty, 2003) and *The Sustainable Intensification of Agriculture* (Pretty and Bharucha). The data on the huge global subsidies of food, agriculture and fishery systems, see Partha Dasgupta (2021).

5. **"Face it, it's all your own fat fault"**: The Boris Johnson quote on obesity is from an article in *The Daily Telegraph* newspaper in 2004.

6. **In Kenya, a remarkable system has been adopted by 250,000 small farmers**: The push-pull mixed agricultural system (*vutu sukumu* in Swahili) from East Africa was developed by Zeyaur Khan, John Pickett and many other colleagues at ICIPE, Nairobi and Rothamsted Research, UK.

7. **By 2020, agriculture worldwide was using 3.5 billion kilogrammes of pesticides each year**: The pesticide use and cost data is from Pretty and Bharucha (*Insects*, 2015) and Pretty (*Science*, 2018). The latest data on the numbers of farmers poisoned each year is reviewed and analysed in Wolfgang Boedeker and colleagues (2020).

8. **Yet, by the 1990s and early 2000s, the insects had largely gone**: on the widespread insect decline and loss data and the disappearance of the moth blizzard (Michael McCarthy first used the term moth snowstorm in the 2015 book *Moth Snowstorm*). See Charlotte Outhwaite and colleagues (2022) for the latest data on insect declines.

9. **The researcher and observer of food systems, Tim Lang**: An excellent analysis of the state of the UK food system is in Tim Lang's *Feeding Britain* (2020). *Food is Culture*, wrote Massimo Montari (2004). Food is as much performed as eaten.

10. **Farmer Managed Natural Regeneration (FMNR) has developed a thousand faces**: The Global Evergreening Alliance chaired by Dennis Garrity provides a collaborative platform to support environmental restoration and regenerative agriculture projects across the world. The goal is to capture 20 Gt (billion tonnes) of carbon annually by 2050. See www.evergreening.org. For more on FMNR, see www.fmnrhub.com.au.

11. **And then I came to a place, where I thought the roads terminated**: The Iroko tree in the next line from Ben Okri is hardwood tree of West Africa and can live for 500 years. See *The Famished Road*.

12. **And anyway, as Karen Lykke Syse of the University of Oslo exclaimed**: See Karen Lykke Syse and Martin Lee Muller M, *Sustainable Consumption and the Good Life* (2016).

13. **In the USA, the food system is US$1770 billion in size**: The costs of NCDs, especially of obesity, are at the Centers for Disease Control website: www.cdc.gov/obesity/adult/causes.html; in World Bank report on Obesity; in UK Chief Medical Officer annual report (2018); in the WHO *Global Status Report on Obesity* (2014). Our papers of 2015 and 2017 contains summaries of the UK health system and wider economic costs of all major NCDs: see Pretty J, Barton J, Bharucha Z P, Bragg R, Pencheon D, Wood C and Depledge M H. 2015. Improving health and well-being independently of GDP: dividends of greener and prosocial economies. *International Journal of Environmental Health Research* 11, 1–26; and Pretty J, Barton J and Rogerson. 2017. Green mind theory: how brain-body-behaviour links into natural and social environments for healthy habits. *International Journal of Environmental Research and Public Health* 14, 706.

14. **There was then a moment of genius for food growthers in the 1980s**: Carlos Monterio developed the term Ultra-Processed Foods: see Bee Wilson's *The Way We Eat Now* (2020). The US clinical trial on the effect of UPF consumption and diets is in Kevin Hall and colleagues (2019). See also Michael Pollan's *In Defense of Food* (2008).

15. **The top 100 fast food brands worldwide had 376,000 outlets in 2019**: The data on food brands and outlets has been gathered from individual annual reports of companies published online. The top 20 brands with 276,000 outlets are Subway, McDonalds, Starbucks, KFC, Burger King, Pizza Hut, Domino's, Dunkin', Baskin Robbins, Hunt Brothers Pizza, Taco Bell, Wendy's, Hardee's, Orange Julius, Papa John's Pizza, Dairy Queen, Little Caesars, Tim Hortons, CNHLS, Sonic Drive-In.

16. **Figure 6.1 Changes in obesity incidence by country:** Country data on obesity and diabetes incidence is drawn from World Bank and WHO datasets.

17. **Ruth Benedict in her anthropological observations on Japan:** There are many cultural insights in her fine 1946 book *The Chrysanthemum and The Sword*.

18. **Wendy Wood suggests that 75% of people think this willpower problem is a personal weakness**: See Wendy Wood's book *Good Habits, Bad Habits* for a review of habit formation, and in particular the deliberate approach of diet companies to keep customers.

19. **Figures 6.3–6.5**: The data on obesity, GDP, carbon emissions and meat consumption are drawn from World Bank, WHO and Our World in Data datasets.

20. **Norman Fischer says of food, "Every gift extends the heart and opens the spirit."** The quote from Norman Fischer on food is from *The World Could Be Otherwise* (2019), from Naomi Klein's on closing the gap from *On Fire*, and from Taiwŏn from Korean Zen: *Garden Chrysanthemum and First Mountain Snow*.

21. **There is one further link to the carbon costs of food: meat**. In the UK, 14% of people are vegetarian, 7% are vegan, and 31% say they eat less meat than they used to. Behaviours are changing. The footprints are instructive: mince beef from intensive livestock systems: 23 kg CO_2eq per kg of meat; plant-based meat analogue: 2.2 kg CO_2eq per kg; quorn/fungal analogue: 0.2 kg CO_2eq per kg.

22. **It was an easy message to remember**: Fruit and vegetable consumption in the UK and USA is from Defra and CDC datasets. Five colours on the plate: Michael Pollan's *In Defense of Food*. Eating the rainbow (colours on the plate) of Okinawan diets and dishes is from Héctor Garcia and Francesc Miralles's *Ikigai* (2017).

23. **The EAT-Lancet Commission in 2019called for a great food transformation**: For more on the EAT-Lancet Commission, see https://eatforum.org/eat-lancet-commission/.

24. **The high consumption of plant variety in South Korea**: The innovative and effective promotion of 100 dishes for the Korean Diet (K-diet) is discussed and shown in Soon Hee Kim and colleagues (2016).

25. **In the late 1960s and 1970s, Barry Popkin identified how nutrition transitions had begun to occur**: Barry Popkin was the first to use the term nutrition transition and has changed thinking about food customs and norms. The Dogen quote is from Kazuaki Tanahashi's *Moon in a Dewdrop*.

26. **In 2020, Yanping Li and colleagues at the Harvard Chan School of Public Health**: The study of 40,000 nurses: see Li and colleagues (2020). John Ji's work on the oldest old in China: see John Ji and colleagues (2019).

27. **Over a 20-year period, Cuba became part of an instructive but unintentional experiment**: The data on the enforced experiment in Cuba on health changes during the 1990s and then after sanctions were listed are from two seminal papers by Manuel Franco and colleagues (2007, 2013).

28. **The importance of the social wrapper to food and health was also demonstrated by remarkable research led by Michael Marmot**: For the remarkable studies of changing health indicators after migration from Japan to the USA and Canada, see Michael Marmot and Leonard Syme (1976), and Boji Huang and colleagues at the University of Hawai'i (1996).

29. **A version of the Hippocratic Oath on food**: Shunryō Masuno is head priest at Kenko-ji temple, and the quote is from his *Zen: The Simple Art of Living* (2019).

30. **Roy Taylor of the University of Newcastle has shown how type 2 diabetes can be reversed**: Roy Taylor's successful programme on extreme calorie restriction to reverse diabetes is in his 2021 book, *Your Simple Guide to Reversing Type 2 Diabetes*.

31. **In the late 1960s, after a six-year stay in Japan**: The lovely story on the gift of the three grapefruits is from Taitetsu Unno's *Shin Buddhism*. The Raizan haiku is from William Scott Wilson.

Chapter 7: The Best Things in Life: How Immersion and Flow Make the World a Better Place

1. **Beginner's Mind: A Trickster Story:** The excellent *The Leaping Hare* by George Ewart Evans and D Thomson covers all aspects of ecological rural and folklore aspects of the hare, and includes this story on the Baby Wasis and the Great Hare. This story also appears in Richard Erdoes and Alfonso Ortiz's *American Indian Myths and Legends* (1984).

2. **The Great Hare was a clever fellow**: For key texts on tricksters, see Lewis Hyde's *Trickster Makes This World*; Richard Erdoes and Alfonso Ortiz's *American Indian Myths and Legends* and *American Indian Trickster Tales*; Italo Calvino's *Italian Folktales*; and Janet Yolen's *Favourite Folktales from Around the World*.

3. **Ruth Benedict wrote of a saying in Japan**: The young child is born happy, it is the task of parents to keep them happy. This is in her 1946 book, *The Chrysanthemum and The Sword*.

4. **Habits are a wonderful thing**: On the formation and breaking of habits: see these aggregator books by Wendy Wood (*Good Habits, Bad Habits*), Mark Williams and Danny Penman (*Mindfulness*), Judson Brewer (*The Craving Mind*); also see the review by Bas Verplanken and Sheina Orbell (2022).

5. **Joan Borysenko calls this the dirty tricks department of the mind**: See Joan Borysenko's *Mending the Body, Mending the Mind*.

6. **Rainer Maria Rilke wrote, "You are not dead yet, it's not too late:"** The Rainer Maria Rilke quote is from *The Poetry of Rainer Maria Rilke* (1982).

7. **Here's an example. Norman Doidge in *The Brain's Way of Healing* described**: unlocking the forced habits of Parkinson's disease, see Norman Doidge's 2014 book on brain plasticity, *The Brain's Way of Healing*.

8. **Catherine Gray has written with great wit of her dispatch of alcohol**: See her personal account in *Sunshine Warm Sober*.

9. **Habits are all about plasticity. This is a term that means the brain changes by doing something with our body**: Daniel Goleman and Richard Davidson in *Altered Traits* (also called *The Science of Meditation*) discuss plasticity and training for new habits.

10. **The journey of a thousand miles starts with a single step, acknowledged Stephen Kotler of the Flow Research Collective**: See Stephen Kotler's *Art of the Impossible* (2021), and Kotler and Wheal's *Stealing Fire* (2017).

11. **"The world is, dust and dirt, Flows away, all is purified**: Otagiki Rengetsu was an Edo poet and nun, and lived from 1791 to 1875. See John Stevens' *Rengetsu: Life and Poetry of Lotus Moon* (2014).

12. **Flow Activities:** The term flow emerged from the long career of Mihalyi Csikzentmihalyi and has lately been advanced by Steve Kotler and the Flow Research Collective. A wide range of spiritual and scientific literature centres on the values of attentiveness or attention. I have used the four terms to cover different aspects of engagement that lead to health benefits: attentive, immersed, focused and flow. You are inside an activity, extraneous thoughts and chatter disappear, and the blue brain is up-regulated.

13. **Andrew Newberg and Mark Waldman of Thomas Jefferson University surveyed 1000 people to explore their varied experiences of enlightenment**: You will find an excellent discussion of enlightenment in *How Enlightenment Changes the Brain* by Andrew Newberg and Mark Waldman (2016), and the outcome of the results of their worldwide survey on the common experiences of enlightenment across many cultures. The authors use the terms "little e" and "big E" to describe differences in outcomes from attentiveness and flow.

14. **Amy Isham, Tim Jackson and colleagues at the University of Surrey have recently shown how flow activities**: For excellent research on how flow and low carbon activities and behaviours work together, see Tim Jackson's *Post Growth* for a summary, and Amy Isham's papers for original research (Isham et al., 2019a, 2019b).

15. **The importance of the lived experiences of the body**: This has been highlighted in Bessel van der Kolk's revealing, shocking and ultimately uplifting book, *The Body*

Keeps the Score (2014). Trauma and abuse is more common that most people think. Some of the solution lie in physical activities that engage the body to shift the brain and mind.

16. **The body and mind are in recovery**: Bessel van der Kolk further notes that abuse and neglect, especially if chronic, makes people feel numb and they shut down. This is one reason why self-harm (such as by skin cutting) makes sufferers feel alive, to feel much better. It is a physical response, and can have an immediate effect. From the outside, it looks non-rational behaviour, but it is rational. Chronically-traumatised people often also feel unsafe in their own body, as the past is painful and alive. There is wordless knowledge (my heart is broken, I was choked up, it made me sick, it made my skin crawl), and sufferers tend to respond to new stresses with migraine, headache, asthma attack, fibromyalgia, chronic fatigue and other chronic body symptoms. Traumatised children have 50 times the rate of asthma than non-traumatised.

17. **As we shall see, these choices and consequences relating to attentiveness, immersion, focus and flow**: See all of Martin Laird's books on flow; the inner castle was the term used by St. Teresa of Ávila.

18. **The light comes from within, wrote Taitetsu Unno**: For Taitetsu Unno, see *Shin Buddhism: Little Bits of Rubble Turn into Gold*.

19. **Mark Twain is said to have written in his later life**: "I am old man and have known a great many troubles, yet most of them never happened." I am grateful to Roger Morris for pointing out that this quote is variously attributed to Mark Twain, Thomas Jefferson, Martin Farquhar Tupper, Seneca, Winston Churchill, James A. Garfield, Thomas Dixon, and Michel de Montaigne. Long ago, Seneca also said, "There is nothing so wretched or foolish as to anticipate misfortunes. What madness it is in your expecting evil before it arrives."

20. **The trouble is, this noise often also has direction as an inner judge**: on thoughts and the noise of mental chatter: see Thich Nhat Hanh's *Silence*, Martin Laird's *An Ocean of Light* and *A Sunlit Absence*, Joan Borysenko's *Mending the Body, Mending the Mind*.

21. **You have to choose something for the body to do**: Transient hypofrontality was a term first used by Arne Dietrich (2003). A wandering mind is an unhappy mind: see the research by Matthew Killingsworth and Daniel Gilbert (2010).

22. **This is also related to what was called the Relaxation Response**: For an example of the huge array of genes switched on and off by body behaviours (the process of epigenetics), see Herbert Benson on the use of the term Relaxation Response, which has now been found to activate more than 4000 genes.

23. **The Idea of a Blue Brain and a Red Brain**: Key aggregator books include Norman Doidge's *The Brain's Way of Healing* (2015), Sharon Begley's *Plastic Mind* (2009), Daniel Levitin's *The Changing Mind* (2020), Daniel Siegel's *Aware* (2018), Judson Brewer's *The Craving Mind* (2017).

24. **Let's look at these structures in more detail**: For more on the split between the functions of the red brain and the blue, see Daniel Kahneman *Thinking Fast and Slow* (2011); Daniel Goleman's *Focus* (2013), Andrew Newberg and Mark Waldman's *How Enlightenment Changes the Brain* (2016).

25. **All humans have pretty much the same starting genetic legacy**: One way to think about these brain structures is to hold up your arm and close your fist. The spinal column is the arm and wrist, the red brain is the lower palm, the mid-brain structures are the thumb, and the four fingers folded over represent the lobes of the cortex. Between the red and the blue are a series of mid-brain or limbic structures in the central region, comprising hippocampus, hypothalamus and thalamus, and amygdala. These are signalling centres, and drivers for emotion, memory formation and bonding. Some of these exist as pairs, the hippocampus and amygdala, but we speak of them in the singular. There are then a large number of hormones and peptides carrying signals between these structures. These neurotransmitters are switches in themselves, often performing more than one job. Some amplify focus and pattern recognition, others produce tighter bonds and heightened cooperation between people. Some define our sense of self.

The most important neurotransmitters include: *Serotonin*: the key to sleep and mood; *Dopamine*: the aid to approach, attention and reward, but when levels fall it makes us feel unhappy; *Norepinephrine*: promotes alertness and arousal; *Acetylcholine*: promotes wakefulness and learning; *Opioids*: reduce pain and buffer stress; *Oxytocin*: plays a key role in social bonding; *Cortisol*: the stress hormone that stimulates the amygdala and inhibits the hippocampus; *Oestrogen*: plays a key in memory; *Adrenalin*: the stress hormone that kicks the red brain into action; *Anandamide*: a fatty acid that causes pleasure.

26. **Loneliness is an example**: On loneliness, see John Caecioppo and Louise Hawkley of the University of Chicago (Cacioppo et al., 2003, Hawkley and Cacioppo, 2007), and Julianne Holt-Lunstad and colleagues at Brigham Young University (2010, 2015, 2017).

27. **A second problem is that we mistakenly take these B-type thoughts to be ourselves, and so become attached to them**: On the types of thoughts and first/second arrows, see this range of Buddhist texts – Matthieu Ricard's *Happiness* (2015), Jack Kornfield's *A Path With Heart* (2002), Shinzen Young's *Science of Enlightenment* (2016), James Kingsland's *Siddhartha's Brain* (2016), Alan Wallace's *The Attention Revolution* (2006), Rick Hanson's *Buddha's Brain* (2009), Daniel Goleman and Richard Davidson's *The Science of Meditation* (also called *Altered Traits*, 2017).

28. **The idea is not to stop thinking, just to limit the B-type thoughts**: You might be wondering about some other big concepts around brain and mind that need explaining. I have deliberately not used the terms consciousness, the unconscious or sub-conscious. This is not to say they are not relevant, just that they might not be necessary here. Steve Hagen said in *The Grand Delusion*, "We haven't a clue as to what constitutes mind or consciousness." We are either attentive to the present, or drawing on memories and habits formed previously during our life course. We act or respond automatically (red brain) or with cognitive effort (blue brain). We sleep, then we're awake. That could be said to be about it.

29. **When Martin Laird says *be the mountain*, or Bruce Lee says *be water***: These quotes on flow are from Martin Laird (2006, 2011, 2019); on Bruce Lee (Lee Jun-fan), see books by Joe Hyams and his daughter Shannon Lee.

30. **Figure 7.1 The enso-circle for the green mind**: The Enso is the black ink semi-completed circle used widely in Zen Buddhism. It is drawn with a single sweep of a brush, and encloses a wide and vast space. It is also never ended to be perfect. The imperfections reflect how life unfolds.

31. **At the centre is the source, the luminous mind surrounded here by the circular form of the brushed enso**: The *enso* of the green mind can also thus be seen as the circle of enlightenment, empty yet full. It also represents the beginner's mind, the state at which we all start at birth. Babies at this age are highly alert and respond to incoming signals. But they do not yet have type B thought responses. So from this empty space comes our whole lives, in a similar way to the universe emerging from nothing at the Big Bang.

32. **It is these four activities and states that encourage well-being**: The brain mechanisms work like this: being attentive activates the parasympathetic nervous system (PNS), and increases neurone growth in the insula, hippocampus and prefrontal cortex, particularly the left prefrontal cortex where feelings of well-being reside. A steady release of dopamine is produced, further increasing well-being. The PNS decreases the stress hormone cortisol, and strengthens activity of the immune system. We feel happier.

33. **This immersion and flow can feel like we are on a high vantage point**: "The *Bhagavad Gita* calls this watchful insight, implying also clarity of mind." See *The Bhagavad Gita* (trans Laurie Patton, 2008).

34. **Carl Jung called the self, "A circle of light"**: For a discussion on Carl Jung, see Joseph Campbell's (1949) *The Hero with a Thousand Faces*.

35. **The Magic Box**: This story comes from Ruth Sawyer's wonderful 1942 book *The Way of the Storyteller*.

36. **The Anglo-Saxon Seafarer said, "I have spent my life on the ice-chilled sea"**: *The Seafarer* is a famed Anglo-Saxon poem from the Exeter Book codex. See translation

in Richard North, Joe Allard and Patricia Gillies: *Longman Anthology of Old English, Old Icelandic and Anglo-Norman Literature.*

37. **We each have available, at any point in life, behaviours that can be called on**. An area of contention centres on contemporary views of genetics. It is a central theme of this book that the environment around us, the web of social, ecological and economic conditions in which we live and grow, that conditions most of who and what we are. The rule of 80:20 is a good guide: studies of identical twins generally show about 20% of the differences between individuals result from genetic differences, and 80% arise from their differing individual life histories from birth. We should take it that most of us start from roughly the same place. We then go on to become individuals fashioned by our emerging and unique lifeways.

No geneticist would say this today, but is has become popular to believe that behaviours, preferences and personalities are mostly caused by innate and thus inherited differences. Some of this worldview came about from socially-conditioned interpretations of Darwinian natural selection, certainly some was guided by the selfish gene. All life, wrote Richard Dawkins in his 1976 book *The Selfish Gene*, is a vehicle for DNA, and so responds to genetic instruction. And because DNA wants to survive, it must act selfishly by instructing its carrying organism to act selfishly too. This played into the hands of neoliberalism when it also wished to suggest that selfishness was the most efficient state for human economies. These aspects of biological determinism demonstrated little for the value of togetherness, and came to be challenged by many.

It undermined notions of free will, it allowed the creep of views that suggested social and racial divides were natural. Moreover, if each of us really is determined only by inherited DNA, then there can be no chance of escape during a lifetime. It's all downhill from birth.

Let us go just back to the start, before this dark night. We humans carry genes that determine our body and brain architectures. After birth, each of us starts in much the same place, with a largely common genetic legacy and a beginner's mind. The environment has not begun its work. We have the same nervous systems and hormones, the same brain structure and inherited capabilities to fight diseases, process foods and reproduce. We then become more unique as we grow and respond to circumstances. We eat, learn habits, bond and learn the rules and norms of our local culture. The science of epigenetics has now also indicated that our body worlds are not unidirectional. Genes are switched on and off by the inner actions of our bodies and by signals from outside. Some genes sleep, others are awoken.

The environment switches genes on and off. Our behaviours do the same. And when environments change, we change too. No organisms live and evolve in static environments. They amend their circumstances, thus shifting the course of their own life success and longer-term evolution. David Krakauer of the Santa Fe Institute has said, "Organisms are builders engaged actively in the planet's construction." Living things modify their environments, trying to make conditions more favourable to survival. Earthworms change the structure and composition of soils by dragging leaves and other organic matter into the soil. Elephants uproot trees, open canopy, create parkland and recycle herbage through their bodies that in turn reduces the incidence of fires. Wild boar create open ground and aid tree germination, and beavers form riverside water meadows and coppice willow to produce more food. The Dawkins view on the selfishness of genes and the primacy of biological determinism was strongly challenged by geneticist Richard Lewontin, paleontologist Stephen Gould and many others: see especially *Not in Our Genes* (1984) by Lewontin, Steven Rose and Leo Kamin. See Nessa Carey's excellent book on *The Epigenetics Revolution* (2011).

38. **But trouble is brewing. The *attentional commons*, a term coined by Matthew Crawford**: Matthew Crawford discusses the Attentional Commons in *Beyond Your Head*, raising important questions around corporate and capitalist enclosures of space we think of as a common. The most extensive invasions and manipulations occur in private spaces such as casinos that are designed to be inviting.

39. **"Wisdom sits in places," said elder Dudley Patterson to Keith Basso**: Keith Basso's two excellent book on the Western Apache and White Mountain Apache are *Wisdom Sits in Places* and *Don't Let the Sun Step Over You*, the account of the life of Eva Tulene Watt.

40. **The Western Apache have a particular phrase when talking about the importance of stories in places**: Places with stories acquire a certain significance in people's lives, and can lead to their protection. A bridge might become a magic vessel. In the 1200s, Dante Alighieri wrote about seeing a young woman called Beatrice on the Ponte Vecchio crossing the Arno River in Florence. Beatrice became his true love (in his eyes), but died young married to a merchant of the city. The famed story travelled down the centuries, and in the Second World War, the bridge was the only one crossing the Arno that was not bombed by Allied forces and also not destroyed by defending German forces. Some say the story of Dante's lost love saved the bridge.

41. **Joe Hyams said to actor and martial artist Bruce Lee**: The Joe Hyams (1979) book on Bruce Lee is *Zen in the Martial Arts*. See also Shannon Lee's (2020) book on her father, Bruce Lee, *Be Water, My Friend*.

Chapter 8: Enoughness: Creative Slowth Is Better Than Infinite Growth

1. **In 2020, ConocoPhillips' Arctic sought approval for the Willow Project**: On the January morning of the Biden inauguration, the Trump administration signed the final permission to allow the Willow Project to proceed. The vital role of national political leadership in preventing planetary climate cascades and breaching of tipping points was later demonstrated when the project was finally not approved by the Biden administration in October 2021. Climate cascades could become more common after the many national fossil fuel dependencies were revealed by the Ukraine war, the imposition of economic sanctions on Russia, and the fracturing of a growing global consensus over actions to prevent the worsening climate crisis. Fossil fuel extraction may thus accelerate as national leaderships elect for new extractive technologies that offer quick returns. These still become stranded assets, as the climate changes further and as green transitions to net zero spread.

2. **At the same time, the largest soft drinks firm in the world**: You can read about Coca-Cola in their published Annual Reports.

3. **This conspicuous consumption, as Thorstein Veblen first called it**: On the first use of conspicuous consumption, see Thorstein Veblen 1899 book *The Theory of the Leisure Class* (for a summary, see *Conspicuous Consumption*, 2005). The piece from Charles Kettering of General Motors can be read in full at https://wwnorton.com/college/history/archive/resources/documents/ch27_02.htm. On the question of enoughness, see Barry Schwartz *The Paradox of Choice* and Robert Frank's *Luxury Fever*. See also my 2013 paper: "The consumption of a finite planet" in *Environmental and Resource Economics*.

4. **Tim Kasser of Knox College in Illinois**: see his *The High Price of Materialism* (2002).

5. **Victoria Husted Medvec and colleagues at Cornell University showed in the mid-1990s that Olympic bronze medallists**: The iconic work by Victoria Husted Medvec and colleagues on Olympic medaling is in Medvec et al. (1995). Bronze medalists are happier than silver, even though all set out to finish as near gold as possible.

6. **When I talked with the Right Livelihood winner, Chilean economist Manfred Max-Neef**: I met Manfred Max-Neef whilst at the International Institute for Environment and Development in London in the late 1980s.

7. **In 2018, 238 scientists called on the EU to abandon GDP**: The calls from several thousand natural and social scientists to abandon GDP as a measure of progress and success: see Domink Wiedenhofer and colleagues (2020) and William Ripple and more than 15,000 others (2017).

8. **"What a glorious luxury it is, to taste life**: The Yoshido Kenko quote is from his 14th-century *Essays on Idleness*.

9. **Look around. Everywhere there are signs of enoughness**: I have used the term *enoughness* to suggest there is a state beyond which we should not need to proceed (for well-being, happiness and contentment). For key further texts, see Tim Jackson's *Beyond Growth* (2021), Gesshin Clare Greenwood's *Just Enough* (2019). Daniel Kahnemen and co-authors write of the Goldilocks point for decisions in *Noise* (2021).

10. **Doris Fuchs and colleagues at the University of Münster**: See the excellent Fuchs and colleagues' book, *Consumption Corridors*. See also Thomas Princen's Sufficiency (2005), and Michael Maniates (in Fuchs et al., 2021). On sufficiency, see the review by Jungell-Michelsson and Heikkurinen (2022). Sufficiency is an end in itself, and a means for action.

11. **A type of magical thinking is also at work**: Michael Maniates (in Fuchs et al., 2019) does not believe individual actions work: "There is scant empirical evidence that individual actions of environmental stewardship lead people to meaningful political action." You would hope, I would hope, this is simply being deliberately cynical to provoke this very action. He further writes, "It is silly to believe a cabal of evil-doers is working overtime to disempower environmentally-concerned publics." Well, Kurt Anderson in *Evil Geniuses* presents another view.

12. **In the 1960s, the Chief Economist to the British Coal Board had been invited on an aid and advice trip to Burma**: You can read E F (Fritz) Schumacher in his prescient *Small is Beautiful* (1973).

13. **The investor and philanthropist John Bogle reported**: The encounter between Kurt Vonnegut and Joseph Heller about enough is recounted in John Bogle's (2010) book *Enough*.

14. **"After my death, what would I claim"**: The Taitetsu Unno quote on a mountain life is from his book, *River of Fire, River of Water* (1998).

15. **What might a stable economy look like that does not breach planetary and health boundaries**: The quotes from Herman Daly are from his fine 1995 book, *Beyond Growth*. Already a generation old, it was both a wise summary of the contemporary predicament and stand well a generation later.

16. **Tim Kasser says this, by way of a recommendation**: This list of activities that will bring a good life appears in his book, *The High Price of Materialism*.

17. **When we cross-pollinate sustainability with the good life**: See Karen Lykke Syse and Martin Lee Muller *Sustainable Consumption and the Good Life* (2016).

18. **The land of always-growth is not going to trickle down**: See Kate Soper's "Towards sustainable flourishing: democracy, hedonism and the politics of prosperity" in Syse and Muller: *Sustainable Consumption and the Good Life* (2016).

19. **An abundant flourishing occurred in Japan's Edo period**: You can read about the Edo Period of Japan (1603–1868) in many texts. On Shinto, see Sokyo Ono' *Shinto* (1962) and Helen Hardacre's *Shinto* (2017); on the growth pilgrimages, see Hardacre again, also Robert Sibley's *The Way of the 88 Temples*; on Edo art and printmaking, see Matthi Ferrer's book on *Hokusai*; on Edo haiku poetry, see the works of Bashō and associated poets of the 1600s and 1700s. The Vicennial Renewal (every 20 years) of the Ise shrines are still hugely popular: on the last occasion in 2013, there were 14 million visitors.

20. **Ben Dooley and Hisako Ueno wrote in The Japan Times in 2020**: Shinese shops are discussed by Ben Dooley and Hisako Ueno in The Japan Times (2020) article, Old Shops.

21. **When Helena Norberg-Hodge first travelled to Ladakh in the 1970s**: You can find out more about the Ladakh in which Helena Norberg-Hodge lived in *Ancient Futures*.

22. **Decoupling has not occurred in ways that can offset the overall carbon and material footprint of consumption**: You can read more of the difficulties and controversies of green growth in Tim Jackson's *Post Growth*, Jason Hickel's *Less is More*, and Naomi Kline's *On Fire*. The review of 11,000 papers on GDP is the paper by Dominik Wiedenhofer and sixteen colleagues (2020). The Club of Rome report of 1972, *The Limits to Growth*, made clear fifty years ago that a total of worldwide consumption was

always likely to swamp green efficiencies made to cut costs a variety of economic sectors. For a recent analysis, see Paul Ekins' analysis in UNEP's Geo-6 Report of 2019. Degrowth is analysed by Inĕs Cosme and Daniel O'Neill in 2017.

23. **If a country invests in renewable energy by solar photovoltaics**: The extensive and remarkable work of the NGO Grameen (and Grameen Shakti) in Bangladesh can be found at https://grameenfoundation.org/.

24. **Another course will be carbon capture and carbon trading**: Michael Mann and Oxford colleagues have been working on identifying 1 Gt projects that could help transform economies by both cutting carbon emissions and creating new sinks.

25. **It's my snow, I think"**: Takari Kikaku (1661–1707) was a haiku poet and pupil of Matsuo Bashō.

26. **And then there might be space for doing not very much**: See Jenny Odell's *How To Do Nothing* (2019).

27. **Personal Growth and Infinity Games:** Infinity games are key to personal growth: see Stephen Kotler's *The Art of the Impossible*. We engage in activities and pursuits (sometimes disparagingly called hobbies) that cannot be perfected. We learn a level of skill, and this deepens the practice, and we find we have more to learn. This can last a lifetime, as we see from the quotes and lives of Pablo Casals and Martha Graham, and the work of sculptor Bernard Lainé. Bill McKibben has described his own infinity game of cross-country skiing in *Falter*: there is always someone ahead and someone behind. The work of Arne Naess and an essay by Alan Drengson can be found in the book, *The Ecology of Wisdom*.

28. **An example of infinity games comes from the anthropologist Wade Davis**: The anthropologist Wade Davis writes of a wide range of cultural aspects of the good life in *The Wayfinders*.

29. **And so, asked Mark Williams and Danny Penman of Oxford**: See Barbara Ehrenreich's *Dancing in the Streets*, and Mark Williams and Danny Penman's very wise question in their book, *Mindfulness*: when did you stop dancing? The Rumi poem is from *Selected Poems* (1995).

30. **Many have written of the tea ceremony in Japan**: On the Japanese tea ceremony, see William Scott Wilson's *The One Taste of Truth*. The brilliant Richard Sennett writes of craft in *The Craftsman*; see also Alexander Langland's *Cræft* and Peter Korn's *Why We Make This*. The Japanese sword ceremonies of manufacture are depicted in the BBC TV programme *Handmade in Japan*.

31. **Alexandra Lamont and Nelinne Ranaweera of Keele University**: On the mental health benefits of knitting and musicians, see Alexandra Lamont and Nelinne Ranaweera (2020).

32. **The potter Bernard Leach and Sōetsu Yanagi, founder of the *mingai-kai* Japanese Craft Society**: The life and work of potter Shōji Hanada is described in *The Unknown Craftsman* by Bernard Potter and Sōetsu Yanagi. The article by Kathryn Wortley on contemporary crafts in Japan is in The Japan Times article: Masters of their Craft (28 Nov 2020).

33. **There is still a remarkable variety of craft traditions in Japan today**. Some are new entrants, some are part of family traditions that have last hundreds of years. Each has this in common: taking time, slowness of production, letting ideas and products season, letting the appearances of pottery, copperware, lacquerware, calligraphy and wood emerge and change. See Uwe Röttgen and Katharina Zetti (2020) *Craftland Japan*. There is Toshihisa Yoshizawa with 350 years of family work as bladesmith, who binds 25 layers of steel to create ripples in the metal. There is Suzuki Morihawa, a 15th generation kettle maker, 400 years of producing just 20 kettles a month. There is Masami Mizuno's copperware which changes colour when heated: "I sometimes burn them to make their colour like that of a used item." Junko Yashino uses seeds and tin powder in lacquerware: "I like to include some imperfections in my works, like a crack, a space to breathe, in otherwise perfectly lacquered surfaces." Kazuro Uganai and family in Nagano have been turning wood for 1000 years, and are called wood grain masters. Isshú and Shukin Moroya create calligraphy, and Shukin says, "When I most get inspired is when my mind is so clear and I'm so alert. I am very aware of what's happening. Each piece is

a journey." Her husband Isshú says: "A journey means, that you have more, when you come back."

34. **Slow can also mean more time to be attentive**: See Victoria Sweet's *Slow Medicine* (2017) and Gavin Francis's *Recovery* (2021). Peter Beresford and Susan Carr are leading researchers and practitioners on new forms of engaged social and health policy: see the edited volume *Social Policy First Hand* (2019).

35. **Musashi was the most famed Samurai of all**: On the life of Musashi, see *The Lone Samurai* by William Scott Wilson (2013). The quote from Saichō is from John Stevens' superb book on the marathon monks of Kyoto, *The Marathon Monks of Mount Hiei*.

36. Keith Basso has written of the Western Apache people in *Wisdom Sits in Places*. Richard Nelson's sensitive account of the Koyukon is in *Make Prayers to the Raven*. Koyukon stories begin with, "Wait, I see something," and Western Apache start with "It happened at . . ."

37. You will find more on stories and sagas in my *Sea Sagas of the North* (2022). There are superb books on the original locations of Iceland sagas and how they look in contemporary times: see W G Collingwood and Stefanson's *A Pilgrimage to the Saga-Steads of Iceland* (1899), Jon Krakauer and David Roberts' *Iceland* (1990), and Richard Fidler and Kári Gíslason's *Sagaland* (2017).

38. The Antoine de Saint-Exupery excerpt is from his wonderful autobiography, *Wind, Sand and Stars* (1939).

Chapter 9: Public Engagement and New Power: The Race to Net Zero

1. **On new power**: see Stephen Timms and Jeremy Heimans (2018) and Jonathan Grant (2021).

2. **It has been well-established that there is a considerable and positive participation premium**: see (Chambers, 1989, 2008; Cernea, 1991; Gibson, 1996; Pretty and Ward, 2001; Beresford and Carr, 2019; Wilson, 2019; Pretty et al., 2020; Beresford, 2021).

3. **This has been called a "deficit-model"**: The "deficit-model" of engagement or communication implies valuable knowledge is transferred to fill a gap for recipients who are assumed not to know something, and thus this should be welcomed by them. See Irwin (1995), Haggar-Johnson et al. (2013) and Wynne (2014).

4. **Yet, paradigms do shift**: On paradigms, see Thomas Kuhn's *The Structure of Scientific Revolutions* and Max Planck's *Scientific Autobiography and Other Papers*

5. **For more on GetUp! Australia**: See the website People, Power, Impact. At www. getup.org.au/.

6. **For more on DonorsChoose**, see Del Valle (2019), and the website at DonorsChoose www.donorschoose.org/.

7. **For more on the NASA Open Innovation platforms and impact**, see Lifshitz-Assaf et al. (2018) and NASA, 2022: Open Innovation. www.nasa.gov/offices/otps/openinnovation. On open innovation at consumer goods giant Proctor and Gamble, see Ozkan (2015).

8. **Wrote Hildegard of Bingen in 1152, "I am a fragile vessel"**: letter to Elisabeth of Schönau. In Hildegard of Bingen, *Selected Writings* (trans Mark Atherton, 2001).

9. **On ensemble awareness and social learning**, see Bawden (1991, 2006, 2011); Pretty (1995); Wilson (2019)

10. **The first typology to express these variations was developed as a "ladder of participation"**: A number of spectrums, ladders and typologies of participation have been developed and refined since Sherry Arnstein in 1969: see Pretty (1995); Haklay (2013); Beresford and Carr (2019); Nielsen (2018); Jünger and Fähnrick (2020); Johnston and Lane (2021); Slotterbach and Lauria (2019).

11. **Scientists must learn to communicate**: On public understanding of science, see Royal Society (1985), and also where there is an assumed gap between science and society, see Miller (2001) and Entradas (2015).

12. **The typology of public engagement:** The sources for this typology are Pretty (2022c), adapted from Arnstein (1969); Bhatnagar and Williams (1992); Pretty (1995); Haklay (2013); Beresford and Carr (2019); Nielsen (2018); Jünger and Fähnrick (2020); Johnston and Lane (2021); Slotterbach and Lauria (2019).

13. **It has been argued that public engagement has become harder precisely because of breakdowns in social capital:** Public engagement has become harder precisely because of breakdowns in social capital: Lesen (2016); Milne (2018).

14. **Paul Mason has observed that "Capitalism is a complex adaptive system . . ."** see Paul Mason (2015): *Post Capitalism.*

15. **We can distinguish three levels of learning:** three levels of learning, in which there is a transition from learning information to meta-learning, and then to epistemic learning: see Bateson (1972); Argyris and Schön (1974); Bawden (1991); Bawden et al. (2007); Ison and Russell (2007); Beresford and Carr (2019).

16. **On the Hawkesbury and Michigan State innovations,** see Richard Bawden (2006, 2011).

17. **A good example of a successful educational public engagement programme is bibliotherapy in prisons:** bibliotherapy in prisons: see Jarjoura and Krumhat (1998); Waxler (2008); Schutt et al. (2013). The **University of Leeds *Writing Back* programme** (2022). Writing Back. https://ahc.leeds.ac.uk/homepage/270/writing_back

18. **Malcolm Green is a storyteller and environmental educator:** On Story Telling about Kittiwakes, see Gersie et al. (2014).

19. **On a nearby branch, a hototogisu cuckoo once was singing:** in Lady Sarashina Nikki, *As I Crossed a Bridge of Dreams* (1008–c1059).

20. **Recent years have seen a flowering of methods, approaches and projects involving citizens as scientists:** Citizen science to engage large numbers of people in gathering unprecedented quantities of good quality data, see Haklay (2013); Heigl et al. (2019); Strasser et al. (2019); Sauermann et al. (2020); Joxhorst et al. (2020).

21. **Citizen science has the power to change behaviours:** See overview by Ryan et al. (2018), and on shaping progress towards achieving the UN's Sustainable Development Goals, see Fritz et al. (2019). **The European Citizen Science Association and US Citizen Science Association:** see ECSA (2022); Citizen Science Association (2022). **The Belgian CurieuzeNeuzen project:** see Kreitzer et al. (2015); de Craemer et al. (2019); van Brussel and Huyse (2019). **The eBird citizen science initiative** from the Cornell Lab of Ornithology and National Audubon Society: see La Sorte and Somville (2020); Sauermann et al. (2020). **Drain detectives in Victoria, Australia:** see Cottam et al. (2021). **The Mildew Mania project** has been run by Curtin University in Western Australia: see Ryan et al. (2018). A study of 28 nature-based citizen science butterfly projects in the USA found citizens involved in the projects has greater knowledge of threats and used their knowledge and commitment to persuade friends, family and co-workers of the value of butterfly and habitat conservation: see Lewandowski and Oberhauser (2016).

22. **The innovative agricultural pesticide-reduction programme in northern Ecuador:** this wonderfully novel project was written about in the book, *The Pesticide Detox:* see chapter by Sherwood et al. (2005). **Science and Technology Backyard Platforms, China:** a fascinating and effective method of building more sustainable agricultural systems. See Zhang et al. (2016). **The MyShake project:** developed by Richard Allen and colleagues at the Berkeley Seismology Lab: see Allen et al. (2020). **Mad studies** seeks to create alternatives to simple medical models of mental ill-health: see Le Francois et al. (2013); Beresford and Russo (2022).

23. **Participatory Budgeting (PB):** a significant social innovation with major positive impacts in cities where it has been implemented: see Touchton and Wampler (2014); Touchton et al. (2017); Wampler et al. (2018).

24. **Place-Based Climate Action Network:** see PCAN (2022).

25. **PatientsLikeMe (PLM):** an online community of 930,000 people: see Wicks et al. (2018); Borentain et al. (2020); PLM (2022). **Buurtzorg Community Health Care, Netherlands:** *see* Kreitzer et al. (2015); Drennan et al. (2018; Leask et al. (2020).

Another example of a very large-scale online community is Avaaz: established in 2007 to use the power of online petitions to affect political decisions worldwide, it has grown to have 70 million members in 194 countries. It promotes global activism on climate change, on human and animal rights, and on corruption, poverty and conflict. Avaaz means "voice" in several European, Middle Eastern and Asian languages. See https://secure.avaaz.org/page/en/.

26. **Grameen Bank Microfinance in Groups**: *see* Rahman (2019); BRAC (2022); Grameen Bank (2022); Proshika (2022). **As we saw earlier, there has been remarkable innovation in rural social capital across the world, particularly in emerging economics**: In Cuba, the Campesino-a-Campesino movement developed out of radical approach to agroecological integration that is redesigning rural systems. It is centred on a Freirean social communication method using adult educational principles. Farmers spread knowledge and technologies to each other through field exchanges, teaching and establishment of cooperatives. There are 100,000 farmers of Campesino-a-Campesino in Cuba: the productivity of this sector has increased by 150% over ten years, and pesticide use is down to 15% of former levels. They are more self-reliant as a result. See Rosset et al. (2011); Freire (1970).

27. **Social Prescribing in the UK**: see Sweet (2017); Beresford and Carr (2019); Holmes et al. (2019); Zaki (2018); Francis (2021). For green SP, see NASP (2019); Pretty and Barton (2020). **Shaping Our Lives**: see Beresford and Carr (2019); Shaping Our Lives (2022).

28. **The Irish Citizens Assembly**: see Muradova et al. (2020); Citizens' Assembly of Ireland (2022).

29. **The famed Ata Island episode, when six Tongan boys**: Rutger Bregman discussed this episode and tells the story of going to meet one the boys as an old man in Australia. He is also critical of Golding, who as a teacher had set boys into competitive projects where the losers knew they would be severely punished. See Bregman's *Human Kind*, and also his *Utopia for Realists*.

30. **I can say at once, that I want nothing to be spared**: This is spoken by one of the principal characters in the famed family saga of Iceland, *The Saga of the People of Laxardal*. Guðrun Osvifsdottir also said one of the most famed phrases in all the sagas, at her death: "Though I treated him worst, I loved him best." See *The Saga of the People of Laxardal* (trans Keneva Kunz, 1997).

Chapter 10: Transformation: Achieving the Low-Carbon Good Lives

1. **The Seafarer** and **The Wanderer** are two of the most highly regarded Anglo-Saxon/Old English poems. Both are in The Exeter Book: the translations of these two quotes are by Richard North, Joe Allard and Patricia Gillies in *Longman Anthology of Old English, Old Icelandic and Anglo-Norman Literature* (2011).

2. **Eva Tulene Watt was a White Mountain Apache born in 1913**: The life of Eva Tulene Watt is told in Keith Basso's *Don't Let the Sun Step Over You*.

3. **The second encyclical of Pope Francis**: The Laudato Si' encyclical can be found at www.vatican.va, or at https://cafod.org.uk/Pray/Laudato-Si-encyclical. The Jack Kornfield quote is from his *After the Ecstasy, The Laundry*.

4. **It is self-creating and self-generating, a system characteristic called autopoiesis**: See Humberto Maturana and Francisco Varela *The Tree of Knowledge* (1987).

5. **"When is a tree?" asks Stuart Walker**, in his book Sustainability and Spirituality (2021). For Aldo Leopold's essay, *Thinking Like a Mountain*, see his book *The Sand County Almanac* (1949). Stuart Walker wisely wrote, "We have been looking for happiness in the wrong place."

6. **How do we collaborate in the creation of diverse regenerative cultures**: see David Wahl, see *Designing Regenerative Cultures* (2021); for Paul Hawken, see *Blessed Unrest* (2007); Thomas Berry: *The Great Work* (1999).

7. **"Under the warm and wide embrace of a summer sun, one and one almost never make two."** See Stuart Walker's *Designing Sustainability* (2014).
8. **Rob Hopkins was onto something**: The great handbook on these transitions is Rob Hopkins' The Transition Handbook. A new movement was launched. There were ideas on practical change, social organisation, small scale technologies, and political influence.
9. **Wise sorcerers are needed**: See Valerie Brown and John Harris, *The Human Capacity for Transformational Change* (2014).
10. **But here is the reality too. Political scientist Ian Budge**: See Ian Budge of the University of Essex and his book, *Kick-Starting Government Action* (2022).
11. **The development of the national K-Diet in South Korea**: See Kim S H et al. Korean diet: characteristics and historical background. *Journal of Ethnic Foods* 3(1), 26–31
12. **One case comes from Oklahoma City in the USA**: This case is discussed by Harold Wilhite in Syse and Muller (2016). But it is worth further noting that adult obesity rates in Oklahoma remain high, with the state at 36% in 2020, the ninth highest rate by state in the USA. If the mayor's programme had not been implemented, it is conceivable that these rates could be worse.
13. **Other changes have arisen from outside national policies**: Further advances on food have come from non-government organisations and international food movements, such as La Vía Campesina with its 180 member groups in 81 countries, and the International Partnership for Satoyama Initiative with 258 member groups in 71 countries. In our survey of the good life, food was valued not as filler or chore. It was celebration, it was shared ceremony with family and friends, it was the time and attentiveness needed to prepare. In a fast world, slow food seems an opportunity for the green mind, to escape the stressors in the default mode network. Slow food, in short, can be health-giving, increasing attentiveness and thus well-being.
14. **The Slow Food movement was set up by Carlo Petrini in Italy**: Slow Food movement: see www.slowfood.com/ and www.slowfood.org.uk/about/about/what-we-do/.
15. **This is the idea of universal basic income**: For discussions of universal basic income principles and schemes, see Ian Budge's *Kick-Starting Government Action* (2022) and Rutger Bregman's *Human Kind* (2020). There are many Universal Basic Income sources online, including The World Bank, Basic Income Hub, Basic Income Conversation, and See also Gerald Huff Fund for Humanity: https://fundforhumanity.org/petition/ and https://fundforhumanity.org/our-projects/. See also Peter Sloman's *Transfer State* (2019), and Emiel de Lange and colleagues' A Global Conservation Basic Income to Safeguard Biodiversity (2022).
16. **Other experiments are beginning to occur on the four-day working week**: Ed Siegel, chief executive of Charity Bank, said it was proud to be one of the first banks in the UK to embrace the four-day week:

    > The 20th-century concept of a five-day working week is no longer the best fit for 21st century business. We firmly believe that a four-day week with no change to salary or benefits will create a happier workforce and will have an equally positive impact on business productivity, customer experience and our social mission.

    See Julia Kollewe, *The Guardian*, 6 June 2022.
17. **The Norwegian Government Pension Fund Global**: One of the best accounts of the importance of the commons since Elinor Ostrom's (1990) *Governing the Commons*, and linking social institutions as natural oil revenue on current account spending and reducing income tax for the wealthy in the 1980s as "The worst economic mistake ever by a British government."
18. **Richard Carmichael has written of behaviour change**: See his paper, Behaviour change, public engagement and net zero (2019). See also further reports of the UK Climate Change Committee online at www.theccc.org.uk/

19. **The cultural historian and priest, Thomas Berry**: See Berry's fine *The Great Work*, written for the new millennium, and anticipating the collective changes that would be needed to reverse the great global crises.

20. **Our primary focus must be on contraction of current carbon emissions, and convergence towards safe space of one tonne per person**: Yet, you might find some muttering about this One Tonne target. In using averages for countries, great differences in current consumption are hidden. The richest 1% in Europe already have carbon footprints of over 50 tonnes per year; the bottom 5% of earners have footprints of about 2 tonnes. The rich have the most to do to change, but then again they also have the resources to deploy. At the same time, there are people how have already made changes in lifestyle and may already be close to One Tonne. This is true for the many people living in low-impact and intentional communities. There is another drawback with this model: it might appear that those countries who have been polluting over decades and centuries are let off the hook. Even without historical reckoning, there is a moral responsibility to step up and take responsibility to act fast and the most.

21. **Not many have taken a coordinated approach to encouraging their citizens**: See Yujie Xue (2022) in south China Morning Post.

22. **We know that the climate crisis is anthropogenic, and so are the solutions. And these will diverge**: A reminder, today's divergence includes 500 million indigenous people, nomads and herders, 100 uncontacted tribes, religious groups, members of intentional low-impact communities. Past divergence brought us 6700 surviving and different languages. It also brought 40+ city-based civilisations that each persisted for more than 1000 years, and then fell. Everything does change- great success can lead to a large fall. Irresistible languages have declined and some disappeared: Akkadian, Sanskrit, Ancient Greek, Latin. The Aramaiss nomads took over the Akkadian empire, bringing an easily learned and transferable alphabetic language, and Akkadian disappeared. The Phoenician sea-trader culture covered the whole of an inland sea, yet no one else learned their language, and eventually all was gone. Was this commerce without culture, and could this be a good definition for today's Capitalocene?

23. **During stability, stories emerge that span great eras**: The oldest surviving written saga is *The Epic of Gilgamesh* of Mesopotamia, its Akkadian manuscripts depicted life some 5000 years ago on tablets of clay. The Greek epic poems were composed by Homer, Virgil and Apollonius of Rhodes, *The Iliad, The Odyssey, The Aeneid, Jason and the Argonauts*, the conclusions already known to audiences as they settled down to each performance. From China came the 2500-year-old Tao stories of life in early eras, as spoken and written by Lao Tzu in *Tao Te Ching*, Chuang Tzu in *the Book of Chuang Tzu*, the folk manual *Seven Taoist Masters*, Luo Guanzhong's epic tale of civil war and collapse of government, and the voices of the ordinary people in *Romance of the Three Kingdoms*. The written Japanese sagas also date back 2000 years with the first *Manyōshū* songs, travel tales in *Journey Along the Sea Road*, *The Tale of Saigyō*, and the later epics, *Tale of the Heike* and *Tale of Ise*. The *Tale of the Heike* of the 1100s was told in verse, accompanied by biwa lutes, sung by blind Buddhist monks. They travelled widely to perform, each was said to hold in memory more than 20,000 lines of tales. From India came the Sanskrit epics of gods and people, *The Mahābhārata*, also written more than 2000 years ago and containing 100,000 couplet lines, and *The Rāmāyaṇa* of 24,000 lines.

24. **Many aspects of Iceland landscapes and land use have remained stable for a millennium**: Today in Iceland you can find the boulders Gisli and his wife Audr scrambled up, where he fought to his last. This is the great rock that Grettir lifted, that is the meadow where Unn the Deep-Minded lost her comb, called Kambsnes ever since. This the farmhouse where Njál and his wife were burned to death, that the nearby farm from which Gunnar the archer could not stay away, the place he loved so much, and where he was killed. Here are the stones outlining the basement where Snorri Sturluson was struck down. This, too, the borg where was an elven church, a further reminder for

protection. Here a Hebridean woman called Thórgunn was showered by a fall of blood during haymaking; over there a Scottish thrall called Nail fought off horse thieves, but was captured and jumped off a cliff; here a russet river flows over amber pools and beds of white pebbles, across which Björn jumped to woo Thuríd. See Jules Pretty (2022). *Sea Sagas of the North.*

Coda: Let's Dance, Together

1. **In Oxford Circus, I was passed a little handwritten note**: This is from Jay Griffith's book, *Why Rebel*, as she headed for a court appearance as a co-defendant following Extinction Rebellion work on the streets of London.
2. **Change is a snare, permanence is a snare**: from Joseph Campbell's still brilliant and seminal *The Hero With a Thousand Faces* (1949).
3. **One day, Chief Henry of the Koyukon people**: The Chief Henry quote is from Richard Nelson's *Make Prayers to the Raven*. The Ieyasu quote about the Japanese cuckoo (hototogisu) is from William Scott Wilson's *Musashi*.
4. **Tokugawa Ieyasu, the first leader of the Edo unification**: This quote about the Japanese cuckoo (hototogisu) is from William Scott Wilson's *Musashi*.
5. **Said the old teacher to Dosho Port**: The shake out your sleeves quote is in Dosho Port's *The Record of Empty Hall* (2021).

DATASETS FOR FIGURES AND TABLES

There are 25 figures in the text. The data for the graphs have been derived from open source datasets held by the World Bank, by United Nations agencies (FAO, WHO, IPCC), by the OECD, by national agencies (Centers for Disease Control, US), by the Global Burden on Disease (Institute of Health Metrics and Evaluation), and by aggregator websites (World Obesity, Our World in Data). I have combined this data into a large interconnected dataset for some 220 countries, covering a wide range of indicators. These include carbon dioxide (total and per capita), all greenhouse gases, GDP per capita (in 2010 constant US$), population, population over 65 years of age, human development index, life expectancy at birth, fertility rates, under-5 mortality, oil consumption, hours worked, meat consumption, calories consumed per day, obesity rates, type 2 diabetes rates, mental ill-health prevalence, all deaths (numbers and causes), world and national happiness data, non-communicable disease burden and deaths, renewable energy production, agricultural yields and livestock numbers, vehicles per capita, advertising spend, GINI index, diet preferences and pesticide use. The data trends are assessed from 1960, where possible, and for carbon emissions from 1800. Further details or explanations can be found in relevant endnotes.

The main sources of data can be found at these website locations:

Global Burden of Disease (Institute of Health Metrics and Evaluation): www. healthdata.org/gbd/

International Panel of Climate Change (IPCC): www.ipcc.ch/

OECD: www.oecd.org

Our World in Data: https://ourworldindata.org/

UNEP: www.unep.org

World Bank: www.worldbank.org

World Database of Happiness (Veenhoven R): https://worlddatabaseofhappiness.
eur.nl/
World Happiness Report: https://worldhappiness.report/
Global Obesity Observatory: https://data.worldobesity.org/

BIBLIOGRAPHY

Addis S. 2008. *Zen Sourcebook*. Hackett Publishing, London.

Addis S and Lombardo S (trans). 1993. *Lao-Tzu: Tao Te-Ching*. Hackelt, Indianapolis/ Cambridge.

Aesop (trans Temple O and Temple R). 1998. *The Complete Fables*. Penguin Classics, London.

Agarwal B. 2018. Can group farms outperform individual family farms? Empirical insights from India. *World Development 108*, 57–73.

Agarwal B and Dorin B. 2019. Group farming in France: Why do some regions have more cooperative ventures than others? *Environment and Planning A: Economy and Space 51*(3), 781–804.

Ahlskog J E, Geda Y E, Graff-Radford N R and Petersen R C. 2011. Physical exercise as a preventive or disease-modifying treatment of dementia and brain aging. *Mayo Clinic Proceedings 86*(9), 876–884.

Aitken R. 1978. *A Zen Wave. Bashō's Haiku and Zen*. Shoemaker and Howard, Washington, DC.

Alderwick H A, Gottlieb L M, Fichtenberg C M and Adler N E. 2018. Social prescribing in the US and England: Emerging interventions to address patients' social needs. *American Journal of Preventive Medicine 54*(5), 715–718.

Alexander S. 2011. Property beyond growth: Toward a politics of voluntary simplicity. In *Property Rights and Sustainability* (pp. 117–148). Brill Nijhoff, Leiden.

Allen O E. 1980. *The Pacific Navigators*. Time-Life, Alexandria, Virginia.

Allen R M, Kong Q and Martin-Short R. 2020. The MyShake platform: A global vision for earthquake early warning. *Pure and Applied Geophysics 177*(4), 1699–1712.

AMS (Academy of Medical Sciences). 2018. *Health of the Public in 2040*. London. At https:// acmedsci.ac.uk/policy/policy-projects/health-of-the-public-in-2040

Andersen K. 2020. *Evil Geniuses*. Ebury Press, London.

Andersen L and Björkman T. 2017. *The Nordic Secret*. Fri Tanke, Stockholm.

Anderson N D, Damianakis T, Kröger E, Wagner L M, Dawson D R, Binns M A, Bernstein S, Caspi E and Cook S L. 2014. The benefits associated with volunteering among seniors: A critical review and recommendations for future research. *Psychological Bulletin 140*(6), 1505.

Antholt C H. 1994. *Getting Ready for the Twenty-First Century: Technical Change and Institutional Modernisation in Agriculture*. World Bank Technical Paper 217. World Bank, Washington, DC.

Apollonius of Rhodes (trans Poochigan A). 2014. *Jason and the Argonauts*. Penguin Classics, London.

Argyris C and Schon D A. 1974. *Theory in Practice: Increasing Professional Effectiveness*. Jossey-Bass, Hoboken, NJ.

Armstrong K. 2005. *A Short History of Myth*. Canongate, Edinburgh.

Arnstein S R. 1969. A ladder of citizen participation. *Journal of the American Institute of Planners 35*(4), 216–224.

Artaraz K, Calestani M and Trueba M L. 2021. Vivir bien/Buen vivir and post-neoliberal development paths in Latin America: Scope, strategies, and the realities of implementation. *Latin American Perspectives 48*(3), 4–16.

Bagadion B U and Korten F F. 1991. Developing irrigators' organisations; a learning process approach. In Cernea M M (ed). *Putting People First*. Oxford University Press, Oxford. 2nd Edition.

Baring-Gould S. 1863 [2007]. *Iceland: Its Scenes and Sagas*. Signal Books, Oxford.

Barraclough E R. 2016. *Beyond the Northlands*. Oxford University Press, Oxford.

Barton J, Bragg R, Pretty J, Roberts J and Wood C. 2016. The wilderness expedition: An effective life course intervention to improve young people's well-being and connectedness to nature. *Journal of Experiential Education 39*(1), 59–72.

Barton J, Bragg R, Wood C and Pretty J. 2017. *Green Exercise: Linking Nature, Health and Well-Being*. Routledge, Oxon.

Barton J, Hine (Bragg) R E and Pretty J. 2009. The health benefits of walking in greenspaces of high natural and heritage value. *Journal of Integrative Environmental Sciences 6*(4), 261–278.

Barton J and Pretty J. 2010. What is the best dose of nature and green exercise for mental health? A meta-study analysis. *Environ Science and Technology 44*, 3947–3955

Bashō M. 1968. *The Narrow Road to the Deep North and Other Travel Sketches* (trans Nobuyuki Yuasa). Penguin, London.

Bashō M. 2008. *The Complete Haiku*. (trans Reichhold J). Kodansha International, Tokyo.

Bashō M (trans Reichhold J). 2008. *Bashō: The Complete Haiku*. Kodansha International, Tokyo.

Bashō M (trans Yuasa N). 1968. *The Narrow Road to the Deep North and Other Travel Sketches*. Penguin, London.

Basso K. 1996. *Wisdom Sits in Places*. University of New Mexico Press, Albuquerque.

Basso K and Eva Tulene Watt. 2004. *Don't Let the Sun Step Over You: A White Mountain Apache Family Life 1860–1975*. University of Arizona Press, Tuscon.

Basu T. 2014. Whatever Happened to Roller Skating? *The Atlantic*, May 13.

Bateson G. 1972 (2000). *Steps to an Ecology of the Mind*. University of Chicago Press, Chicago and London.

Bawden R. 1991. Systems thinking and practice in agriculture: Keynote address, 85th annual meeting, American dairy science association, Raleigh, NC, June 1990. *Journal of Dairy Science 74*, 2362–2373.

Bawden R. 2006. Have we *Ever* been critically engaged? In Fear F, Rosaen C L, Bawden R J and Foster-Fishman P G (eds). *Coming to Critical Engagement*. University Press of America, Lanham, MD.

Bawden R J. 2011. Epistemic aspects of social ecological conflict. In Wright D, Camden-Pratt C and Hill S (eds). *Social Ecology: Applying Ecological Understandings to Our Lives and Our Planet*. Hawthorn Press, Glos.

Bawden R J, McKenzie B and Packham R. 2007. Moving beyond the academy: A commentary on extra-mural initiatives in systemic development. *Systems Research and Behavioural Science: The Official Journal of the International Federation for Systems Research 24*(2), 129–141.

BBC TV. 2017. *Handmade in Japan* (Swordmakers). BBC, London.

Begley S. 2009. *Plastic Mind*. Robinson, New York.

Beling A E, Cubillo-Guevara A P, Vanhulst J and Hidalgo-Capitán A L. 2021. Buen vivir (Good Living): A "glocal" genealogy of a Latin American utopia for the world. *Latin American Perspectives 48*(3), 17–34.

Belk R W and Pollay R W. 1987. The good life in twentieth century US advertising. *Media Information Australia 46*(1), 51–57.

Benedict R. 1946 (2019). *The Chrysanthemum and the Sword*. Albatross, Naples.

Benor D, Harrison J Q and Baxter M. 1984. *Agricultural Extension: The Training and Visit System*. The World Bank, Washington, DC.

Beresford M W. 1954. *The Lost Villages of England*. Lutterworth Press, London.

Beresford P. 2019. Public participation in health and social care: Exploring the co-production of knowledge. *Frontiers in Sociology 3*, 41.

Beresford P. 2021. *Participatory Ideology: From Exclusion to Involvement*. Policy Press, Bristol.

Beresford P and Carr S. 2019. *Social Policy First Hand*. Policy Press, Bristol.

Beresford P and Russo J. 2022. *Routledge International Handbook on Mad Studies*. Routledge, Oxon.

Berkes F. 1999. *Sacred Ecology*. Taylor and Francis, Philadelphia.

Berkes F. 2020. *Advanced Introduction to Community-Based Conservation*. Edward Elgar, Cheltenham.

Berkes F, Colding J and Folke C (eds). 2008. *Navigating Social-Ecological Systems: Building Resilience for Complexity and Change*. Cambridge University Press, Cambridge.

Berman M. 2015. *Neurotic Beauty: An Outside Looks at Japan*. Water Street Press, Healdsburg, CA.

Berry T. 1999. *The Great Work*. Bell Tower, New York.

Berry W. 1997. *The Unsettling of America*. Sierra Club Books, San Francisco.

Bhagavad Gita (trans Patton L). 2008. Penguin, London.

Bharucha Z P, Mitjans S B and Pretty J. 2020. Towards redesign at scale through zero budget natural farming in Andhra Pradesh, India. *International Journal of Agricultural Sustainability 18*(1), 1–20.

Bharucha Z P and Pretty J. 2010. The role and importance of wild foods in agricultural systems. *Philosophical Transactions of the Royal Society of London B 365*, 2912–2926.

Bhasin M K, Dusek J A, Chang B H, Joseph M G, Denninger J W, Fricchione G L, Benson H and Libermann T A. 2013. Relaxation response induces temporal transcriptome changes in energy metabolism, insulin secretion and inflammatory pathways. *PLoS One 8*(5), e62817.

1Bhatnagar B and Williams A C (eds). 1992. *Participatory Development and the World Bank: Potential Directions for Change*. World Bank Discussion Papers 183. World Bank, Washington, DC.

Bird Rose D. 1996. *Nourishing Terrains*. Australian Heritage Commission, Canberra.

Bird Rose D. 2000. *Dingo Makes Us Human*. Cambridge University Press, Cambridge.

Blackburn J. 2019. *Time Song: Searching for Doggerland*. Jonathan Cape, London.

Blamires H. 1969. *Word Unheard: A Guide through Eliot's Four Quartets*. Methuen, London.

Blofield J. 1959 [1972]. *The Wheel of Life*. Rider, London.

Bloomfield D. 2017. What makes nature-based interventions for mental health successful? *British Journal of Psychology International 14*(4), 82–85.

Blythe R. 1988. *Divine Landscapes*. Canterbury Press, Norwich.

Boedeker W, Watts M, Clausing P and Marquez E. 2020. The global distribution of acute unintentional pesticide poisoning: Estimations based on a systematic review. *BMC Public Health 20*(1), 1–19.

Bogle J. 2010. *Enough*. Wiley, London.

Booker C. 2004. *The Seven Basic Plots*. Continuum, London.

Book of Chuang Tzu (trans Palmer M). 1996 [2006]. Penguin Classics, London.

Borentain S, Nash A I, Dayal R and DiBernardo A. 2020. Patient-reported outcomes in major depressive disorder with suicidal ideation: A real-world data analysis using Patient-sLikeMe platform. *BMC Psychiatry 20*(1), 1–11.

Borgonovi F. 2008. Doing well by doing good: The relationship between formal volunteering and self-reported health and happiness. *Social Science & Medicine 66*(11), 2321–2334.

Borgonovi F. 2010. A life-cycle approach to the analysis of the relationship between social capital and health in Britain. *Social Science & Medicine 71*(11), 1927–1934.

Borysenko J. 2007. *Mending the Body, Mending the Mind*. Perseus Books, Philadelphia.

Bourdieu P. 1986. The forms of capital. In Bourdieu P and Richardson J G (eds). *Handbook of Theory and Research for the Sociology of Education*. Richardson, New York, NY.

Boyle D and Simms A. 2009. *The New Economics: A Bigger Picture*. Earthscan, London.

BRAC. 2022. *BRAC Microfinance Programmes*. At www.brac.net

Bragdon J. 2021. *Economies That Mimic Life*. Routledge, Oxon.

Bragg R, Atkins G and Leck C. 2017. *Good Practice in Social Prescribing for Mental Health*. NECR 228. Natural England, London.

Bratman G N, Anderson C B, Berman M G, Cochran B, De Vries S, Flanders J, Folke C, Frumkin H, Gross J J, Hartig T and Kahn P H. 2019. Nature and mental health: An ecosystem service perspective. *Science Advances 5*(7), eaax0903.

Bratman G N, Hamilton J P and Daily G C. 2012. The impacts of nature experience on human cognitive function and mental health. *Annals of New York Academy of Sciences 1249*, 118–136.

Bratton S P. 2021. *Religion and the Environment*. Routledge, Oxon.

Braun A and Duveskog D. 2009. *The Farmer Field School Approach: History, Global Assessment and Success Stories*. IFAD, Rome.

Bregman R. 2014. *Utopia for Realists*. Bloomsbury, London.

Bregman R. 2020. *Human Kind*. Bloomsbury, London.

Breuning G. 2016. *Habits of a Healthy Brain*. Adams, Avon MA.

Brewer J. 2017. *The Craving Mind*. Yale University Press, New Haven.

Bringhurst R. 1999 [2011]. *A Story as Sharp as a Knife: The Classical Haida Mythtellers and Their World*. Douglas and McIntyre, Madeira Park.

Brits L T. 2016. *Hygge: The Danish Art of Living Well*. Penguin, London.

Brown V and Harris J. 2014. *The Human Capacity for Transformational Change*. Routledge, London.

Budge I. 2022. *Kick-Starting Government Action against Climate Change*. Routledge, Oxon.

Bunch R. 2018 [2012]. *Restoring the Soil*. Canadian Foodgrains Bank, Winnipeg, 2nd Edition.

Bunker S, Coates C, Dennis, J and Ho J. 2014. *Low Impact Living Communities*. Diggers and Dreamers Publications, London.

Buric I, Farias M, Jong J, Mee C and Brazil I A. 2017. What is the molecular signature of mind: Body interventions? A systematic review of gene expression changes induced by meditation and related practices. *Frontiers in Immunology 8*, 670.

Burrows M S and Sweeney J M (eds). 2017. *Meister Eckhart's Book of Secrets*. Hampton Roads, Charlottesville.

Butcher C A (trans). 2018. *The Cloud of Unknowing*. Shambhala, Boulder, CO.

Cacioppo J T, Hawkley L C and Berntson G G. 2003. The anatomy of loneliness. *Current Directions in Psychological Science 12*, 71–74.

Calvino I. 1980 (2002). *Italian Folktales*. Penguin, London.

Campbell A. 2020. *Living Better: How I Learned to Survive Depression*. John Murray, London.

Campbell A, Alexandra J and Curtis D. 2017. Reflections on four decades of land restoration in Australia. *Rangeland Journal 39*, 405–416.

Campbell J. 1949 [2008]. *The Hero With a Thousand Faces*. New World Library, Novatto, CA.

Campbell J. 2003. *Myths of Light*. New World Library, Novato.

Campbell J. 2004. *Pathways to Bliss*. New World Library, Novato.

Cardillo J. 2003. *Be Like Water*. Grand Central, New York.

Carey N. 2011. *The Epigenetics Revolution*. Columbia University Press, New York.

Carmichael R. 2019. *Behaviour Change, Public Engagement and Net Zero*. UK Climate Change Committee, London.

Carolan M. 2016. Adventurous food futures: Knowing about alternatives is not enough, we need to feel them. *Agriculture and Human Values 33*(1), 141–152.

Carreño-Calderón A. 2021. Living well and health practices among Aymara People in Northern Chile. *Latin American Perspectives 48*(3), 69–81.

Caspi A, Moffitt T E, Newman D L and Silva P A. 1996. Behavioral observations at age 3 years predict adult psychiatric disorders: Longitudinal evidence from a birth cohort. *Archives of General Psychiatry 53*(11), 1033–1039.

CCC. 2020. *Reducing UK Emissions: 2020 Progress Report*. Committee on Climate Change, UK government, London.

Cederlöf G. 2016. Low-carbon food supply: The ecological geography of Cuban urban agriculture and agroecological theory. *Agriculture and Human Values 33*(4), 771–784.

Cernea M M. 1987. Farmer organisations and institution building for sustainable development. *Regional Development Dialogue 8*, 1–24.

Cernea M M. 1991. *Putting People First*. Oxford University Press, Oxford, 2nd Edition.

Chambers R. 1989. *Farmer First*. IT Publications, Rugby.

Chambers R. 2008. *Revolutions in Development Inquiry*. Earthscan, London.

Charities Aid Foundation. 2022. *World Giving Index 2021*. At www.cafonline.org/about-us/publications/2021-publications/caf-world-giving-index-2021

Chatterjee H J, Camic P M, Lockyer B, Thomson L J. 2018. Non-clinical community interventions: A systematised review of social prescribing schemes. *Arts & Health 10*(2), 97–123.

Chatwin B. 1998. *Songlines*. Picador, London.

Chen S, Kuhn M, Prettner K and Bloom D E. 2019. The global macroeconomic burden of road injuries: Estimates and projections for 166 countries. *The Lancet Planetary Health 3*(9), e390–e398.

Chitiwere T. 2018. *Sustainable Communities and Green Lifestyles*. Routledge, Oxon.

Christie D. 2013. *Blue Sapphire of the Mind*. Oxford University Press, Oxford.

Citizens' Assembly of Ireland. 2022. At www.citizensassembly.ie/en/

Citizen Science Association. 2022. *The Power of Citizen Science*. At https://citizenscience.org/

Clark A E, Fleche S, Layard R, Powdthavee N and Ward G. 2018. *The Origins of Happiness*. Princeton University Press, Princeton.

Clark A E and Georgellis Y. 2013. Back to baseline in Britain: Adaptation in the British household panel survey. *Economica 80*(319), 496–512.

Clark D M. 2018. Realizing the mass public benefit of evidence-based psychological therapies: The IAPT program. *Annual Review of Clinical Psychology 14*.

Clark J K, Bean M, Raja S, Loveridge S, Freedgood J and Hodgson K. 2017. Cooperative extension and food system change: Goals, strategies and resources. *Agriculture and Human Values 34*(2), 301–316.

Clunies Ross M. 2010. *The Old Norse-Icelandic Saga.* Cambridge University Press, Cambridge.

CMO (Chief Medical Officer). 2013. *Chief Medical Officer's Annual Report 2012: Our Children Deserve Better: Prevention Pays.* UK Government, London.

CMO (Chief Medical Officer). 2018. *Annual Report 2018: Better Health Within Reach.* UK Government, London.

Colegate I. 2002. *A Pelican in the Wilderness.* Harper Collins, London.

Coleman J. 1990. *Foundations of Social Theory.* Harvard University Press, Boston.

Collingwood W G and Stefánson J. 1899. *A Pilgrimage to the Saga-Steads of Iceland.* Reprinted by Viking Society for Northern Research (Edgeler M (ed), 2015). Raubling, Germany.

Cooperatives UK. 2018. *The Cooperative Economy 2018.* At reports.uk.coop/economy2018

Coplan K. 2020. *Live Sustainably Now: A Low Carbon Visions of the Good Life.* Columbia University Press, New York.

Cosme I, Santos R and O'Neill D W. 2017. Assessing the degrowth discourse: A review and analysis of academic degrowth policy proposals. *Journal of Cleaner Production 149*, 321–334.

Costanza R, Atkins P W, Bolton M, Cork S, Grigg N J, Kasser T and Kubiszewski I. 2017. Overcoming societal addictions: What can we learn from individual therapies? *Ecological Economics 131*, 543–550.

Cottam D, McGuire C, Mossop D, Davis G, Donlen J, Friend K, Lewis B, Boucher E, Kirubakaran H, Goulding R and Jovanovic D. 2021. Drain detectives: Lessons learned from citizen science monitoring of beach drains. *Citizen Science: Theory and Practice 6*(1).

Cousineau P. 1998. *The Art of Pilgrimage.* Conari Press, San Francisco.

Couturier A. 2017. *The Abundance of Less.* North Atlantic Books, Berkeley, CA.

Covey S R. 1991. *The Seven Habits of Highly Effective People.* Covey Leadership Center, Provo, UT.

Crawford M. 2015. *The World beyond Your Head.* Penguin, London.

Crisp B F and Kelly M J. 1999. The socioeconomic impacts of structural adjustment. *International Studies Quarterly 43*(3), 533–552.

Cron L. 2012. *Wired for Story.* Ten Speed Press, London.

Crook J and Low J. 1997 (2012). *Yogins of Ladakh.* Motilal Banardsidass Publ, New Delhi.

Crossley-Holland K. 1980. *The Penguin Book of Norse Myths.* Penguin, London.

Crossley-Holland K. 1999. *The Anglo-Saxon World: An Anthology.* Oxford World's Classics. Oxford University Press, Oxford.

Crum S J. 1994. *The Road on Which We Came: A History of the Western Shoshone.* University of Utah Press, Salt Lake City.

Csikszentmihalyi M. 2002. *Flow: The Psychology of Happiness.* Rider, London.

Cummings V, Jordan P and Zvelebil M (eds). 2014. *The Oxford Handbook of the Archaeology and Anthropology of Hunter-Gatherers.* Oxford University Press, Oxford.

Cunliffe B. 2012. *Britain Begins.* Oxford University Press, Oxford.

Curry O S, Mullins D A and Whitehouse H. 2019. Is it good to cooperate? Testing the theory of morality-as-cooperation in 60 societies. *Current Anthropology 60*(1), 47–69.

Daly H. 1996. *Beyond Growth.* Beacon Press, Boston.

Danne, D, Snowdon D A and Friesen W V. 2001. Positive emotions in early life and longevity: Findings from the nun study. *Journal of Personality and Social Psychology 80*(5), 804–813.

Dasgupta P. 2021. *The Economics of Biodiversity.* UK Treasury, London.

Davis R. 2019. *Lawrence Elkins: Consensus All-American 1963 and 1964.* Brownwood, TX.

Davis W. 2002. *Light at the Edge of the World*. Douglas and McIntyre, Vancouver.

Davis W. 2009. *The Wayfinders: Why Ancient Wisdom Matters*. Anansi, Toronto.

Dawson J. 2006. *Ecovillages: New Frontiers for Sustainability*. Green Books, Totnes.

de Craemer S, Vercauteren J, Fierens F, Lefebre W, Hooyberghs H and Meysman F. 2019. *CurieuzeNeuzen: Monitoring Air Quality Together with 20.00 0 Citizens*. 40th AIVC-8th TightVent-6th Venticool Conference. At https://venticool.eu/

Defra. 2019. *The 25 Year Environment Plan*. Department for Environment, Food and Rural Affairs, London.

de Lange E, Sze J, Allan J, Atkinson S C, Booth H, Fletcher R, Khanyari M and Saif O. 2022. A global conservation basic income to safeguard biodiversity. *OSF Preprints*, April 6. doi:10.31219/osf.io/nvpfh

Del Giorno J M, Hall E E, O'Leary K C, Bixby W R and Miller P C. 2010. Cognitive function during acute exercise: A test of the transient hypofrontality theory. *Journal of Sport and Exercise Psychology 32*(3), 312–323.

Del Valle G. 2019. DonorsChoose and other crowdfunding sites are coming under scrutiny. *VoxMedia*, March 29.

Denevan W M. 2001. *Cultivated Landscapes of Native Amazonia and the Andes*. Oxford University Press, Oxford.

de Saint-Exupery A. 1939. *Wind, Sand and Stars*. Penguin, London.

Deshimaru T. 1982. *The Zen Way to the Martial Arts*. Arkana, London.

DHSC. 2018. *Prevention is Better than Cure*. Department of Health and Social Care, London.

Diamond J. 2005. *Collapse: How Societies Choose to Fail or Survive*. Penguin, London.

Dias M de O and Teles A. 2018. Agriculture cooperatives in Brazil and the Importance for economic development. *International Journal of Business Research and Management 9*(2).

Dickinson E. 2016. *The Collected Poems of Emily Dickinson*. Digireads Publishing, Overland Park, KS.

Dietrich A. 2003. Functional neuroanatomy of altered states of consciousness: The transient hypofrontality hypothesis. *Consciousness and Cognition 12*(2), 231–256.

Dietrich A. 2006. Transient hypofrontality as a mechanism for the psychological effects of exercise. *Psychiatry Research 145*(1), 79–83.

Dillard A. 1990. *Pilgrim at Tinker Creek*. (In *Three by Annie Dillard*, 2001). Harper Perennial, New York.

Doidge N. 2015. *The Brain's Way of Healing*. Allen Lane, London.

Dorling D. 2020. *Slowdown*. Yale University Press, New Haven.

Dorling D and Koljonen A. 2020. *Finntopia*. Agenda Publishing, Newcastle.

Drennan V M, Calestani M, Ross F, Saunders M and West P. 2018. Tackling the workforce crisis in district nursing: Can the Dutch Buurtzorg model offer a solution and a better patient experience? A mixed methods case study. *BMJ Open 8*(6), e021931.

Duhigg C. 2012. *The Power of Habit*. Random House, London.

Dusek J A, Otu H, Wohlhueter A L, Bhasin M, Zerbini L F, Joseph M G, Benson H and Libermann T A. 2008. Genomic counter-stress changes induced by the relaxation response. *PLoS One 3*(7), e2576.

Easterlin R A. 1974. Does economic growth improve the human lot? Some empirical evidence. In *Nations and Households in Economic Growth*. Academic Press, Cambridge, MA, pp. 89–125.

ECSA. 2022. *European Citizen Science Association*. At https://ecsa.citizen-science.net/

Ehrenreich B. 2007. *Dancing in the Streets*. Granta, London.

Ehrlich G. 2001. *This Cold Heaven*. Harper Collins, London.

Einarsson N. 2009. From good to eat to good to watch: whale watching, adaptation and change in Icelandic fishing communities. *Polar Research 28*(1), 129–138.

Einarsson N. 2011a. *Culture, Conflict and Crises in Icelandic Fisheries*. Uppsala Studies in Cultural Anthropology 48. Uppsala University, Sweden.

Einarsson N. 2011b. Fisheries governance and social discourse in post-crisis Iceland: Responses to the UN Human Rights Committee's views in case 1306/2004. *The Yearbook of Polar Law Online 3*, 479–515.

Ekins P and Gupta J. 2019. Perspective: A healthy planet for healthy people. *Global Sustainability 2*.

Elder Edda (trans Orchard A). 2011. Penguin, London.

Ellis Davidson H R. 1964. *Gods and Myths of Northern Europe*. Penguin, London.

Ellis Davidson H R. 1969. *Scandinavian Mythology*. Newnes Books, Middlesex.

Elwood P, Galante J, Pickering J, Palmer S, Bayer A, Ben-Shlomo Y, Longley M and Gallagher J. 2013. Healthy lifestyles reduce the incidence of chronic diseases and dementia: Evidence from the Caerphilly Cohort Study. *PLoS One* 8(12), e81877.

Entradas M. 2015. Science and the public: The public understanding of science and measurements. *Portuguese Journal of Social Science* 14(1), 71–85.

Epic of Gilgamesh (trans George A). 1998. Penguin Classics, London.

Erdoes R and Ortiz A. 1984. *American Indian Myths and Legends*. Pantheon Books, New York.

Erdoes R and Ortiz A. 1998. *American Indian Trickster Tales*. Penguin, London.

Ernle, Lord (Prothero R E). 1912. *English Farming: Past and Present*. Longman Green, London.

Esch T, Fricchione G L and Stefano G B. 2003. The therapeutic use of the relaxation response in stress-related diseases. *Medical Science Monitoring* 9(2), RA23–RA34.

Everrit D. 2008. *Don't Sleep, There Are Snakes*. Profile, London.

Ewart Evans G and Thomson D. 1972. *The Leaping Hare*. Faber and Faber, London.

FAO. 2016a. *Farmer Field School Guidance Document*. FAO, Rome.

FAO. 2016b. *Forty Years of Community-Based Forestry*. FAO, Rome.

FAO. 2019. *Farmers Taking the Lead: Thirty Years of Famer Field Schools*. FAO, Rome.

Fear F A, Bawden R J, Rosaen C L and Foster-Fishman P G. 2002. A model of engaged learning: Frames of reference and scholarly underpinnings. *Journal of Higher Education Outreach and Engagement* 7(3), 55–68.

Fear F A, Rosaen C L, Bawden R J and Foster-Fishman P G. 2006. *Coming to Critical Engagement*. University Press of America, Lanham, MD.

Feldman C and Kuyken W. 2019. *Mindfulness*. Guilford Press, New York and London.

Fidler R and Gíslason K. 2017. *Saga Land*. ABC Books, Sydney.

Fields in Trust. 2019. *Revaluing Parks and Green Space*. At www.fieldsintrust.org/revaluing

First Poems in English (trans Alexander M). 1966 [2008]. Penguin Classics, London.

Fischer N. 2008. *Sailing Home*. North Atlantic Books, Berkeley, CA.

Fischer N. 2019. *The World Could Be Otherwise*. Shambhala, Boulder, CO.

Fitton N, Ejerenwa C P, Bhogal A, Edgington P, Black H, Lilly A, Barraclough D, Worrall F, Hillier J and Smith P. 2011. Greenhouse gas mitigation potential of agricultural land in Great Britain. *Soil Use and Management* 27(4), 491–501.

Ford J I and Blacker M M. 2011. *The Book of Mu*. Wisdom Publications, Boston.

Forero J E. 2021. Buen vivir as an alternative development model: Ecuador's bumpy road toward a post-extractivist society. *Latin American Perspectives* 48(3), 227–244.

Forrer M. 2008. *Hokusai*. Prestel, Munich.

Forster T, Kentikelenis A E, Reinsberg B, Stubbs T H and King L P. 2019. How structural adjustment programs affect inequality: A disaggregated analysis of IMF conditionality, 1980–2014. *Social Science Research* 80, 83–113.

Forster T, Kentikelenis A E, Stubbs T H and King L P. 2020. Globalization and health equity: The impact of structural adjustment programs on developing countries. *Social Science & Medicine* 267, 112496.

Francis G. 2021. *Recovery*. Wellcome, London.

Franco M, Bilal U, Orduñez P, Benet M, Morejón A, Caballero B, Kennelly J F and Cooper R S. 2013. Population-wide weight loss and regain in relation to diabetes burden and cardiovascular mortality in Cuba 1980–2010: Repeated cross sectional surveys and ecological comparison of secular trends. *BMJ 346*.

Franco M, Ordunez P, Caballero B, Tapia Granados J A, Lazo M, Bernal J L, Guallar E and Cooper R S. 2007. Impact of energy intake, physical activity, and population-wide weight loss on cardiovascular disease and diabetes mortality in Cuba, 1980–2005. *American Journal of Epidemiology 166*(12), 1374–1380.

Frank R H. 1999. *Luxury Fever*. Princeton University Press, Princeton.

Frankl V. 1959 (2011). *Man's Search for Meaning*. Rider, London.

Freedman P. 2007. *Food: The History of Taste*. Thames and Hudson, London.

Freidberg S. 2020. Assembled but unrehearsed: Corporate food power and the 'dance' of supply chain sustainability. *The Journal of Peasant Studies 47*(2), 383–400.

Freire P. 1970. *Pedagogy of the Oppressed*. Penguin, London.

French P. 2003. *Tibet, Tibet*. Penguin, London.

Friedman H S and Martin L R. 2011. *The Longevity Project*. Hay House, London.

Fritz S, See L, Carlson T, Haklay M, Oliver J L, Fraisl D, Mondardini R, Brocklehurst M, Shanley L A, Schade S and Wehn U. 2019. Citizen science and the United Nations sustainable development goals. *Nature Sustainability 2*(10), 922–930.

Frumkin, H. 2005. The health of places, the wealth of evidence. In *Urban Place, Reconnecting with the Natural World*. The Massachusetts Institute of Technology Press, Cambridge, MA, pp. 253–269.

Fuchs D, Sahakian M, Gumbert T, Di Giulio A, Maniates M, Lorek S and Graf A. 2021. *Consumption Corridors: Living a Good Life within Sustainable Limits*. Routledge, Oxon.

Funakoshi. 1975. *Karate-Dō: My Way of Life*. Kodansha, New York.

Gaiman N. 2017. *Norse Mythology*. Bloomsbury, London.

Gallagher W. 2009. *Rapt. Attention and Focused Life*. Penguin, London.

Garcia H and Miralles F. 2017. *Ikigai: The Japanese Secret to a Long and Happy Life*. Cornerstone, London.

Gardner B, Lally P and Wardle J. 2012. Making health habitual: The psychology of 'habit-formation' and general practice. *British Journal of General Practice 62*, 664–666.

Geoffrey of Monmouth (trans Thorpe L). 1966. *The History of the Kings of Britain*. Penguin Classics, London.

Ger G. 1997. Human development and humane consumption: Well-being beyond the 'Good Life'. *Journal of Public Policy & Marketing 16*(1), 110–125.

Gerber J F. 2020. Degrowth and critical agrarian studies. *The Journal of Peasant Studies 47*(2), 235–264.

Gersie A, Nanson A and Schieffelin E (eds). 2014. *Storytelling for a Greener World*. Hawthorn Press, Stroud.

GetUp! 2022. *People, Power, Impact*. At www.getup.org.au/

Gibson T. 1996. *The Power in Our Hands*. Jon Carpenter, Charlbury, Oxon.

Giller K E, Hijbeek R, Andersson J A and Sumberg J. 2021. Regenerative agriculture: An agronomic perspective. *Outlook on Agriculture 50*(1), 13–25.

Gilligan S and Dilts R. 2009. *The Hero's Journey*. Crown House Publishing, Carmarthan.

Giradet H. 2015. *Creating Regenerative Cities*. Routledge, Oxon.

Gisli Sursson's Saga (trans Regal M S). 1997. Penguin Classics, London.

Global Ecovillage Network. 2022. At https://ecovillage.org/

Goldberg E. 2021. *Three Simple Lines*. New World Library, Novato.

Goldschmidt W. 1946. *As You Sow: Three Studies into the Social Consequences of Agribusiness.* Allanheld, Osmunn and Co, Montclair, NJ.

Goldschmidt W. 1978. Large-scale farming and the rural social structure. *Rural Sociology* *43*(3), 362.

Goleman D. 2013. *Focus: The Hidden Driver of Excellence.* Bloomsbury, London.

Goleman D and Davidson R J. 2017. *Altered Traits: Science Reveals How Meditation Changes Your Mind, Brain, and Body.* Avery, New York [also called The Science of Meditation. Penguin, London].

Goyal M, Singh S, Sibinga E M, Gould N F, Rowland-Seymour A, Sharma R, Berger Z, Sleicher D, Maron D D, Shihab H M, Ranasinghe P D. 2014. Meditation programs for psychological stress and well-being: A systematic review and meta-analysis. *JAMA Internal Medicine* *174*(3), 357–368.

Grameen Bank. 2022. *Grameen Bank Microfinance Programmes.* At www.grameen.com

Grant J. 2021. *The New Power University.* Pearson Education, London.

Graton L and Scott A. 2016. *The 100-Year Life.* Bloomsbury, London.

Gray C. 2021. *Sunshine Warm Sober.* Aster, London.

Gray J. 1998. *False Dawn.* The New Press, New York.

Greenwood C. 2019. *Just Enough.* New World Library, Novato CA.

Greer J M. 2019. *The Long Descent.* Founders House Publishing. At https://foundershouse-books.com/

Griffiths J. 2006. *Wild: An Elemental Journey.* Hamish Hamilton, London.

Griffiths J. 2021. *Why Rebel.* Penguin, London.

Hagen S. 2020. *The Grand Delusion.* Wisdom, Somerville, MA.

Hagger-Johnson G E, Hegarty P, Barker M and Richards C. 2013. Public engagement, knowledge transfer and impact validity. *Journal of Social Issues* *69*(4), 664–683.

Haklay M. 2013. Citizen science and volunteered geographic information: Overview and typology of participation. In *Crowdsourcing Geographic Knowledge*, Springer, Berlin, pp. 105–122.

Halifax J. 1993. *Fruitful Darkness.* Grove Press, New York.

Hall K D, Ayuketah A, Brychta R, Cai H, Cassimatis T, Chen K Y, Chung S T, Costa E, Courville A, Darcey V and Fletcher L A. 2019. Ultra-processed diets cause excess calorie intake and weight gain: An inpatient randomized controlled trial of ad libitum food intake. *Cell Metabolism* *30*(1), 67–77.

Hammill S and Seaton J P. 2004. *The Poetry of Zen.* Shambhala, Boston.

Hanbury-Tenison R. 2006. *The Seventy Great Journeys.* Thames and Hudson, London.

Hanson R. 2013. *Hardwiring Happiness.* Harmony, Easton, PA.

Hanson R and Mendius R. 2009. *Buddha's Brain: The Practical Neuroscience of Happiness.* New Harbinger Publishing, Oakland.

Haraway D. 2015. Anthropocene, capitalocene, plantationocene, chthulucene: Making kin. *Environmental Humanities* *6*(1), 159–165.

Hardacre H. 2017. *Shinto.* Oxford University Press, Oxford.

Hare B and Woods V. 2020. *Survival of the Friendliest.* One World, London.

Harris D. 2014. *10% Happier.* Yellow Kite, London.

Harris S. 2014. *Waking Up.* Black Swan, London.

Hawken P. 2007. *Blessed Unrest.* Penguin, London.

Hawkley L C and Cacioppo J T. 2007. Aging and loneliness: Downhill quickly? *Current Dir in Psych Science* *16*(4), 187–191.

Haywood J. 2005. *Penguin Historical Atlas of Ancient Civilisation.* Penguin, London.

Heaney S. 1999. *Beowulf.* Faber and Faber, London.

Heaney S. 2002. *Finders Keepers.* Faber and Faber, London.

Heigl F, Kieslinger B, Paul K T, Uhlik J and Dörler D. 2019. Opinion: Toward an international definition of citizen science. *Proceedings of the National Academy of Sciences 116*(17), 8089–8092.

Henriksen G. 2007. *I Dreamed the Animals*. Berghahn Books, New York and Oxford.

Herrigel E. 1953. *Zen in the Art of Archery*. Penguin, London.

Hesiod (trans West M L). 1988. *Theogony and Works and Days*. Oxford Classics, Oxford.

Hickel J. 2020a. *Less Is More*. William Heinemann, London.

Hickel J. 2020b. What does degrowth mean? A few points of clarification. *Globalizations*, 1–7.

Hildegard of Bingen (trans Atherton M). 2001. *Selected Writings*. Penguin Classics, London.

Hill S. 1985. Redesigning the food system for sustainability. *Alternatives 12*, 32–36.

Hinton D (trans and ed). 2002. *Mountain Home: The Wilderness Poetry of Ancient China*. Anvil Press Poetry, London.

Hoey L and Sponseller A. 2018. 'It's hard to be strategic when your hair is on fire': Alternative food movement leaders' motivation and capacity to act. *Agriculture and Human Values 35*(3), 595–609.

Holmes L, Cresswell K, Williams S, Parsons S, Keane A, Wilson C, Islam S, Joseph O, Miah J, Robinson E and Starling B. 2019. Innovating public engagement and patient involvement through strategic collaboration and practice. *Research Involvement and Engagement 5*(1), 1–12.

Holt-Lunstad J, Robles T F and Sbarra D A. 2017. Advancing social connection as a public health priority in the United States. *American Psychologist 72*(6), 517.

Holt-Lunstad J, Smith T B, Baker M, Harris T and Stephenson D. 2015. Loneliness and social isolation as risk factors for mortality: A meta-analytic review. *Perspectives on Psychological Science 10*(2), 227–237.

Holt-Lunstad J, Smith T B and Layton J B. 2010. Social relationships and mortality risk: A meta-analytic review. *PLoS Med 7*(7), e1000316.

Homer (trans Rieu E V). 1950. *The Iliad*. Penguin Classics, London.

Homer (trans Rieu E V and Rieu D C H). 1946 [2003]. *The Odyssey*. Penguin Classics, London.

Hopkins R. 2008. *The Transition Handbook*. Green Books, Totnes.

Huang B, Rodriguez B L, Burchfiel C M, Chyou P H, Curb J D and Yano K. 1996. Acculturation and prevalence of diabetes among Japanese-American men in Hawai'i. *American Journal of Epidemiology 144*(7), 674–681.

Hung Ying-Ming (trans Wilson). 2009. *Living the Good Life: Master of the Three Ways*. Shambhala, Boulder.

Hunter J. 1999. *The Last of the Free*. Mainstream, Edinburgh.

Hunter J. 2014. Rights-based land reform in Scotland: Making the case in the light of international experience. *Community Land Scotland*. At https://www.communitylandscotland.org.uk/

Hutton R. 2013. *Pagan Britain*. Yale University Press, New Haven and London.

Hyams J. 1979. *Zen in the Martial Arts*. Penguin Putnam, New York.

Hyde L. 1979 [2012]. *The Gift*. Canongate Books, Edinburgh.

Hyde L. 1998. *Trickster Makes This World*. Canongate Books, Edinburgh.

Hynes W J and Doty W G. 1993. *Mythical Trickster Figures*. University of Alabama Press, Tuscaloosa.

Ibn Fadlān (trans Lunde P and Stone C). 2016. *Ibn Fadlān and the Land of Darkness*. Penguin Classics, London.

IIED. 2021. *Indigenous Peoples' Food Systems Hold the Key to Feeding Humanity*. IIED and Kew Gardens, London.

ILO. 2011. *Manual on the Measurement of Volunteering*. International Labour Organisation, Geneva.

Imamura F, O'Connor L, Ye Z, Mursu J, Hayashino Y, Bhupathiraju S N and Forouhi N G. 2015. Consumption of sugar sweetened beverages, artificially sweetened beverages, and fruit juice and incidence of type 2 diabetes: Systematic review, meta-analysis, and estimation of population attributable fraction. *BMJ 351*.

Ingram J (trans). 1823. *The Anglo-Saxon Chronicle*. Red and Black, Florida.

Insel T R and Young L J. 2001. The neurobiology of attachment. *Nature Reviews Neuroscience 2*(2), 129–136.

Institute for Global Environmental Strategies, Aalto University and D-mat ltd. 2018. *1.5-Degree Lifestyles: Targets and Options for Reducing Lifestyle Carbon Footprints*. At www.iges.or.jp/en/pub/15-degrees-lifestyles-2019/en

IPBES. 2019. *Summary for Policymakers of the Global Assessment Report on Biodiversity and Ecosystem Services of the Intergovernmental Science-Policy Platform on Biodiversity and Ecosystem Services*. Díaz S, Settele J, Brondízio E S, Ngo H T, Guèze M, Agard J, Arneth A, Balvanera P, Brauman K A, Butchart S H M, Chan K M A, Garibaldi L A, Ichii K, Liu J, Subramanian S M, Midgley G F, Miloslavich P, Molnár Z, Obura D, Pfaff A, Polasky S, Purvis A, Razzaque J, Reyers B, Roy Chowdhury R, Shin Y J, Visseren-Hamakers I J, Willis K J and Zayas C N (eds). IPBES secretariat, Bonn, Germany, 56pp.

IPCC. 2007. *Fourth Assessment Report: Climate Change 2007*. IPCC, Switzerland.

IPCC. 2013. *Fifth Assessment Report: Climate Change 2013*. IPCC, Switzerland.

IPCC. 2019a. *Climate Change and Land*. IPCC, Geneva.

IPCC. 2019b. *Land: An IPCC Special Report on Climate Change, Desertification, Land Degradation, Sustainable Land Management, Food Security, and Greenhouse Gas Fluxes in Terrestrial Ecosystems*. IPCC, Geneva.

IPCC. 2022. *Sixth Assessment Report: Climate Change 2022*. IPCC, Switzerland.

iPES-Food. 2016. *From Uniformity to Diversity*. International Panel of Experts on Sustainable Food Systems, Brussels.

iPES Food. 2021. *A Long Food Movement*. At www.ipes-food.org/pages/LongFoodMovement

Irvine W. 2009. *A Guide to the Good Life: The Ancient Art of Stoic Joy*. Oxford University Press, Oxford.

Irwin A. 1995. *Citizen Science: A Study of People, Expertise and Sustainable Development*. Routledge, London.

Isham A, Gatersleben B and Jackson T. 2019a. Flow activities as a route to living well with less. *Environment and Behavior 51*(4), 431–461.

Isham A, Mair S and Jackson T. 2019b. Wellbeing and productivity: A review of the literature. In *Report for the Economic and Social Research Council*. University of Surrey, Guildford.

Ison R and Russell D (eds). 2007. *Agricultural Extension and Rural Development: Breaking Out of Knowledge Transfer Traditions*. Cambridge University Press, Cambridge.

Ivanova D, Barrett J, Wiedenhofer D, Macura B, Callaghan M and Creutzig F. 2020. Quantifying the potential for climate change mitigation of consumption options. *Environmental Research Letters 15*(9), 093001.

Ivanova D, Vita G, Steen-Olsen K, Stadler K, Melo P C, Wood R and Hertwich E G. 2017. Mapping the carbon footprint of EU regions. *Environmental Research Letters 12*(5), 054013.

Ivanova D, Vita G, Wood R, Lausselet C, Dumitru A, Krause K, Macsinga I and Hertwich E G. 2018. Carbon mitigation in domains of high consumer lock-in. *Global Environmental Change 52*, 117–130.

Ivanova D and Wood R. 2020. The unequal distribution of household carbon footprints in Europe and its link to sustainability. *Global Sustainability 3*.

IWGIA (ed Mamo D). 2022. *The Indigenous World 2022*. 850pp. At www.iwgia.org/en/

Jackson J. 2016. *Black Elk: The Life of American Visionary*. Farrar, Straus and Giroux, New York.

Jackson T. 2009. *Prosperity without Growth*. Earthscan, London.

Jackson T. 2021. *Post Growth: Life after Capitalism*. Polity, Cambridge.

Jahnke R, Larkey L, Rogers C, Etnier J. and Lin F. 2010. A comprehensive review of health benefits of qigong and tai chi. *American Journal of Health Promotion* 24(6), e1–e25.

Jahren H. 2020. *The Story of More*. Fleet, London.

Japan Times (Ben Dooley and Hisako Ueno). 2020. Old Shops. December 15, Tokyo.

Japan Times (Kathryn Wortley). 2020. Masters of Their Craft. November 28, Tokyo.

Jarjoura G R and Krumholz S T. 1998. Combining bibliotherapy and positive role modelling as an alternative to incarceration. *Journal of Offender Rehabilitation* 28(1–2), 127–139.

Ji J S, Zhu A, Bai C, Wu C D, Yan L, Tang S, Zeng Y and James P. 2019. Residential greenness and mortality in oldest-old women and men in China: A longitudinal cohort study. *The Lancet Planetary Health* 3(1), e17–e25.

Jodha N S. 1990. *Rural Common Property Resources: A Growing Crisis*. IIED Gatekeeper Series 24. IIED, London.

John, F L D and Erdoes R. 1972. *Lame Deer: Sioux Medicine Man*. Davis-Poynter, London.

Johnston K A and Lane A B. 2021. Communication with intent: A typology of communicative interaction in engagement. *Public Relations Review* 47(1), 101925.

Joxhorst T, Vrijsen J, Niebuur J and Smidt N. 2020. Cross-cultural validation of the motivation to change lifestyle and health behaviours for dementia risk reduction scale in the Dutch general population. *BMC Public Health* 20(1), 1–9.

Jungell-Michelsson J and Heikkurinen P. 2022. Sufficiency: A systematic literature review. *Ecological Economics* 195, 107380.

Jünger J and Fähnrich B. 2020. Does really no one care? Analyzing the public engagement of communication scientists on Twitter. *New Media & Society* 22(3), 387–408.

Kaage E. 2017. *Silence in the Age of Noise*. Penguin, London.

Kahneman D. 2011. *Thinking, Fast and Slow*. Macmillan, London.

Kahneman D, Sibony O and Sunstein C. 2021. *Noise*. William Collins, London.

Kallis G. 2011. In defence of degrowth. *Ecological Economics* 70(5), 873–880.

Kane S. 1998 (2010). *Wisdom of the Mythtellers*. Broadview Press, Peterborough, Ontario.

Kaplan C P, Erickson P I and Juarez-Reyes M. 2002. Acculturation, gender role orientation, and reproductive risk-taking behavior among Latina adolescent family planning clients. *Journal of Adolescent Research* 17(2), 103–121.

Kaplan S. 1995. The restorative benefits of nature: Toward an integrative framework. *Journal of Environmental Psychology* 15(3), 169–182.

Kasser T. 2002. *The High Price of Materialism*. MIT Press, Cambridge, MA.

Kelly R L. 1995. *The Foraging Spectrum: Diversity in Hunter-Gatherer Lifeways*. Smithsonian, Washington, DC.

Kenkō Y and Chomei K (trans McKinney M). 2013 *Essays in Idleness and Hōjōku*. Penguin, London.

Kerr A. 2020. *The Heart Sutra*. Allen Lane, London.

Kettering C F. 1929. Keep the consumer dissatisfied. *Nation's Business* 17(1), 30–31.

Khan Z R, Midega C A O, Hooper A and Pickett J A. 2016. Push-pull: Chemical ecology-based integrated pest management technology. *Journal of Chemical Ecology* 42(7), 689–697.

Killingsworth M A and Gilbert D T. 2010. A wandering mind is an unhappy mind. *Science* 330, 932.

Kim S H, Kim M S, Lee M S, Park Y S, Lee H J, Kang S A, Lee H S, Lee K E, Yang H J, Kim M J and Lee Y E. 2016. Korean diet: characteristics and historical background. *Journal of Ethnic Foods* 3(1), 26–31.

Kimmerer R W. *Braiding Sweetgrass*. Milkweed, Minnesota.

King F H. 1911. *Farmers of Forty Centuries*. Rodale Press, Pennsylvania.

King L A and Napa C K. 1998. What makes a life good? *Journal of Personality and Social Psychology* 75(1), 156.

Kingsland J. 2016. *Siddhartha's Brain*. Robinson, London.

Kirschner H. 2021. *Water, Wood and Wild Things*. Viking, London.

Klein N. 2015. *This Changes Everything*. Penguin, London.

Klein N. 2019. *On Fire*. Allen Lane, London.

Kleine S S and Baker S M. 2004. An integrative review of material possession attachment. *Academy of Marketing Science Review* 1(1), 1–39.

Kline D. 1990. *Great Possessions*. Wooster Book Company, Wooster Ohio.

Knorr and WWF-UK. 2019. *Fifty Foods for Healthier People and Healthier Planet*. At www.wwf.org.uk/updates/wwf-and-knorr-launch-future-50-foods

Kolbert E. 2021. *Under a White Sky*. Bodley Head, London.

Kondoh K. 2015. The alternative food movement in Japan: Challenges, limits, and resilience of the teikei system. *Agriculture and Human Values* 32, 143–153.

Korean Zen (ed Haight I). 2010. *Garden Chrysanthemum and First Mountain Snow*. White Pine Press, Buffalo.

Korn P. 2018. *Why We Make Things and Why It Matters*. Vintage, London.

Kornfield J. 2000. *After the Ecstasy, the Laundry*. Random House, London.

Kornfield J. 2002. *A Path with Heart*. Rider, London.

Kotler S. 2021. *The Art of the Impossible*. Harper Wave, New York.

Kotler S and Wheal J. 2017. *Stealing Fire*. HarperCollins, London.

Kozar R, Galang E, Alip A, Sedhain J, Subramanian S and Saito O. 2019. Multi-level networks for sustainability solutions: The case of the International Partnership for the Satoyama Initiative. *Current Opinion in Environmental Sustainability* 39, 123–134.

KPMG and Landcare Australia. 2021. *Well-Being Benefits of Participating in Landcare*. At https://landcareaustralia.org.au/wellbeing-report/

Krakauer J and Roberts D. 1990. *Iceland: Land of the Sagas*. Villard, New York.

Kraybill D B. 1989 (2001). *The Riddle of Amish Culture*. Johns Hopkins University Press, Baltimore.

Kreitzer M J, Monsen K A, Nandram S and De Blok J. 2015. Buurtzorg Nederland: A global model of social innovation, change, and whole-systems healing. *Global Advances in Health and Medicine* 4(1), 40–44.

Kubiszewski I, Costanza R, Anderson S and Sutton P. 2017. The future value of ecosystem services: Global scenarios and national implications. *Ecosystem Services* 26, 289–301.

Kuhn T. 1965. *The Structure of Scientific Revolutions*. University of Chicago Press, Chicago.

Kuyken W, Hayes R, Barrett B, Byng R, Dalgleish T, Kessler D, Lewis G, et al. 2015. Effectiveness and cost-effectiveness of mindfulness-based cognitive therapy compared with maintenance antidepressant treatment in the prevention of depressive relapse or recurrence (PREVENT): A randomised controlled trial. *The Lancet* 386(9988), 63–73.

Laird M. 2006. *Into the Silent Land*. Oxford University Press, Oxford.

Laird M. 2011. *A Sunlit Absence*. Oxford University Press, Oxford.

Laird M. 2019. *An Ocean of Light*. Oxford University Press, Oxford.

Lal R. 2022. Reducing carbon footprints of agriculture and food systems. *Carbon Footprints* 1(1), 3.

Lally P, Van Jaarsveld C H, Potts H W and Wardle J. 2010. How are habits formed: Modelling habit formation in the real world. *European Journal of Social Psychology 40*, 998–1009.

Lamont A and Ranaweera N A. 2020. Knit one, play one: Comparing the effects of amateur knitting and amateur music participation on happiness and wellbeing. *Applied Research in Quality of Life 15*(5), 1353–1374.

Lancet. 2019. Planetary health: From concept to decisive action. *The Lancet 3*, e402.

Lancet. 2020. *WHO-UNICEF-Lancet Children's Commission*. The Lancet, London.

Lane J. 2002. *Spirit of Silence*. Green Books, Dartington.

Lang T. 2020. *Feeding Britain*. Penguin, London.

Lang T and Rayner G. 2012. Ecological public health: The 21st century's big idea? An essay by Tim Lang and Geoff Rayner. *BMJ 345*.

Langlands A. 2017. *Cræft*. WW Norton, New York.

Lao Tzu (trans Lau D). 2000. *Tao Te Ching*. Penguin Classics, London.

Lao Tzu (trans Mitchell S). 1988. *Tao Te Ching: An Illustrated Journey*. Frances Lincoln, London.

Lappé F M. 2021. *Diet for a Small Planet* (50th Anniversary Edition). Ballantine Books, New York.

La Sorte F A and Somville M. 2020. Survey completeness of a global citizen-science database of bird occurrence. *Ecography 43*(1), 34–43.

Lauria M and Slotterback C S (eds). 2021. *Learning from Arnstein's Ladder*. Routledge, Oxon.

Layard R. 2020. *Can We Be Happier?* Pelican, London.

Leask C F, Bell J and Murray F. 2020. Acceptability of delivering an adapted Buurtzorg model in the Scottish care context. *Public Health 179*, 111–117.

Lee H S, Duffey K J and Popkin B M. 2012. South Korea's entry to the global food economy: Shifts in consumption of food between 1998 and 2009. *Asia Pacific Journal of Clinical Nutrition 21*(4), 618.

Lee I-M and Paffenbarger R S. 2000. Associations of light, moderate, and vigorous intensity physical activity with longevity the Harvard Alumni Health Study. *American Journal of Epidemiology 151*, 293–299.

Lee J Y, Jun N R, Yoon D, Shin C and Baik I. 2015. Association between dietary patterns in the remote past and telomere length. *European Journal of Clinical Nutrition 69*(9), 1048–1052.

Lee R B. 1979. *The !Kung San*. Cambridge University Press, Cambridge.

Lee R B and Daly R. 1999. *Cambridge Encyclopedia of Hunter-Gatherers*. Cambridge University Press, Cambridge.

Lee S (Sam). 2020. *The Nightingale*. Century, London.

Lee S (Shannon). 2020. *Be Water, My Friend*. Rider, London.

Le Francois B A, Menzies R and Reaume G (eds). 2013. *Mad Matters: A Critical Reader in Canadian Mad Studies*. Canadian Scholars Press, Toronto ON.

Leopold A. 1949 [2020]. *A Sand County Almanac, and Sketches Here and There*. Penguin, London.

Lesen A E. 2016. *Scientists, Experts, and Civic Engagement: Walking a Fine Line*. Routledge, Oxon.

Lethbridge T C. 1950. *Herdsmen and Hermits*. Bowes, London.

Levey J M. 2015. *Mindfulness, Meditation and Mind Fitness*. Conari Press, San Francisco.

Levin T (with Suzukei V). 2006. *Where Rivers and Mountains Sing*. Indiana University Press, Bloomington and Indianapolis.

Levitin D. 2020. *The Changing Mind: A Neuroscientist's Guide to Ageing Well*. Penguin, London.

Lewandowski E and Oberhauser K. 2016. Butterfly citizen science projects support conservation activities among their volunteers. *Citizen Science: Theory and Practice 1*(1).

248 Bibliography

Lewis D. 1972. *We, the Navigators: The Ancient Art of Landfinding in the Pacific*. University of Hawaii Press, Honolulu.

Lewontin R C, Rose S and Kamin L. 1984. *Not in Our Genes: Biology, Ideology and Human Nature*. Pantheon, New York.

Li Y S, Liu C F, Yu W P, Mills M E C and Yang B H. 2020. Caring behaviours and stress perception among student nurses in different nursing programmes: A cross-sectional study. *Nurse Education in Practice 48*, 102856.

Lifshitz-Assaf H, Tushman M L and Lakhani K R. 2018. A study of NASA scientists shows how to overcome barriers to open innovation. *Harvard Business Review*, May 29.

Linn V. 2015. *The Buddha on Wall Street*. Windhorse, Cambridge.

Li Po and Tu Fu (trans Cooper A). 1973. *Poems*. Penguin Classics, London.

Litfin K T. 2014. *Ecovillages: Lessons for Sustainable Community*. Polity Press, Cambridge.

Liu X, Clark J, Siskind D, Williams G M, Byrne G, Yang J L and Doi SA. 2015. A systematic review and meta-analysis of the effects of Qigong and Tai Chi for depressive symptoms. *Complementary Therapies in Medicine 23*(4), 516–534.

Lobao L M. 1990. *Locality and Inequality: Farm and Industry Structure and Socioeconomic Conditions*. SUNY Press, New York.

Lopez B. 1998. *About This Life*. Harvill, London.

Lopez B. 2019. *Horizon*. Bodley Head, London.

Lorimer E O. 1939. *Language Hunting in the Karakoram*. George Allen and Unwin, London.

Luo Guanzhong (trans Palmer M). 2018. *The Romance of the Three Kingdoms*. Penguin Classics, London.

Lynton R P. 1961. *Tide of Learning: The Aloka Experience*. Routledge, London.

Macfarlane R. 2007. *The Wild Places*. Granta, London.

Macfarlane R. 2014. *Landmarks*. Hamish Hamilton, London.

Magnason A S. 2020. *On Time and Water*. Serpent's Tail, London.

Magnuson J. 2013. *The Approaching Great Transformation*. Policy Press, University of Bristol, Bristol.

Mahābhāra (trans Smith J D). 2009. Penguin Classics, London.

MARA (Ministry of Agriculture and Rural Affairs, China) (2019). *Farmer Cooperatives Have Become the Backbone for Rural Vitalization*. At www.moa.gov.cn/ztzl/70zncj/201909/t20190916_6327995.htm

Marmot M G, Friel S, Bell R, Houweling T A, Taylor S and Commission on Social Determinants of Health. 2008. Closing the gap in a generation: Health equity through action on the social determinants of health. *The Lancet 372*(9650), 1661–1669.

Marmot M G and Syme S L. 1976. Acculturation and coronary heart disease in Japanese-Americans. *American Journal of Epidemiology 104*(3), 225–247.

Marseglia A, Wang H X, Rizzuto D, Fratiglioni L and Xu W. 2019. Participating in mental, social, and physical leisure activities and having a rich social network reduce the incidence of diabetes-related dementia in a cohort of Swedish older adults. *Diabetes Care 2*(2), 232–239.

Mason P. 2015. *Post-Capitalism*. Penguin, London.

Massey C. 2017. *Call of the Reed Warbler*. Chelsea Green, White River Junction, VT.

Masuno S. 2019. *Zen: The Simple Art of Living*. Michael Joseph, London.

Maturana H R and Varela F J. 1987. *The Tree of Knowledge: The Biological Roots of Human Understanding*. New Science Library/Shambhala Publications, Boulder, CO.

McCarthy C. 2009. *Out of the Marvellous*. RTÉ Television, Dublin.

McCarthy M. 2015. *Moth Snowstorm*. John Murray, London.

McCloughan P, Batt W H, Costine M and Scully D. 2011. Second European quality of life survey. *Participation in Volunteering and Unpaid Work: European Foundation for the Improvement of Living and Working Conditions*, Brussels.

McIntosh A. 2020. *Riders on the Storm*. Berlin, Edinburgh.

McKee R. 1999. *Story*. Methuen, London.

McKibben B. 2019. *Falter*. Headline, London.

McMichael P. 2013. *Food Regimes and Agrarian Questions*. Fernwood, Halifax.

McPhee J. 1989. *The Control of Nature*. Farrar, Strauss and Giroux, New York.

McPherson, D (ed). 2017. *Spirituality and the Good Life: Philosophical Approaches*. Cambridge University Press, Cambridge.

Meadows D. 2008. *Thinking in Systems: A Primer*. Chelsea Green Publishing, Vermont.

Meadows D, Randers J and Meadows D. 2004. *Limits to Growth: The 30-Year Update*. Chelsea Green Publishing, Vermont.

Medvec V H, Madey S F and Gilovich T. 1995. When less is more: Counterfactual thinking and satisfaction among Olympic medalists. *Journal of Personality and Social Psychology* 69(4), 603–610.

Merton T. 1961 (1999). *Mystics and Zen Masters*. Farrar, Straus and Giroux, New York.

Micha R, Peñalvo J L, Cudhea F, Imamura F, Rehm C D and Mozaffarian D. 2017. Association between dietary factors and mortality from heart disease, stroke, and type 2 diabetes in the United States. *JAMA* 317(9), 912–924.

Milne R G (ed). 2018. *Civic Engagement*. Routledge, Oxon.

Miller F (ed). 2018. *Ecovillages around the World*. Findhorn Press, Rochester, VT.

Miller M. 2018. *Circe*. Bloomsbury, London.

Miller S. 2001. Public understanding of science at the crossroads. *Public Understanding of Science* 10(1), 115–120.

Miltenberger R G, Fuqua R W and Woods D W. 1998. Applying behavior analysis to clinical problems: Review and analysis of habit reversal. *Journal of Applied Behavior Analysis* 31(3), 447–469.

Mindfulness Institute. 2021. *Implementing Mindfulness in Schools*. Mindfulness Institute, London.

Mitchell S (trans). 1982. *Selected Poetry of Rainer Maria Rilke*. Random House, London.

Mitchell R J and Popham F. 2008. Effect of exposure to natural environment on health inequalities: An observational population study. *The Lancet* 372(9650), 1655–1660.

Mitchell R J, Richardson E A, Shortt N K and Pearce J R. 2015. Neighborhood environments and socioeconomic inequalities in mental well-being. *American Journal of Preventive Medicine* 49(1), 80–84.

Montari M. 2004. *Food is Culture*. Columbia University Press, New York.

Monteiro C A, Cannon G, Levy R B, Moubarac J C, Louzada M L, Rauber F, Khandpur N, Cediel G, Neri D, Martinez-Steele E and Baraldi L G. 2019. Ultra-processed foods: what they are and how to identify them. *Public Health Nutrition* 22(5), 936–941.

Monteiro C A, Moubarac J C, Levy R B, Canella D S, da Costa Louzada M L and Cannon G. 2018. Household availability of ultra-processed foods and obesity in nineteen European countries. *Public Health Nutrition* 21(1), 18–26.

Moore E O. 1981. A prison environment's effect on health care service demands. *Journal of Environmental Systems* 11(1), 17–34.

Moore J W. 2018. The capitalocene part II: Accumulation by appropriation and the centrality of unpaid work/energy. *The Journal of Peasant Studies* 45(2), 237–279.

Morris J N and Raffle P A B. 1954. Coronary heart disease in transport workers: A progress report. *British Journal of Industrial Medicine* 11(4), 260–264.

Muradova L, Walker H and Collis F. 2020. Climate change communication and public engagement in interpersonal deliberative settings: Evidence from the Irish citizens' assembly. *Climate Policy* 20(10), 1322–1335.

Murphy S. 2014. *Mending the Earth, Mending the World*. Counterpoint, Berkeley.

Myths of Mesopotamia (trans Dalley S). 1989. *Creation, the Flood, Gilgamesh and Others.* Oxford World's Classics. Oxford University Press, Oxford.

Naess A. 2008 (2016). *The Ecology of Wisdom.* Penguin, London.

NASA. 2022. *Open Innovation.* At www.nasa.gov/offices/otps/openinnovation

NASP. 2019. *National Academy for Social Prescribing 2020–2023 Strategic Plan.* NASP, London.

Natural England. *Good Practice in Social Prescribing for Mental Health: The Role of Nature-Based Interventions.* Natural England Commissioned Report NECR228, London, 2017.

Nayak P K. 2021. *Making Commons Dynamic.* Routledge, Oxon.

NEA. 2011. *National Ecosystem Assessment.* Defra, London.

Nef. 2013. *The Economic Benefits of Ecominds.* New Economics Foundation, London.

Neima A. 2021. *The Utopians.* Picador, London.

Newberg A and Waldman M. 2016. *How Enlightenment Changes the Brain.* Hay House, Carlsbad, CA.

Nielsen R K. 2018. No one cares what we know: Three responses to the irrelevance of political communication research. *Political Communication 35*(1), 145–149.

Niles J O, Brown S, Pretty J, Ball A S and Fay J. 2002. Potential carbon mitigation and income in developing countries from changes in use and management of agricultural and forest lands. *Philosophical Transactions of the Royal Society of London. Series A 360*(1797), 1621–1639.

Nisbet E K, Zelenski J M and Grandpierre Z. 2019. Mindfulness in nature enhances connectedness and mood. *Ecopsychology 11*(2), 81–91.

Njál's Saga (trans Cook R). 1997. Penguin Classics, London.

Norberg-Hodge H. 1991 (2000). *Ancient Futures.* Rider, London.

Norgaard R B. 2019. Economism and the Econocene: A coevolutionary interpretation. *Real-World Economic Review 114.*

North R, Allard J and Gillies P. 2011. *Longman Anthology of Old English, Old Icelandic and Anglo-Norman Literature.* Longman, London.

Odell J. 2019. *How to Do Nothing.* Melville House, Brooklyn, NY.

O'Driscoll D. 2009. *Stepping Stones: Interviews with Seamus Heaney.* Faber and Faber, London.

Ogden D. 2013. *Dragons, Serpents and Slayers in the Classical and Early Christian Worlds.* Oxford University Press, Oxford.

Okakura K. 1906 (2016). *The Book of Tea.* Penguin Classics, London.

Okri B. 1990. *The Famished Road.* Vintage, London.

Oliver M. 2014. *Upstream.* Penguin Press, New York.

O'Neill D W, Dietz R and Jones N (eds). 2010. *Enough Is Enough: Ideas for a Sustainable Economy in a World of Finite Resources.* Report of the Steady State Economy Conference. Centre for the Advancement of the Steady State Economy and Economic Justice for All, Leeds.

O'Neill D W, Fanning A L, Lamb W F and Steinberger J K. 2018. A good life for all within planetary boundaries. *Nature Sustainability 1*(2), 88–95.

Ono S. 1962. *Shinto.* Tuttle, Tokyo.

Oreskes N and Conway E M. 2010. *Merchants of Doubt.* Bloomsbury, New York.

Orkneyinga Saga (trans Palsson H and Edwards P). 1978. Penguin Classics, London.

O'Rourke B. 2010. *Finding Your Hidden Treasure.* Darton, Longman and Todd, London.

Orr D W. 1992. *Ecological Literacy.* State University of New York Press, New York.

Ostler N. 2005. *Empires of the World.* Harper Perennial, London.

Ostrom E. 1990. *Governing the Commons: The Evolution of Institutions for Collective Action.* Cambridge, Cambridge University Press.

Oswald Y, Owen A and Steinberger J K. 2020. Large inequality in international and intranational energy footprints between income groups and across consumption categories. *Nature Energy 5*(3), 231–239.

Ouellette J A and Wood W. 1998. Habit and intention in everyday life: The multiple processes by which past behaviour predicts future behaviour. *Psychological Bulletin 124*(1), 54.

Outhwaite C L, McCann P and Newbold T. 2022. Agriculture and climate change are reshaping insect biodiversity worldwide. *Nature 605*, 97–102.

Owens S. 2020. *The Spirit of Place*. Thames and Hudson, London.

Ozkan N N. 2015. An example of open innovation: P&G. *Procedia-Social and Behavioral Sciences 195*, 1496–1502.

Panlasigui S, Spotswood E, Beller E and Grossinger R. 2021. Biophilia beyond the building: applying the tools of urban biodiversity planning to create biophilic cities. *Sustainability 13*(5), 2450.

Park N and Peterson C. 2009. Achieving and sustaining a good life. *Perspectives on Psychological Science 4*(4), 422–428.

Peat D. 1994. *Blackfoot Physics*. Fourth Estate, London.

Perelman M and Shea K P. 1972. The big farm. *Environment: Science and Policy for Sustainable Development 14*(10), 10–15.

Peters J. 2019. *The Art of Japanese Living*. Sumersdale, London.

Peterson C, Park N and Seligman M E. 2005. Orientations to happiness and life satisfaction: The full life versus the empty life. *Journal of Happiness Studies 6*(1), 25–41.

Piketty T and Saez E. 2014. Inequality in the long run. *Science 344*(6186), 838–843.

Place-Based Climate Action Network (PCAN). 2022. At https://pcancities.org.uk/

Planck M. 1950 (1968). *Scientific Autobiography and Other Papers*. Philosophical Library, New York.

Plieninger T, Kohsaka R, Bieling C, Hashimoto S, Kamiyama C, Kizos T, Penker M, Kieninger P, Shaw B J, Sioen G B, Yoshida Y and Saito O. 2018. Fostering biocultural diversity in landscapes through place-based food networks: A 'solution scan' of European and Japanese models. *Sustain Science 13*, 219–233.

PLM. 2022. *Patients Like Me*. At www.patientslikeme.com/

Plotkin B. 2003. *Soulcraft*. New World Library, Novato, CA.

Pollan M. 2008. *In Defense of Food: An Eater's Manifesto*. Penguin, London.

Pongsiri M J, Bickersteth S, Colón C, et al. 2019. Planetary health: From concept to decisive action. *The Lancet Planetary Health 3*(10), e402–e404.

Port D. 2021. *The Record of Empty Hall: One Hundred Classic Koans*. Shambhala, Boulder, CO.

Pretty J. 1995. Participatory learning for sustainable agriculture. *World Development 23*(8), 1247–1263.

Pretty J. 2003. Social capital and the collective management of resources. *Science 302*, 1912–1915.

Pretty J. 2007. *The Earth Only Endures*. Earthscan, London.

Pretty J. 2011. *This Luminous Coast*. Full Circle Editions, Saxmundham.

Pretty J. 2013. The consumption of a finite planet: Well-being, convergence, divergence and the nascent green economy. *Environmental and Resource Economics 55*(4), 475–499.

Pretty J. 2014. *The Edge of Extinction*. Cornell University Press, Ithaca, NY.

Pretty J. 2018. Intensification for redesigned and sustainable agricultural systems. *Science 362*, eaav0294.

Pretty J. 2022a. *Sea Sagas of the North: Travels and Tales at Warming Waters*. Hawthorn Press, Stroud.

Pretty J. 2022b. *The Low Carbon Good Life*. CPPE Paper 2, University of Essex, Colchester.

Pretty J. 2022c. *How Public Engagement Improves Lives: Principles, Typology and Evidence*. CPPE Paper 1, University of Essex, Colchester.

Pretty J, Attwood S, Bawden R, Van den Berg H, Bharucha Z, Dixon J . . . . Yang P (29 authors). 2020. Assessment of the growth in social groups for sustainable agriculture and land management. *Global Sustainability 3*, E23.

Pretty J and Barton J. 2020. Nature-based interventions and mind: Body interventions: Saving public health costs whilst increasing life satisfaction and happiness. *International Journal of Environmental Health Research 17*, 7769.

Pretty J, Barton J, Bharucha Z P, Bragg R, Pencheon D, Wood C and Depledge M H. 2015. Improving health and well-being independently of GDP: Dividends of greener and prosocial economies. *International Journal of Environmental Health Research 11*, 1–26.

Pretty J, Benton T G, Bharucha Z P, Dicks L, Butler Flora C, Hartley S, Lampkin N, Morris C, Pierzynski G, Prasad P V V, Reganold J, Rockström J, Smith P, Thorne P and Wratten S. 2018. Global assessment of agricultural system redesign for sustainable intensification. *Nature Sustainability 1*, 441–446.

Pretty J, Barton J and Rogerson M. 2017. Green mind theory: How brain-body-behaviour links into natural and social environments for healthy habits. *International Journal of Environmental Research and Public Health 14*, 706.

Pretty J and Bharucha Z P. 2015. Integrated pest management for sustainable intensification of agriculture in Asia and Africa. *Insects 6*(1), 152–182.

Pretty J and Bharucha Z P. 2018. *The Sustainable Intensification of Agriculture*. Routledge, Oxon.

Pretty J, Guijt I, Thompson J and Scoones I. 1995. *Participatory Learning and Action: A Trainers Guide*. IIED, London.

Pretty J, Peacock J, Sellens M and Griffin M. 2005. The mental and physical health outcomes of green exercise. *International Journal of Environmental Health Research 15*(5), 319–337.

Pretty J and Ward H. 2001. Social capital and the environment. *World Development 29*(2), 209–227.

Princen T. 2005. *The Logic of Sufficiency*. MIT Press, Cambridge, MA.

Project Drawdown. 2020. *The Drawdown Review*. At www.drawdown.org/drawdown-review

Proshika. 2022. *Proshika Microfinance Programmes*. At www.proshika.org

Pryor F. 2019. *The Fens*. Head of Zeus, London.

Putnam R. 1995. Bowling alone: America's declining social capital. *Journal of Democracy 6*(1), 65–78.

Rahman A. 2019. *Women and Microcredit in Rural Bangladesh: An Anthropological Study of Grameen Bank Lending*. Routledge, Oxon.

Ramsden A and Hollingsworth S. 2013. *The Storyteller's Way*. Hawthorn Press, Stroud.

Rao G. 2019. Familiarity does not breed contempt: Generosity, discrimination, and diversity in Delhi schools. *American Economic Review 109*(3), 774–809.

Raworth K. 2017. *Doughnut Economics: Seven Ways to Think Like a 21st Century Economist*. Random House, London.

Reid A. 2002. *The Shaman's Coat: A Native History of Siberia*. Phoenix, London.

Ricard M. 2015. *Happiness: A Guide to Developing Life's Most Important Skill*. Atlantic Books, London.

Rinaudo T. 2021. *The Forest Underground*. ISCAST, Forest Hill, Victoria.

Ripple W J, Wolf C, Newsome T M, Galetti M, Alamgir M, Crist E, Mahmoud M I, Laurance W F and 15,364 Scientist Signatories from 184 Countries. 2017. World scientists' warning to humanity: a second notice. *BioScience 67*(12), 1026–1028.

Rizzuto D, Orsini N, Qiu C, Wang H X and Fratiglioni L., 2012. Lifestyle, social factors, and survival after age 75: Population based study. *BMJ 345*.

Rockefeller Foundation. 2021. *True Cost of Food*. Rockefeller Foundation, New York.

Rockström J and 28 other co-authors. 2009. A safe operating space for humanity. *Nature 461*(7263), 472–475.

Rockström J, Williams J, Daily G, Noble A, Matthews N, Gordon L, Wetterstrand H, DeClerck F, Shah M, Steduto P, Fraiture C, Hatibu N, Unver O, Bird J, Sibanda L and

Smith J. 2017. Sustainable intensification of agriculture for human prosperity and global sustainability. *Ambio 46*, 4–17.

Rogerson M, Wood C, Pretty J, Schoenmakers P, Bloomfield D and Barton J. 2020. Regular doses of nature: The efficacy of green exercise interventions for mental well-being. *International Journal of Environmental Research and Public Health 17*(5), 1526.

Rohr R. 2012. *Falling Upward*. SPCK, London.

Rojas-Rueda D, Nieuwenhuijsen M J, Gascon M, Perez-Leon D and Mudu P. 2019. Green spaces and mortality: A systematic review and meta-analysis of cohort studies. *The Lancet Planetary Health 3*(11), e469–e477.

Rosenthal N E. 2011. *Transcendence*. Hay House, Carlsbad, CA.

Rosling H. 2018. *Factfulness*. Sceptre, London.

Ross M C. 2010. *The Cambridge Introduction to the Old Norse-Icelandic Saga*. Cambridge University Press, Cambridge.

Rosset P M, Machín Sosa B, Roque Jaime A M and Ávila Lozano D R. 2011. The campesino-to-campesino agroecology movement of ANAP in Cuba: Social process methodology in the construction of sustainable peasant agriculture and food sovereignty. *Journal of Peasant Studies 38*, 161–191.

Röttgen U and Zetti K. 2020. *Craftland Japan*. Thames and Hudson, London.

Royal Society. 1985. *Public Understanding of Science*. The Royal Society, London.

Royal Society. 2012. *People and the Planet*. Royal Society, London.

Rumi (trans Banks L). 1995. *Selected Poems*. Penguin Classics, London.

Ryan S F, Adamson N L, Aktipis A, Andersen L K, Austin R, Barnes L, Beasley M R, Bedell K D, Briggs S, Chapman B, Cooper C B and 31 more authors. 2018. The role of citizen science in addressing grand challenges in food and agriculture research. *Proceedings of the Royal Society B 285*(1891), 20181977.

Ryokan T (trans Stevens J). 1993. *Dewdrops on a Lotus Leaf*. Shambhala, Boulder.

Saga of King Hrolf Kraki (trans Byock J L). Penguin Classics, London.

Saga of the People of Laxardal (trans Kunz K). 1997. Penguin Classics, London.

Sagas of the Volsungs (trans Byock J). 1990. Penguin Classics, London.

Sagas of the Warrior Poets (trans Eiriksson L). 1997 [2002]. Penguin Classics, London.

Saint Teresa (trans Cohen J M). 1957. *The Life of Saint Teresa of Avila by Herself*. Penguin Classics, London.

Sánchez-Bayo F and Wyckhuys K A. 2021. Further evidence for a global decline of the entomofauna. *Austral Entomology 60*(1), 9–26.

Sarashina Nikki (also known as Lady Sarashina) (trans Ivan Morris). 1971. *As I Crossed a Bridge of Dreams*. Penguin Classics, London.

Sauermann H, Vohland K, Antoniou V, Balázs B, Göbel C, Karatzas K, Mooney P, Perelló J, Ponti M, Samson R and Winter S. 2020. Citizen science and sustainability transitions. *Research Policy 49*(5), 103978.

Sawyer R. 1942 (1962). *The Way of the Storyteller*. Bodley Head, London.

Schoen M. 2013. *Your Survival Instinct Is Killing You*. Plume, New York.

Schor T K. 2017. *A Step Away from Paradise*. City Lion Press and Penguin, India.

Schumacher E F. 1973 (2011). *Small Is Beautiful*. Vintage, London.

Schutt R K, Deng X and Stoehr T. 2013. Using bibliotherapy to enhance probation and reduce recidivism. *Journal of Offender Rehabilitation 52*(3), 181–197.

Schwartz B. 2004. *The Paradox of Choice*. Harper, New York.

Schwarz D W. 1987. *Breaking Through*. Green Books, Dartington.

Schwarz D W. 1998. *Living Lightly*. Jon Carpenter, Charlbury.

Scialabba N E H and Müller-Lindenlauf M. 2010. Organic agriculture and climate change. *Renewable Agriculture and Food Systems 25*(2), 158–169.

Scott A and Grafton L. 2020. *The New Long Life*. Bloomsbury, London.

Sennett R. 2008. *The Craftsman*. Penguin, London.

Shanahan D F, Bush R, Gaston K J, Lin B B, Dean J, Barber E and Fuller R A. 2016. Health benefits from nature experiences depend on dose. *Scientific Reports 6*, 28551.

Shaping Our Lives. 2022. *Inclusive Involvement Matters*. At https://shapingourlives.org.uk/

Sheldrake M. 2020. *Entangled Life*. Vintage, London.

Sherwood S, Cole D, Crissman C and Paredes M. 2005. From pesticides to people: Improving ecosystem health in the northern Andes. In Pretty J (ed). *The Pesticide Detox*. Earthscan, London.

Shi Z, Zhang T, Byles J, Martin S, Avery J C and Taylor A W. 2015. Food habits, lifestyle factors and mortality among oldest old Chinese: The Chinese Longitudinal Healthy Longevity Survey (CLHLS). *Nutrients 7*(9), 7562–7579.

Shire W. 2022. *Bless the Daughter: Raised by a Voice in Her Head*. Chatto and Windus, London.

Shukman H. 2019. *One Blade of Grass*. Yellow Kite, London.

Sibley R. 2013. *The Way of the 88 Temples*. University of Virginia Press, Charlottesville.

Sidelsky R and Sidelsky E. 2012. *How Much Is Enough?* Penguin, London.

Siegel D. 2018. *Aware*. Scribe, Melbourne and London.

Silke R M. 2015. *The Poetry of Rainer Maria Silke*. Create Space, Scotts Valley.

Silko L M. 2010. *The Turquoise Ledge*. Penguin, London.

Simard S. 2021. *Finding the Mother Tree*. Allen Lane, London.

Simms A and Smith J. 2008. *Do Good Lives Have to Cost the Earth?* Constable, London.

Singer P. 2015. *The Most Good You Can Do*. Yale University Press, New Haven.

Skinner J D. 2017. *Practical Zen*. Singing Dragon, London.

Sloman P. 2019. *Transfer State: The Idea of a Guaranteed Income and the Politics of Redistribution in Modern Britain*. Oxford University Press, Oxford.

Slotterback C S and Lauria M. 2019. Building a foundation for public engagement in planning: 50 years of impact, interpretation, and inspiration from Arnstein's Ladder. *Journal of the American Planning Assoc 85*(3), 183–187.

Smiley J. 2000. *The Sagas of the Icelanders*. Penguin Classics, London.

Smith J. 2019. The Aristotelian good life and virtue theory. *International Journal of Business and Social Science 10*(1), 6–12.

Smith L G, Jones P J, Kirk G J, Pearce B D and Williams A G. 2018. Modelling the production impacts of a widespread conversion to organic agriculture in England and Wales. *Land Use Policy 76*, 391–404.

Smith P. 2013. Delivering food security without increasing pressure on land. *Global Food Security 2*(1), 18–23.

Smith P, Bhogal A, Edgington P, Black H, Lilly A, Barraclough D, Worrall F, Hillier J and Merrington, G. 2010. Consequences of feasible future agricultural land-use change on soil organic carbon stocks and greenhouse gas emissions in Great Britain. *Soil Use and Management 26*(4), 381–398.

Smith P, Milne R, Powlson D S, Smith J U, Falloon P and Coleman K. 2000. Revised estimates of the carbon mitigation potential of UK agricultural land. *Soil Use and Management 16*(4), 293–295.

Snyder G. 1979 [2007]. *He Who Hunted Birds in His Father's Village*. Shoemaker Hoard, Berkeley.

Snyder G. 1990. *The Practice of the Wild*. Shoemaker Hoard, Washington, DC.

Snyder G. 1995. *A Place in Space*. Counterpoint, New York.

Solnit R. 2009. *A Paradise Built in Hell*. Viking, London.

Soper K. 2016. Towards sustainable flourishing: Democracy, hedonism and the politics of prosperity. In Syse K L and Muller M (eds). *Sustainable Consumption and the Good Life*. Routledge, Oxon.

Stafford R, Chamberlain B, Clavey L, Gillingham P K, McKain S, Morecroft M D, Morrison-Bell C and Watts O (eds). 2021. *Nature-Based Solutions for Climate Change in the UK: A Report by the British Ecological Society*. London, UK. At www.britishecologicalsociety. org/nature-basedsolutions

Standing G. 2019. *Plunder of the Commons*. Penguin, London.

Stanford P. 2021. *Pilgrimage*. Thames and Hudson, London.

Steptoe A and Wardle J. 2012. Enjoying life and living longer. *Arch of Internal Medicine* 172(3), 273–275.

Sternberg E. 2009. *Healing Spaces*. Belknap, Harvard.

Stevens J (trans and ed). 1993. *Dewdrops on a Lotus Leaf: Zen Poems of Ryōkan*. Shambhala, Boston.

Stevens J. 2007. *Zen Bow, Zen Arrow*. Shambhala, Boulder.

Stevens J. 2013. *The Marathon Monks of Mount Hiei*. Echo Point Books, Brattleboro, VT.

Stevens J. 2014. *Rengetsu: Life and Poetry of Lotus Moon*. Echo Point, Brattleboro, VT.

Stone B. 1959. *Sir Gawain and the Green Knight*. Penguin, London.

Strasser B, Baudry J, Mahr D, Sanchez G and Tancoigne E. 2019. 'Citizen science'? Rethinking science and public participation. *Science & Technology Studies 32*, 52–76.

Sturlusson S. 1966. *King Harald's Saga*. Penguin Classics, London.

Sturlusson S (trans Byock J). 2005. *The Prose Edda*. Penguin Classics, London.

Sturlusson S (trans Magnusson M and Pálsson K). 1966 [2002]. *The Elder Edda*. Penguin Classics, London.

Sun W, Watanabe M, Tanimoto Y, Shibutani T, Kono R, Saito M, Usuda K and Kono K. 2007. Factors associated with good self-rated health of non-disabled elderly living alone in Japan: A cross-sectional study. *BMC Public Health 7*(1), 297.

Sundar B. 2017. Joint forest management in India: An assessment. *International Forestry Review 19*(4), 495–511.

Suzuki D T. 1938 [2010]. *Zen and Japanese Culture*. Princeton University Press, New York.

Suzuki S. 1970 [2009]. *Zen Mind, Beginner's Mind*. Weatherhill, Boston and London.

Sweet V. 2017. *Slow Medicine*. Riverhead Books, New York.

Syse K L and Muller M L (eds). 2016. *Sustainable Consumption and the Good Life*. Routledge, Oxon.

Tabb P. 2021. *Biophilic Urbanism*. Routledge, Oxon.

Tale of the Heike (trans Tyler R). 2012. Penguin Classics, London.

Tales of the Ise (trans MacMillan P). 2016. Penguin Classics, London.

Tanahashi K (ed). 1985. *Moon in a Dewdrop: Writings of Zen Master Dogen*. North Point Press, New York.

Tarrant J. 1998. *The Light Inside the Dark*. Harper Collins, London.

Tarrant J. 2004. *Bring Me the Rhinoceros*. Shambhala, Boulder.

Taylor R. 2021. *Your Simple Guide to Reversing Type 2 Diabetes*. Short Books, London.

Temple R. 1986. *The Genius of China*. Andre Deutsch, London.

Theobald E J, Ettinger A K, Burgess H K, DeBey L B, Schmidt N R, Froehlich H E, Wagner C, HilleRisLambers J, Tewksbury J, Harsch M A and Parrish J K. 2015. Global change and local solutions: Tapping the unrealized potential of citizen science for biodiversity research. *Biological Conservation 181*, 236–244.

Theocritus of Syracuse (trans Verity A). 2002. *Idylls*. Oxford University Press, Oxford.

Thich Nhat Hanh. 2015. *Silence*. Rider, London.

Thompson E P. 1975. *Of Whigs and Hunters*. Penguin, London.

Timms H and Heimans J. 2018. *#newpower*. Picador, London.

Tirivayi N, Nennen L, Tesfaye W and Ma Q. 2018. The benefits of collective action: Exploring the role of forest producer organizations in social protection. *Forest Policy and Economics 90*, 106–114.

Tolkein J R R. 1964. *Tree and Leaf.* George Allen and Unwin, London.

Tornstamd L. 2011. Maturing into gerotranscendence. *Journal of Transpersonal Psychology* 43(2).

Touchton M, Sugiyama N B and Wampler B. 2017. *Democracy at Work: Moving Beyond Elections to Improve Well-Being.* Department of Political Science, Boise State University.

Touchton M and Wampler B. 2014. Improving social well-being through new democratic institutions. *Comparative Political Studies* 47(10), 1442–1469.

Ueshiba K. 1992. *A Life in Aikido.* Kodansha, New York.

UKRI. 2019. *Citizen Science.* At www.ukri.org/news/citizen-science-awards-to-put-public-at-heart-of-key-research/

Ulrich R S. 1984. View through a window may influence recovery from surgery. *Science* 224(4647), 420–421.

UNEP. 2011. *Towards a Green Economy: Pathways to Sustainable Development and Eradication of Poverty.* UNEP, Nairobi.

United Skates (Directors Tina Brown and Dyana Winkler). 2019. Storyville Documentary. BBC, London.

University of Leeds. 2022. *Writing Back.* At https://ahc.leeds.ac.uk/homepage/270/writing_back

Unno T. 1998. *River of Fire, River of Water.* Doubleday, New York.

Unno T. 2002. *Shin Buddhism: Bits of Rubble Turn into Gold.* Doubleday, New York.

USDA. 2018. *The State of the Co-Op Economy.* US Department of Agriculture, Washington, DC.

Vaillant G E. 2002. *Aging Well.* Little, Brown and Co, Boston.

Vainshtein S. 1980. *Nomads of South Siberia.* Cambridge University Press, Cambridge.

Van Brussel S and Huyse H. 2019. Citizen science on speed? Realising the triple objective of scientific rigour, policy influence and deep citizen engagement in a large-scale citizen science project on ambient air quality in Antwerp. *Journal of Environmental Planning and Management* 62(3), 534–551.

van den Berg H, Ketelaar J W, Dicke M and Fredrix M. 2020. Is the farmer field school still relevant? Case studies from Malawi and Indonesia. *NJAS-Wageningen Journal of Life Sciences* 92, 100329.

van den Berg H, Phillips S, Dicke M and Fredrix M. 2020. Impacts of farmer field schools in the human, social, natural and financial domain: a qualitative review. *Food Security* 12(6), 1443–1459.

van der Kolk B. 2014. *The Body Keeps the Score.* Penguin, London.

Vardakoulias, O. 2013. *The Economic Benefits of Ecominds.* MIND, London.

Veblen T. 2005. *Conspicuous Consumption.* Penguin, London.

Verma S, Sonkar V K, Kumar A and Roy D. 2019. Are farmer producer organizations a boon to farmers? The evidence from Bihar, India. *Agricultural Economics Research Review* 32, 123–137.

Verplanken B and Melkevik O. 2008. Predicting habit: The case of physical exercise. *Psychology of Sport and Exercise* 9(1), 15–26.

Verplanken B and Orbell S. 2022. Attitudes, habits and behaviours. *Annual Review of Psychology* 73.

Virgil (trans Knight W F J). 1956. *The Aeneid.* Penguin Classics, London.

Vogler C. 1998 [2007]. *The Writer's Journey.* Michael Wiese, Studio City, CA.

Waddington H, Snilstveit B, Hombrados J, Vojtkova M, Phillips D, Davies P and colleagues. 2014. Farmer field schools for improving farming practices and farmer outcomes in low- and middle-income countries: A systematic review. *Campbell Systematic Reviews* 10, 1–335.

Wahl D C. 2021. *Designing Regenerative Cultures.* Triarchy Press, Bridport.

Waldinger R. 2015. What makes a good life? Lessons from the longest study on happiness. *The Harvard Study of Adult Development. Retrieved 28*(8), 2017.

Walker S. 2011. *The Spirit of Design.* Earthscan, Oxon.

Walker S. 2014. *Designing Sustainability.* Routledge, Oxon.

Walker S. 2021. Design and Spirituality. Routledge, Oxon.

Wallace B A. 2006. *The Attention Revolution.* Wisdom, Somerville, MA.

Waller M. 2016. *The Life of Merlin.* Amberley, Stroud.

Wampler B, McNulty S and Touchton M. 2018. Participatory budgeting: Spreading across the globe. *Transparency & Accountability Initiative.* At https://www.transparency-initiative.org/

Wang C, Li K, Choudhury A and Gaylord S. 2019. Trends in yoga, Tai Chi, and Qigong use among US adults, 2002–2017. *American Journal of Public Health 109*(5), 755–761.

Wang H (with Libby T). 2021. *Searching for Centre.* Littlewood Press, Issaquah, PA.

Wang Wei (trans Robertson G). 1973. *Poems.* Penguin Classics, London.

Ward Thompson C, Silveirinha de Oliveira E, Tilley S, Elizalde A, Botha W, Briggs A, Cummins S, Leyland A H, Roe J J, Aspinall P and Brookfield K. 2019. Health impacts of environmental and social interventions designed to increase deprived communities' access to urban woodlands: A mixed-methods study. *Public Health Research 7*(2), 1–172.

Waxler R P. 2008. Changing lives through literature. *PMLA 123*(3), 678–683.

Wayne P M and Fuerst M L. 2013. *The Harvard Medical School Guide to Tai Chi.* Shambhala, Boston.

Weller M. 2014. *The Battle for Open: How Openness Won and Why It Doesn't Feel Like Victory.* Ubiquity Press, London.

Wester P, Mukherjee A, Mishra A and Shrestha A B (eds). 2019. *Hindu Kush Himalaya Assessment.* ICIMOD and Springer Open, Kathmandu.

WHO. 2016. *Urban Green Space and Health.* WHO, Geneva.

WHO. 2018. *A Healthy Diet Sustainably Produced.* WHO/NMH/NHD 18.18 Technical Report, Copenhagen.

WHR. (eds Helliwell J F, Layard R and Sachs J D). 2019. *World Happiness Report 2019.* Sustainable Development Solutions Network, New York.

WHR. (eds Helliwell J F, Layard R, Sachs J D, De Neve J-E). 2020. *World Happiness Report 2020.* Sustainable Development Solutions Network, New York.

Wicks P, Thorley E M, Simacek K, Curran C and Emmas C. 2018. Scaling PatientsLikeMe via a 'generalized platform' for members with chronic illness: Web-based survey study of benefits arising. *Journal of Medical Internet Research 20*(5), e9909.

Wiedenhofer D, Virág D, Kalt G, Plank B, Streeck J, Pichler M, Mayer A, Krausmann F, Brockway P, Schaffartzik A and Fishman T. 2020. A systematic review of the evidence on decoupling of GDP, resource use and GHG emissions, part I: Bibliometric and conceptual mapping. *Environmental Research Letters 15*(6), 063002.

Wiking M. 2016. *The Little Book of Hygge.* Penguin, London.

Wiking M. 2017. *The Key to Happiness.* Penguin, London.

Wiking M (ed). 2020. *Copenhagen: Beyond Green.* Happiness Research Institute, Copenhagen.

Wilhite H. 2016. The problem of habits for a sustainable transition. In Syse K L and Muller M (eds). *Sustainable Consumption and the Good Life.* Routledge, Oxon.

Wilkinson R and Pickett K. 2009. *The Spirit Level.* London and New York, Penguin.

Wilkinson R and Pickett K. 2016. *The Inner Level.* Allen Lane, London.

Willcox B J, Willcox D C and Ferrucci, L. 2008. Secrets of healthy aging and longevity from exceptional survivors around the globe: Lessons from octogenarians to supercentenarians. *The Journals of Gerontology Series A: Biological Sciences and Medical Sciences 63*(11), 1181–1185.

Willcox D C, Scapagnini G and Willcox B J. 2014. Healthy aging diets other than the Mediterranean: A focus on the Okinawan diet. *Mechanisms of Ageing and Development 136*, 148–162.

Williams M and Penman D. *Mindfulness*. Piatkus, London.

Wilson B. 2020. *The Way We Eat Now*. Basic Books, New York.

Wilson J. 1998. *The Earth Shall Weep: A History of Native America*. Grove Press, New York.

Wilson P A. 2019. *The Heart of Community Engagement*. Routledge, Oxon.

Wilson S. 2012. *One Taste of Truth*. Shambhala, Boston.

Wilson W S. 2013. *The Lone Samurai: The Life of Miyamoto Musashi*. Shambhala, Boston.

Wilson W S. 2014. *The Swordsman's Handbook*. Shambhala, Boston.

Wilson W S. 2015. *Walking the Kiso Road*. Shambhala, Boulder, CO.

Wirth J M. 2017. *Mountains, Rivers, and the Great Earth: Reading Gary Snyder and Dōgen in an Age of Ecological Crisis*. Suny Press, New York.

Wissler C and Duval D C. 1908 [2007]. *Mythology of the Blackfoot Indians*. University of Nebraska Press, Lincoln.

Wittman H and Blesh J. 2017. Food sovereignty and Fome Zero: Connecting public food procurement programmes to sustainable rural development in Brazil. *Journal of Agrarian Change 17*, 81–105.

Wolsko C, Lindberg K and Reese R. 2019. Nature-based physical recreation leads to psychological well-being: Evidence from five studies. *Ecopsychology 11*(4), 222–235.

Wood A M, Froh J J and Geraghty A W. 2010. Gratitude and well-being: A review and theoretical integration. *Clinical Psychology Review 30*(7), 890–905.

Wood C, Pretty J and Griffin M. 2015. A case: Control study of the health and well-being benefits of allotment gardening. *Journal of Public Health 1*–9.

Wood W. 2019. *Good Habits, Bad Habits*. MacMillan, London.

World Bank (eds Shekar and Popkin). 2020. *Obesity: Health and Economic Consequences of an Impending Global Challenge*. World Bank Publications, Washington, DC.

World Happiness Report (eds Helliwell J F, Layard R, Sachs J D, De Neve J-E). 2020. *World Happiness Report 2020*. Sustainable Development Solutions Network, New York.

World Health Organization. 2014. *Global Status Report on Non-Communicable Diseases 2014* (No. WHO/NMH/NVI/15.1). World Health Organization.

Worpole K. 2021. *No Matter How Many Skies Have Fallen*. Little Toller, Dorset.

Wynne B. 2014. Further disorientation in the hall of mirrors. *Public Understanding of Science 23*(1), 60–70.

Xue Y J. 2022. China introduces guidance to encourage low carbon behaviours. *South China Morning Post*, May 10.

Yanaga S and Leach B. 1972 (2013). *The Unknown Craftsman*. Kodansha, New York.

Yekti M I, Schultz B, Norken I N, Gany A H A and Hayde L. 2017. Learning from experiences of ancient Subak schemes for participatory irrigation system management in Bali. *Irrigation and Drainage 66*(4), 567–576.

Yeshe Losal Rinpoche. 2020. *From a Mountain in Tibet*. Penguin Life, London.

Yolen J. 1986. *Favourite Folktales from around the World*. Pantheon Books, New York.

Yorke J. 2013. *Into the Woods*. Penguin, London.

Yoshino G. 1937 [1982]. *How Do You Live?* Penguin, London.

Young S. 2016. *The Science of Enlightenment*. Sounds True, Boulder.

Yunkaporta T. 2019. *Sand Talk*. Text Publishing, Melbourne.

Zak P J, Stanton A A and Ahmadi S. 2007. Oxytocin increases generosity in humans. *PLoS One 2*(11), e1128.

Zaki J. 2018. *The War for Kindness*. Robinson, London.

Zhang R. 2019. Quest for a pathway to human good life in the Chinese cultural context. *Counselling Psychology Quarterly 32*(3–4), 516–528.

Zhang W, Cao G, Li X, Zhang H, Wang C, Liu Q, Chen X, Cui Z, Shen J, Jiang R and Mi G. 2016. Closing yield gaps in China by empowering smallholder farmers. *Nature 537*(7622), 671–674.

Zhou Q, Deng X, Wu F, Li Z and Song W. 2017. Participatory irrigation management and irrigation water use efficiency in maize production: Evidence from Zhangye City, Northwestern China. *Water 9*(11), 822.

INDEX

Printed in the United States
by Baker & Taylor Publisher Services

Printed in the United States
by Baker & Taylor Publisher Services